Mathematical Methods for Physics

Mathematical Methods for Physics

45th anniversary edition

H.W. Wyld

CRC Press
Taylor & Francis Group
Boca Raton London New York

CRC Press is an imprint of the
Taylor & Francis Group, an **informa** business

First edition published 2021
by CRC Press
6000 Broken Sound Parkway NW, Suite 300, Boca Raton, FL 33487-2742

and by CRC Press
2 Park Square, Milton Park, Abingdon, Oxon, OX14 4RN

© 2021 Taylor & Francis Group, LLC

CRC Press is an imprint of Taylor & Francis Group, LLC

ISBN: 9780367477080 (hbk)
ISBN: 9780367479732 (pbk)
ISBN: 9781003037460 (ebk)

Typeset in Computer Modern font
by Cenveo Publisher Services

Contents

List of Figures   xiii

List of Tables   xix

Editor's Preface to the 45th Anniversary Edition   xxi

Preface to the First Edition   xxiii

SECTION I Homogeneous Boundary Value Problems and Special
    Functions

CHAPTER   1 ▪ The Partial Differential Equations of Mathematical Physics 3

1.1   INTRODUCTION   3
1.2   HEAT CONDUCTION AND DIFFUSION   3
1.3   QUANTUM MECHANICS   5
1.4   WAVES ON STRINGS AND MEMBRANES   5
1.5   HYDRODYNAMICS AND AERODYNAMICS   7
1.6   ACOUSTIC WAVES IN A COMPRESSIBLE FLUID   8
1.7   IRROTATIONAL FLOW IN AN INCOMPRESSIBLE FLUID   9
1.8   ELECTRODYNAMICS   11
       1.8.1   Time Independent Phenomena   12
       1.8.2   Vacuum Equations   13
       1.8.3   General Case   14
1.9   SUMMARY   15
PROBLEMS   17

CHAPTER   2 ▪ Separation of Variables and Ordinary Differential Equations 21

2.1   INTRODUCTION   21
2.2   SEPARATION OF VARIABLES   21
2.3   RECTANGULAR COORDINATES (x, y, z) 22
2.4   CYLINDRICAL COORDINATES (r, θ, z) 23

2.5   SPHERICAL COORDINATES (r, θ, ϕ) 25

2.6   SERIES SOLUTIONS OF ORDINARY DIFFERENTIAL EQUATIONS: PRELIMINARIES   27

2.7   EXPANSION ABOUT A REGULAR SINGULAR POINT   30

2.8   STURM-LIOUVILLE EIGENVALUE PROBLEM   34

2.9   FOURIER SERIES AND INTEGRALS   39

2.10 NUMERICAL SOLUTION OF ORDINARY DIFFERENTIAL EQUATIONS   41

PROBLEMS   45

CHAPTER   3 ■ Spherical Harmonics and Applications   51

3.1   INTRODUCTION   51

3.2   SERIES SOLUTION OF LEGENDRE'S EQUATION - LEGENDRE POLYNOMIALS   52

3.3   PROPERTIES OF LEGENDRE POLYNOMIALS   55

3.4   THE SECOND SOLUTION $Q_l(x)$ OF LEGENDRE'S EQUATION   60

3.5   ASSOCIATED LEGENDRE POLYNOMIALS   62

3.6   SPHERICAL HARMONICS   65

3.7   THE SPHERICAL HARMONICS ADDITION THEOREM   68

3.8   MULTIPOLE EXPANSIONS   71

3.9   LAPLACE'S EQUATION IN SPHERICAL COORDINATES   75

3.9.1   Interior Problem I, $r \leq a$ with $\psi(a, \theta, \varphi) = u(\theta, \varphi)$ given   76

3.9.2   Interior Problem II, $r \leq a$ with $\left.\frac{\partial \psi}{\partial r}\right|_{r=a} = v(\theta, \varphi)$ given   76

3.9.3   Exterior Problem, $r \geq a$ with $\psi(a, \theta, \varphi) = u(\theta, \varphi)$ given   77

3.9.4   Exterior Problem, $r \geq a$ with $\left.\frac{\partial \psi}{\partial r}\right|_{r=a} = v(\theta, \varphi)$ given   77

3.9.5   Region Between Two Spheres, $a \leq r \leq b$ with $\psi(a, \theta, \varphi) = u(\theta, \varphi)$ and $\psi(b, \theta, \varphi) = v(\theta, \varphi)$ given   77

3.9.6   Notes on Solving Other Boundary Conditions on Regions Between Two Spheres   78

3.10 CONDUCTING SPHERE IN A UNIFORM EXTERNAL ELECTRIC FIELD   78

3.11 FLOW OF AN INCOMPRESSIBLE FLUID AROUND A SPHERICAL OBSTACLE   79

PROBLEMS   81

CHAPTER   4 ■ Bessel Functions and Applications   85

4.1   INTRODUCTION   85

4.2   SERIES SOLUTIONS OF BESSEL'S EQUATION; BESSEL FUNCTIONS   85

4.3   NEUMANN FUNCTIONS   87

4.4   SMALL ARGUMENT AND ASYMPTOTIC EXPANSIONS   91

4.5   BESSEL FUNCTIONS OF IMAGINARY ARGUMENT   92

4.6   LAPLACE'S EQUATION IN CYLINDRICAL COORDINATES   94

4.7   INTERIOR OF A CYLINDER OF FINITE LENGTH   94

4.8   THE STURM-LIOUVILLE EIGENVALUE PROBLEM AND APPLICATION OF THE EXPANSION THEOREM 96

4.9   INTERIOR OF A CYLINDER OF FINITE LENGTH - CONTINUED   97

4.10 EXTERIOR OF AN INFINITELY LONG CYLINDER   98

4.11 CYLINDER IN AN EXTERNAL FIELD   99

4.12 SPACE BETWEEN TWO INFINITE PLANES   100

4.13 FOURIER BESSEL TRANSFORMS   101

4.14 SPACE BETWEEN TWO INFINITE PLANES - CONTINUED   102

PROBLEMS   104

CHAPTER   5 ▪ Normal Mode Eigenvalue Problems   107

5.1   INTRODUCTION   107

5.2   REDUCTION OF THE DIFFUSION EQUATION AND WAVE EQUATION TO AN EIGENVALUE PROBLEM   107

5.3   THE VIBRATING STRING   111

5.4   THE VIBRATING DRUMHEAD   112

5.5   HEAT CONDUCTION IN A CYLINDER OF FINITE LENGTH   114

5.6   PARTICLE IN A CYLINDRICAL BOX (QUANTUM MECHANICS)   115

5.7   NORMAL MODES OF AN ACOUSTIC RESONANT CAVITY   116

5.8   ACOUSTIC WAVE GUIDE   117

PROBLEMS   120

CHAPTER   6 ▪ Spherical Bessel Functions and Applications   123

6.1   INTRODUCTION   123

6.2   FORMULAS FOR SPHERICAL BESSEL FUNCTIONS IN TERMS OF ELEMENTARY FUNCTIONS   124

6.3   EIGENVALUE PROBLEM AND APPLICATION OF THE EXPANSION THEOREM   128

6.4 EXPANSION OF PLANE AND SPHERICAL WAVES IN SPHERICAL COORDINATES   130

6.5 THE EMISSION OF SPHERICAL WAVES   133

6.6 SCATTERING OF WAVES BY A SPHERE   136

PROBLEMS   141

Summary of Part I   145

SECTION II Inhomogeneous Problems, Green's Functions, and Integral Equations

CHAPTER 7 ■ Dielectric and Magnetic Media   157

7.1 INTRODUCTION   157

7.2 MACROSCOPIC ELECTROSTATICS IN THE PRESENCE OF DIELECTRICS   157

7.3 BOUNDARY VALUE PROBLEMS IN DIELECTRICS   163

7.3.1 Free Charge Distribution ρ_F Embedded in an Infinite Uniform Dielectric with a Constant Dielectric Constant ε 163

7.3.2 Point Charge in Front of a Semi-infinite Dielectric   164

7.3.3 Dielectric Sphere in a Uniform External Electric Field   167

7.4 MAGNETOSTATICS AND THE MULTIPOLE EXPANSION FOR THE VECTOR POTENTIAL   169

7.5 MAGNETIC MEDIA   172

7.6 BOUNDARY VALUE PROBLEMS IN MAGNETIC MEDIA   177

7.6.1 Uniformly Magnetized Sphere, $\mathbf{M}$ Given   177

7.6.2 Magnetic Sphere in a Uniform External Magnetic Field   180

7.6.3 Long Straight Wire Carrying Current I Parallel to a Semi-infinite Slab of Material of Permeability μ 182

PROBLEMS   185

CHAPTER 8 ■ Green's Functions: Part One   189

8.1 INTRODUCTION   189

8.2 ORDINARY DIFFERENTIAL EQUATIONS   189

8.3 GENERAL THEORY, VARIOUS BOUNDARY CONDITIONS   190

8.4 THE BOWED STRETCHED STRING   193

8.5 EXPANSION OF GREEN'S FUNCTION IN EIGENFUNCTIONS   196

8.6 POISSON'S EQUATION   198

8.7   POISSON'S EQUATION FOR ALL SPACE   200

8.8   ELECTROSTATICS WITH BOUNDARY CONDITIONS ON
       SURFACES AT FINITE DISTANCES – THE IMAGE METHOD   200

8.9   EXPANSION OF THE GREEN'S FUNCTION FOR THE INTERIOR OF
       A SPHERE IN SERIES   202

8.10   THE HELMHOLTZ EQUATION – THE FORCED DRUMHEAD   204

8.11   EIGENFUNCTION EXPANSION OF THE GREEN'S FUNCTION FOR
       THE HELMHOLTZ EQUATION   208

       PROBLEMS   210

CHAPTER   9 ■ Green's Functions: Part Two   215

9.1   INTRODUCTION   215

9.2   THE HELMHOLTZ EQUATION FOR INFINITE REGIONS,
       RADIATION, AND THE WAVE EQUATION; SINUSOIDAL TIME
       DEPENDENCE   215

9.3   GENERAL TIME DEPENDENCE   218

9.4   THE WAVE EQUATION   220

9.5   THE WAVE EQUATION FOR ALL SPACE, NO BOUNDARIES AT
       FINITE DISTANCES   223

9.6   FIELD DUE TO A POINT SOURCE   227

       9.6.1   Point Source Moving with Constant Velocity, $v < c$ 229

       9.6.2   Point Source Moving with Constant Velocity, $v > c$ 230

9.7   THE DIFFUSION EQUATION   232

9.8   THE DIFFUSION EQUATION FOR ALL SPACE, NO BOUNDARIES
       AT FINITE DISTANCES   235

       PROBLEMS   238

CHAPTER   10 ■ Integral Equations   241

10.1   INTRODUCTION   241

10.2   QUANTUM THEORY OF SCATTERING   241

10.3   TYPES OF INTEGRAL EQUATIONS   244

       10.3.1   First Kind   244

       10.3.2   Second Kind   244

       10.3.3   Volterra   244

       10.3.4   Eigenvalue Problem   245

10.4   INTEGRAL EQUATIONS WITH SEPARABLE KERNELS   245

10.5   CONVOLUTION INTEGRAL EQUATIONS   248
10.6   ITERATION – LIOUVILLE-NEUMANN SERIES   249
10.7   NUMERICAL SOLUTION   251
10.8   FREDHOLM'S FORMULAS   254
10.9   CONDITIONS FOR VALIDITY OF FREDHOLM'S FORMULAS   258
10.10 HILBERT-SCHMIDT THEORY   260
PROBLEMS   265

Section III Complex Variable Techniques

Chapter 11 ▪ Complex Variables; Basic Theory   273

11.1   INTRODUCTION   273
11.2   ANALYTIC FUNCTIONS; THE CAUCHY-RIEMANN EQUATIONS   274
11.3   POWER SERIES   280
11.4   MULTIVALUED FUNCTIONS; CUTS; RIEMANN SHEETS   286
11.5   CONTOUR INTEGRALS; CAUCHY'S THEOREM   298
11.6   CAUCHY'S INTEGRAL FORMULA   305
11.7   TAYLOR AND LAURENT EXPANSIONS   307
11.8   ANALYTIC CONTINUATION   313
PROBLEMS   318

Chapter 12 ▪ Evaluation of Integrals   323

12.1   INTRODUCTION   323
12.2   THE RESIDUE THEOREM   323
12.3   RATIONAL FUNCTIONS $(-\infty, \infty)$ 325
12.4   EXPONENTIAL FACTORS; JORDAN'S LEMMA   326
12.5   INTEGRALS ON THE RANGE $(0, \infty)$ 330
12.6   ANGULAR INTEGRALS   333
12.7   TRANSFORMING THE CONTOUR   334
12.8   PARTIAL FRACTION AND PRODUCT EXPANSIONS   335
PROBLEMS   340

Chapter 13 ▪ Dispersion Relations   343

13.1   INTRODUCTION   343
13.2   PLEMELJ FORMULAS; DIRAC'S FORMULA   343

13.3   DISCONTINUITY PROBLEM   345

13.4   DISPERSION RELATIONS; SPECTRAL REPRESENTATIONS   347

13.5   EXAMPLES   354

13.6   INTEGRAL EQUATIONS WITH CAUCHY KERNELS   359

PROBLEMS   363

CHAPTER 14 ▪ Special Functions   365

14.1   INTRODUCTION   365

14.2   THE GAMMA FUNCTION   365

14.3   ASYMPTOTIC EXPANSIONS; STIRLING'S FORMULA   368

14.4   THE HYPERGEOMETRIC FUNCTION   372

14.5   LEGENDRE FUNCTIONS   380

14.6   BESSEL FUNCTIONS   386

14.7   ASYMPTOTIC EXPANSIONS FOR BESSEL FUNCTIONS   393

PROBLEMS   400

CHAPTER 15 ▪ Integral Transforms in the Complex Plane   403

15.1   INTRODUCTION   403

15.2   THE CALCULATION OF GREEN'S FUNCTIONS BY FOURIER
       TRANSFORM METHODS   404

       15.2.1   The Helmholtz Equation   404

       15.2.2   The Wave Equation   407

       15.2.3   The Klein-Gordon Equation   409

15.3   ONE-SIDED FOURIER TRANSFORMS; LAPLACE TRANSFORMS   414

15.4   LINEAR DIFFERENTIAL EQUATIONS WITH CONSTANT
       COEFFICIENTS   419

15.5   INTEGRAL EQUATIONS OF CONVOLUTION TYPE   420

15.6   MELLIN TRANSFORMS   421

15.7   PARTIAL DIFFERENTIAL EQUATIONS   422

15.8   THE WIENER-HOPF METHOD   424

       15.8.1   Potential Given on Semi-Infinite Plate   425

       15.8.2   Diffraction by a Knife Edge   430

PROBLEMS   439

Bibliography   443

Index   447

List of Figures

1.1	Closed volume with differential surface normal	4
1.2	Transverse displacement of string under tension	5
1.3	Rectilinear coordinates for surface element of area	6
1.4	Closed volume for Gauss's law	8
1.5	Irrotational flow remains irrotational over time	11
2.1	System of cylindrical coordinates	24
2.2	System of spherical coordinates	25
2.3	Dirac delta function centered at a	39
3.1	Angle between two directions in spherical coordinates	68
3.2	Rotation of coordinate axis	69
3.3	Vectors to observation point, $\mathbf{r}$, and to sources, $\mathbf{r}'$, for multipole expansion	71
4.1	Finite length cylinder	94
4.2	Cross-sectional view of infinitely long cylinder in flow	99
4.3	Two infinite planes a distance b apart	100
5.1	Closed volume and associated surface for boundary conditions	108
5.2	Normal modes of oscillating string	111
5.3	Lowest order normal modes and eigenfunctions for drumhead	113
5.4	Finite length cylinder for heat conduction problem	114
5.5	Cylindrical box for sound waves	116
5.6	Cylindrical acoustic wave guide	117
6.1	Waves scattered by a sphere	136
7.1	Conceptual diagram of charge density distribution of an atom	158
7.2	Conceptual diagram of atomic charge near boundary in dielectric	159
7.3	Vectors for density of charge at boundary	159
7.4	Pillbox region across an interface for dielectric	162

7.5   Loop crossing an interface for dielectric   163

7.6   Point charge in vacuum a distance a from dielectric   164

7.7   Potential in vacuum from charge, q, and image charge, q' 165

7.8   Potential in dielectric from effective charge, q'' 165

7.9   Boundary and geometry for surface charge density on dielectric   167

7.10   Dielectric sphere in uniform field   167

7.11   Polarization charge on surface of dielectric sphere   169

7.12   Vectors for region of charge distribution, $\mathbf{r}'$, and point of observation, $\mathbf{r}$ 171

7.13   Planar current loop   172

7.14   Current loops at boundary of material   173

7.15   Conceptualization of atom position and current density   173

7.16   Vectors for density of current at boundary   174

7.17   Pillbox across the interface for magnetic media   176

7.18   Loop at boundary of magnetic media   176

7.19   Uniformly magnetized sphere   177

7.20   Pillbox on surface of magnetized sphere   178

7.21   Magnetization, $\mathbf{M}$, and surface normal of magnetized sphere   179

7.22   Surface density of magnetic poles   180

7.23   Magnetic sphere in uniform magnetic field   180

7.24   Hysteresis curve of B vs. H 181

7.25   Infinitely long wire a distance a from semi-infinite slab   182

7.26   Field in vacuum from current, I and image current I' 182

7.27   Field in slab from effective current, I'' 183

7.28   Geometry of boundary conditions at slab   183

8.1   Image charge near a conducting sphere   200

8.2   Potential within the sphere   201

8.3   Two-dimensional drumhead in polar coordinates   205

9.1   Arbitrary closed volume for the wave equation   221

9.2   Particle path relative to observer at $\mathbf{r}$ 228

9.3   Wake of wave as sum of two wavelets   231

9.4   Point at leading edge of wake   232

9.5   Arbitrary closed volume for diffusion equation   233

9.6   Delta function at two different times   236

10.1   Set of points with associated weights   251

10.2   Set of points equally spaced   254

11.1   Point, z, on the complex plane   275
11.2   Circle of convergence about point z_0 282
11.3   Convergence at point z_1 within circle of convergence of radius r 283
11.4   Two branches, w_1 and w_2 on the complex plane for $z = re^{i\theta}$ 286
11.5   Circuit loops paths in complex plane   287
11.6   Regions including and excluding the origin   288
11.7   Progression of region enlarging yet avoiding branch point at $z = 0$ 289
11.8   Cut choice along negative x axis   290
11.9   Cut choice along positive x axis   290
11.10 Riemann sheets connected along branch line   291
11.11 Cross section view of cut crossings   291
11.12 Cross section view of Riemann surface with multiple sheets   292
11.13 Two arbitrary cuts for two branch points   293
11.14 Cut along line between two branch points   293
11.15 Curves around two branch points   294
11.16 Curves for function with two cuts and four branch points   295
11.17 Multiple passes through branch point   296
11.18 Cut along negative real axis for $\log z$ 296
11.19 Infinite sheets attached at branch point   297
11.20 Points z_1 and z_2 near the cut on negative real axis   298
11.21 Contour of integration in complex plane   298
11.22 Circular contour in complex plane about point ζ 299
11.23 Path of integration from z_1 to z_2 in complex plane   300
11.24 Example paths from z_1 to z_2 in a region R 301
11.25 Closed contour enclosing an area S in the complex plane   302
11.26 Annular region R with unknown behavior of function within shaded
       area   303
11.27 Separate curves in the annular region, R 303
11.28 Combining contours leaves two concentric contours in annular region   304
11.29 Contours around multiple excluded shaded regions   304
11.30 Integral from z_0 to z in complex plane   305
11.31 Counterclockwise contour, C, in region R 306
11.32 Circle centered at z_0 in a region, R 308
11.33 Convergence in largest circle centered about z_0 in region R 309
11.34 Annular region for an analytic function   309

11.35 Contour within annular region 310

11.36 Remaining contours after removing contiguous portions 310

11.37 Forbidden region within annulus which includes a singularity 311

11.38 Convergence in a region outside a singularity, z_0 312

11.39 Coinciding points within circle neighborhoods as illustration of analytic continuation in a region 313

11.40 Subregion of overlapping regions 314

11.41 Circle chain method 314

11.42 Circle chain around a singularity 315

11.43 Contour in t plane avoiding $t = 0$ 317

12.1 Contour with poles within contours 323

12.2 Semicircle closed contour of radius, R, with pole, ia 325

12.3 Semicircle contour of radius R without pole 326

12.4 Jordan's Lemma: contour for upper half-plane 327

12.5 Jordan's Lemma: contour for lower half-plane 327

12.6 Combined contours for semicircle with displaced contour around pole on axis 329

12.7 Cut along positive real axis 330

12.8 Contour avoiding positive x axis 331

12.9 Transforming positive x axis, C_1, to a preferred contour, $C_2 + C_3$ 334

12.10 Contour including the poles of $1/\sin \pi z$ 335

12.11 Contour choice to exclude a pole on or near the real axis 335

12.12 Contour, Γ, expanding to infinity while avoiding poles z_1 and z_2 336

12.13 Contour, C_1, in complex l plane 338

12.14 Deformed contour, C_1, parallel to imaginary axis avoiding poles 339

12.15 Contour, C_2, remains with sum of the residue poles 339

13.1 Singularity above contour 345

13.2 Singularity below contour 345

13.3 Contour in upper half-plane 347

13.4 Pole a along x axis for function defined for x between $-b$ and c 349

13.5 Contour with semicircular indentation about a pole 349

13.6 Singularity at a with cuts on axis for less than $-b$ and greater than c 350

13.7 Region of convergence for points z_1 and z_1^* 351

13.8 Contour going around poles and cuts 352

13.9 Distorting the contour 353

13.10 Diagram to use with evaluating Example (1) – equation (13.57)   355

13.11 Cut along negative real axis for $x < -b$ for Example (2)   355

13.12 Cut along real axis between -1 and +1 for Example (3)   356

13.13 Contour deformed around a point x on the real axis   358

14.1 Contour looped around the origin   366

14.2 Contour choice and cuts in t plane   381

14.3 Angle analysis for equation (14.98)   382

14.4 Contour used with second solution of Legendre's equation, $Q_l(z)$ 383

14.5 Angle analysis for equation 14.107   384

14.6 Contour squeezed along x axis while including the poles at ± 1 384

14.7 Contours for use with equation (14.122)   387

14.8 Shaded areas identify allowable and useful contour regions   388

14.9 Shaded regions for generally allowed contours   388

14.10 Contour choice for the Bessel function $J_\lambda(z)$ 389

14.11 Contour choice for the Bessel function $J_{-\lambda}(z)$ 390

14.12 Shifting of contour $-L$ by π to obtain $-L'$ 391

14.13 New contours, L_1 and L'_1, from combining contours L and $-L'$ 391

14.14 Contour for use with equation (14.147)   393

14.15 Deforming the contour until a saddle point is determined   394

14.16 Contour used in equation (14.167) for $H_\lambda^{(1)}(z)$ 396

14.17 Contour direction and associated phase arguments   397

15.1 Consideration of contour, C_1, around the poles, $\pm\lambda$ 406

15.2 Poles displaced with contour along real k axis   406

15.3 Consideration of contour, C_2, around the poles, $\pm\lambda$ 407

15.4 Contour choice above the poles for ω plane   408

15.5 Contour choice below the poles for ω-plane   409

15.6 Contour above the poles in the ω plane for equation (15.35)   409

15.7 Geometry of cut from $-m$ to $+m$ in reference to equation (15.43)   411

15.8 Contour shifted to wrap just around the cut line from $-m$ to $+m$ 411

15.9 Cut from change of variables from ω plane to z plane   412

15.10 Change of variables for the integration of equation (15.35)   413

15.11 Contour for use with equation (15.56)   413

15.12 Contour used with equation (15.81)   416

15.13 Expansion of contour to the upper and lower half-planes   417

15.14 Function $f(t)$ from equation (15.86) only defined for $t > 0$ 418

15.15 Semi-infinite sheet with potential $\phi(x, y)$ 425

15.16 Geometry for treating $|k|$ as an analytic function with cuts in the imaginary plane for equation (15.144)   427

15.17 Plane wave incident upon a knife edge   430

15.18 Geometry configuration for equation (15.169)   431

15.19 Contour for use with equation (15.187)   435

15.20 Shaded regions indicate allowable contours for equation (15.189)   436

15.21 Reflected plane wave from knife edge at angle φ 437

List of Tables

15.1   Potential for ranges of x for semi-infinite plate   426

15.2   Potential for ranges of x for knife edge   432

15.3   Table of Laplace transforms for Problem 15.2   439

Editor's Preface to the 45th Anniversary Edition

It has been forty-five years since the initial printing of H. W. Wyld's *Mathematical Methods for Physics*. It has been nearly thirty years since I was introduced to his text in my undergraduate studies in physics at Colorado Mesa University (previously Mesa State College). The text was chosen to provide an introduction to solving partial differential equations and boundary value problems. It was also to be a useful reference for continued studies. Not long ago I was reviewing boundary value problems and picked up the first edition again. In my review I rediscovered what a great reference this is and sought out CRC Press to pursue an updated version. Together we are pleased to provide this anniversary edition.

Changes to the text are minor, maintaining the voice and lecture style of William Wyld. The corrections came from a complete review: checking the consistency of equation references and verifying equations. This edition modernizes the typesetting and corrects minor typos. With the update in typesetting, key or summary equations are boxed and proofs are called out(ending with $\square$) where suitable. Further changes in the updated edition include breaking Chapter 8 on Green's functions into two chapters, number equations by chapter, and including captions and references in the text for figures and tables. The following table indicates notation used throughout.

Description	Notation Example
scalars and variables	$a, b, c, x, x', r, \theta, \varphi$
functions	$f(x), g(x), G(x, x'), \psi(r, \theta, \varphi)$
vectors	$\mathbf{r}, \mathbf{A}, \mathbf{B}, \mathbf{H}, \mathbf{E}, \mathbf{D}$
unit vectors	$\hat{\mathbf{i}}, \hat{\mathbf{j}}, \hat{\mathbf{k}}, \hat{\mathbf{r}}, \hat{\theta}, \hat{\varphi}$
div, grad, curl, Laplacian	$\nabla \cdot, \nabla, \nabla \times, \nabla^2$

It has been a pleasure to work with the publishers to make available this anniversary edition with a focus on the appearance of the material while retaining the original content. I reviewed against the original material many times and worked through the math and derivations. Any remaining issues are mine. With updated formatting and style, it is my hope that this text by William Wyld will continue to be a valued resource for learning and an excellent reference.

Gary D. Powell, Ph.D.

Preface to the First Edition

This book is a written version of the lecture course I have given over a number of years to first-year graduate students at the University of Illinois on the subject of mathematical methods of physics. The course (and the book) are intended to provide the students with the basic mathematical background which they will need to perform typical calculations in classical and quantum physics. The level is intermediate; the usual undergraduate course in advanced calculus should provide an adequate prerequisite and would even provide some overlap (e.g., Fourier series) with the subjects covered in the present work. The treatment is limited to certain standard topics in classical analysis; no attempt is made to cover the method of characteristics, Hilbert space, or group theoretical methods. What I have tried to do is provide a short readable textbook from which the average physics or engineering student can learn the most important mathematical tools he will need in his professional career. The physics which lies behind the mathematical problems is all explained in some detail, so that the treatment should be intelligible also to pure mathematicians and might even provide an introduction to some of the advanced texts by mathematicians on the subject.

The mathematical methods sequence, as presently constituted at the University of Illinois, consists of three half-semester courses, i.e., all together 3/4 of an academic year. Together with a fourth half-semester course in classical mechanics, these courses provide a basis for more advanced work in electrodynamics, quantum mechanics, particle, nuclear, and solid state physics. The subject matter of the three parts, intentionally kept independent, and the corresponding chapters in the present book, are:

I Homogeneous Boundary Value Problems and Special Functions. 1-6

II Inhomogeneous Problems, Green's Functions, and Integral Equations. 7-9

III Complex Variable Techniques. 10-14

The low level of mathematical rigor which is customarily found in the writing of physicists will also be found in the present work. I feel that students seriously concerned with rigor should consult the mathematicians. I have, however, attempted to give, at appropriate spots, page references to works in which rigorous mathematical proofs and accurately worded theorems can be found.

Finally, I want to record here my debt to Mary Ostendorf for the excellent job she did in typing the manuscript.

H. W. Wyld

I

Homogeneous Boundary Value Problems
and
Special Functions

CHAPTER 1

The Partial Differential Equations of Mathematical Physics

1.1 INTRODUCTION

A large fraction of classical, and also quantum, physics uses a common type of mathematics. Certain partial differential equations occur over and over again in different fields. The methods of solution of these equations and the special functions which arise are thus generally useful tools which should be known to all physicists. The purpose of this book is to provide a guide to the study of this part of mathematics and to show how it is used in various physical applications.

Similar courses are offered in mathematics departments. There, one is usually concerned with the rigorous logical development of the mathematics. The student interested in such matters should and must go to the mathematicians. Here we will minimize the rigor and concentrate on a rough and ready approach to applications.

We start by reviewing the physical basis of the various equations we wish to solve.

1.2 HEAT CONDUCTION AND DIFFUSION

The flow of heat through a medium can be described by a flux vector $\mathbf{F}$, whose direction gives the direction of the heat flow and whose magnitude gives the magnitude of the heat flow in $\text{cal/cm}^2/\text{s}$. This vector $\mathbf{F}$ is related to the gradient of the temperature T by the thermal conductivity K of the medium:

$$\mathbf{F} = -K\boldsymbol{\nabla}T. \tag{1.1}$$

We also introduce the specific heat c and the density ρ of the medium.

In terms of these quantities we can write two different but equal expressions for the rate of change with time t of the heat Q in a volume V represented in Figure 1.1:

$$\frac{dQ}{dt} = \int_V d^3x \; c\rho\frac{\partial T}{\partial t} = -\int_S d\mathbf{A} \cdot \mathbf{F}$$

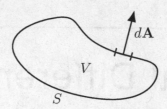

Figure 1.1: Closed volume with differential surface normal

$$\frac{dQ}{dt} = \int_S d\mathbf{A} \cdot K\boldsymbol{\nabla}T$$

$$= \int_V d^3x \, \boldsymbol{\nabla} \cdot (K\boldsymbol{\nabla}T). \tag{1.2}$$

Here the last step follows from the mathematical identity known as Gauss's theorem,

$$\int_S d\mathbf{A} \cdot \mathbf{F} = \int_V d^3x \, \boldsymbol{\nabla} \cdot \mathbf{F}, \tag{1.3}$$

valid for any vector field $\mathbf{F}$. Since the volume V is arbitrary, we obtain from (1.2) the relation

$$c\rho\frac{\partial T}{\partial t} = \boldsymbol{\nabla} \cdot (K\boldsymbol{\nabla}T), \tag{1.4}$$

or, if K is a constant,

$$\boxed{\boldsymbol{\nabla}^2 T = \frac{1}{\kappa}\frac{\partial T}{\partial t}} \tag{1.5}$$

with $1/\kappa = (c\rho)/K$ and

$$\boldsymbol{\nabla}^2 = \boldsymbol{\nabla} \cdot \boldsymbol{\nabla} = \frac{\partial^2}{\partial x^2} + \frac{\partial^2}{\partial y^2} + \frac{\partial^2}{\partial z^2}. \tag{1.6}$$

A similar equation is obtained for processes involving the diffusion of particles. If $n(\mathbf{r}, t)$ is the concentration of particles (number/cm^3), the flux of particles is given by

$$\mathbf{F} = -C\boldsymbol{\nabla}n,$$

where C is a constant. We can then write two expressions for the rate of change with time of the number N of particles in a volume V:

$$\frac{dN}{dt} = \int_V d^3x \, \frac{\partial n}{\partial t} = -\int_S d\mathbf{A} \cdot \mathbf{F} = C\int_V d^3x \, \boldsymbol{\nabla} \cdot (\boldsymbol{\nabla}n). \tag{1.7}$$

Since the volume V is arbitrary, we obtain

$$\boxed{\boldsymbol{\nabla}^2 n = \frac{1}{C}\frac{\partial n}{\partial t}.} \tag{1.8}$$

The heat conduction equation (1.5), or diffusion equation (1.8), is a standard equation of mathematical physics. In the important special case of no time dependence, T or n independent of time, we obtain Laplace's equation:

$$\boxed{\nabla^2 T = 0 \quad \text{or} \quad \nabla^2 n = 0.}$$
(1.9)

For the less restrictive special case of exponential time dependence, $T(\mathbf{r}, t) = e^{-\kappa k^2 t} u(\mathbf{r})$ or $n(\mathbf{r}, t) = e^{-C k^2 t} u(\mathbf{r})$, we obtain the Helmholtz equation:

$$\boxed{(\nabla^2 + k^2) u(\mathbf{r}) = 0.}$$
(1.10)

1.3   QUANTUM MECHANICS

The potential V in Schrödinger's equation,

$$\boxed{-\frac{\hbar^2}{2m} \nabla^2 \psi + V\psi = i\hbar \frac{\partial \psi}{\partial t},}$$
(1.11)

makes each quantum mechanics problem a special case. If $V = 0$, we find

$$\boxed{\nabla^2 \psi = -i \frac{2m}{\hbar} \frac{\partial \psi}{\partial t},}$$
(1.12)

which is the diffusion equation with an imaginary diffusion constant. If we assume an exponential time dependence,

$$\psi(\mathbf{r}, t) = u(\mathbf{r}) e^{-iEt/\hbar}, \quad E = \frac{\hbar^2 k^2}{2m},$$
(1.13)

we again find the Helmholtz equation (1.10).

1.4   WAVES ON STRINGS AND MEMBRANES

Consider a string of mass per unit length σ, under tension T, stretched along the x axis and then subjected to a transverse displacement $u(x, t)$, illustrated in Figure 1.2. Newton's second law, applied to the transverse motion for a small length dx of

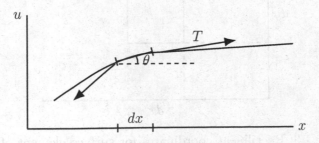

Figure 1.2: Transverse displacement of string under tension

the string, assumes the form

$$(T\sin\theta)_{x+dx} - (T\sin\theta)_x = \sigma \, dx \frac{\partial^2 u}{\partial t^2}. \tag{1.14}$$

For a <u>sufficiently small displacement</u> u, such that $\partial u/\partial x \ll 1$, we can approximate $\sin\theta \approx \tan\theta = \partial u/\partial x$ and the transverse component of the force in the string is given by

$$T\sin\theta \approx T\frac{\partial u}{\partial x}. \tag{1.15}$$

Newton's second law (1.14) then simplifies to

$$\frac{\partial}{\partial x}\left(T\frac{\partial u}{\partial x}\right) = \sigma\frac{\partial^2 u}{\partial t^2}. \tag{1.16}$$

If there is no longitudinal motion of the string, $T = $ constant and we obtain the one-dimensional wave equation,

$$\boxed{\frac{\partial^2 u}{\partial x^2} = \frac{1}{c^2}\frac{\partial^2 u}{\partial t^2},} \tag{1.17}$$

where

$$c = \sqrt{\frac{T}{\sigma}} \tag{1.18}$$

is the velocity of waves on the string. We note again that the linear equation (1.17) depends on the small amplitude assumption $\partial u/\partial x \ll 1$; without this assumption we would obtain a very complicated nonlinear equation.

For a stretched membrane (a drumhead) we can carry through a similar argument in two dimensions. The mass density σ is now a mass/unit area, the tension T is a force/unit length, and the transverse displacement $u(x, y, t)$ depends on two spatial variables x and y, as well as the time t. Newton's second law, applied to a small square surface element of area $dx \, dy$, as shown in Figure 1.3, gives

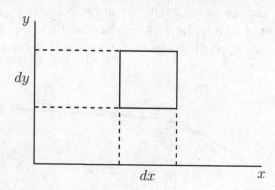

Figure 1.3: Rectilinear coordinates for surface element of area

$$\left[\left[T\frac{\partial u}{\partial x}\right]_{x+dx,y} - \left[T\frac{\partial u}{\partial x}\right]_{x,y}\right]dy + \left[\left[T\frac{\partial u}{\partial y}\right]_{x,y+dy} - \left[T\frac{\partial u}{\partial y}\right]_{x,y}\right]dx$$

$$= T\left[\frac{\partial^2 u}{\partial x^2} + \frac{\partial^2 u}{\partial y^2}\right]dx\,dy = \sigma\frac{\partial^2 u}{\partial t^2}dx\,dy, \qquad (1.19)$$

or

$$\boxed{\boldsymbol{\nabla}^2 u = \frac{\partial^2 u}{\partial x^2} + \frac{\partial^2 u}{\partial y^2} = \frac{1}{c^2}\frac{\partial^2 u}{\partial t^2},} \qquad (1.20)$$

with

$$c = \sqrt{\frac{T}{\sigma}}. \qquad (1.21)$$

The equation (1.20) is the wave equation in two dimensions, and (1.21) gives the velocity of waves on the membrane.

1.5 HYDRODYNAMICS AND AERODYNAMICS

We consider a small volume element dV of fluid and apply Newton's second law, $\mathbf{F} = m\mathbf{a}$, to its motion. If ρ is the density of fluid and $\mathbf{v}$ its velocity, we have

$$m\mathbf{a} = \rho\,dV\frac{d\mathbf{v}}{dt}. \qquad (1.22)$$

In general $\rho = \rho(\mathbf{r}, t)$ and $\mathbf{v} = \mathbf{v}(\mathbf{r}, t)$ are functions of position and time. As the volume element dV moves, $\mathbf{r}$ changes and we find

$$\frac{d\mathbf{v}}{dt} = \frac{\partial\mathbf{v}}{\partial t} + \frac{dx}{dt}\frac{\partial\mathbf{v}}{\partial x} + \frac{dy}{dt}\frac{\partial\mathbf{v}}{\partial y} + \frac{dz}{dt}\frac{\partial\mathbf{v}}{\partial z} = \frac{\partial\mathbf{v}}{\partial t} + (\mathbf{v}\cdot\boldsymbol{\nabla})\mathbf{v}. \qquad (1.23)$$

On the other side of Newton's second law we have the forces. One force is the pressure. We shall assume here that the pressure p is a scalar and thus eliminate from consideration many of the phenomena in an elastic solid. With a scalar pressure $p(\mathbf{r}, t)$ the force on dV is

$$[p(x, y, z, t) - p(x + dx, y, z, t)]\,dy\,dz\,\hat{\mathbf{i}}$$
$$+ [p(x, y, z, t) - p(x, y + dy, z, t)]\,dx\,dz\,\hat{\mathbf{j}}$$
$$+ [p(x, y, z, t) - p(x, y, z + dz, t)]\,dx\,dy\,\hat{\mathbf{k}}$$
$$= -\boldsymbol{\nabla}p\,dV. \qquad (1.24)$$

Thus Newton's second law assumes the form

$$\rho\left[\frac{\partial\mathbf{v}}{\partial t} + (\mathbf{v}\cdot\boldsymbol{\nabla})\mathbf{v}\right] = -\boldsymbol{\nabla}p. \qquad (1.25)$$

For a viscous fluid there are additional terms

$$\mu\boldsymbol{\nabla}^2\mathbf{v} + (\mu + \lambda)\boldsymbol{\nabla}(\boldsymbol{\nabla}\cdot\mathbf{v}) \qquad (1.26)$$

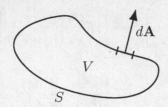

Figure 1.4: Closed volume for Gauss's law

on the right side of (1.25). We shall have no further use in this course for such terms, which are of course discussed in detail in textbooks and treatises on fluid mechanics.

In addition to Newton's second law (1.25) we need the continuity equation which is the mathematical statement of the conservation of matter. The flux vector for matter flow is $\rho\mathbf{v}$. Using this flux vector in a derivation similar to that in Section 1.2, where Figure 1.4 illustrates the enclosed volume and surface normal, we find

$$\frac{d}{dt}\int_V d^3x\,\rho = \int_V d^3x\,\frac{\partial\rho}{\partial t}$$

$$= -\int_S d\mathbf{A}\cdot\rho\mathbf{v}$$

$$= -\int_V d^3x\,\boldsymbol{\nabla}\cdot(\rho\mathbf{v}), \tag{1.27}$$

or, since the volume V is arbitrary,

$$\frac{\partial\rho}{\partial t} + \boldsymbol{\nabla}\cdot(\rho\mathbf{v}) = 0. \tag{1.28}$$

In the next two sections we apply our general results, Newton's second law in the form of (1.25) and the continuity equation (1.28), to two relatively simple common cases.

1.6 ACOUSTIC WAVES IN A COMPRESSIBLE FLUID

We consider small oscillations on a uniform background $\rho_0 = $ const., $p_0 = $ const. Thus $\rho = \rho_0 + \rho'$ with $\rho' << \rho_0$, and $p = p_0 + p'$ with $p' << p_0$. The amplitude of the oscillations is assumed to be sufficiently small that $\mathbf{v}$ is very small compared to the sound velocity (to be derived shortly). With these assumptions we can linearize (1.25) and (1.28), i.e., drop nonlinear terms like $(\mathbf{v}\cdot\boldsymbol{\nabla})\mathbf{v}$ and $\rho'\mathbf{v}$ which are quadratic in the quantities assumed small. Newton's second law (1.25) becomes

$$\rho_0\frac{\partial\mathbf{v}}{\partial t} = -\boldsymbol{\nabla}p' \tag{1.29}$$

and the continuity equation (1.28) becomes

$$\frac{\partial\rho'}{\partial t} + \rho_0\boldsymbol{\nabla}\cdot\mathbf{v} = 0. \tag{1.30}$$

If the pressure is assumed to be a function only of the density, with other variables held constant, the pressure and density fluctuations will be related:

$$p' = \Delta p = \frac{\partial p}{\partial \rho}\Delta\rho = \frac{\partial p}{\partial \rho}\rho' \equiv c^2\rho'. \tag{1.31}$$

Which other variable should be held constant? Newton assumed isothermal fluctuations so that

$$pV = nRT = \text{const.}$$

$$p \propto \frac{1}{V} \propto \rho$$

$$c^2 \equiv \frac{\partial p}{\partial \rho} = \frac{p}{\rho}. \tag{1.32}$$

Laplace observed that the sound oscillations are so rapid that no heat can be transferred so that one should use the adiabatic derivative:

$$PV^\gamma = \text{const.}$$

$$p \propto \frac{1}{V^\gamma} - \rho^\gamma$$

$$c^2 \equiv \frac{\partial p}{\partial \rho} = \gamma\frac{p}{\rho}. \tag{1.33}$$

As we see below c turns out to be the sound velocity and the correct result (1.33) for c^2 is larger by a factor $\gamma =$ the specific heat ratio c_P/c_V than the isothermal formula (1.32).

Taking the divergence of (1.29) and eliminating $\mathbf{v}$ and p' by (1.30) and 1.6.3. we obtain

$$\boxed{\nabla^2\rho' = \frac{1}{c^2}\frac{\partial^2\rho'}{\partial t^2}} \tag{1.34}$$

or, if we use (1.31) again,

$$\boxed{\nabla^2 p' = \frac{1}{c^2}\frac{\partial^2 p'}{\partial t^2}.} \tag{1.35}$$

The pressure and density fluctuations thus satisfy the three-dimensional wave equation with c the velocity of the waves. Again we find that a linear equation is obtained only for small amplitude phenomena. The nonlinear terms lead to various interesting phenomena such as shock waves, which we cannot cover here.

1.7   IRROTATIONAL FLOW IN AN INCOMPRESSIBLE FLUID

For an incompressible fluid $\rho = \text{const.}$ and the continuity equation (1.28) reduces to

$$\nabla \cdot \mathbf{v} = 0. \tag{1.36}$$

<u>If</u> we can further assume irrotational flow, i.e.,

$$\nabla \times \mathbf{v} = 0, \tag{1.37}$$

then we obtain a simple equation describing the flow. For (1.37) implies, according to a general and well-known result of vector analysis, that $\mathbf{v}$ is the gradient of some scalar function:

$$\mathbf{v} = -\nabla\phi. \tag{1.38}$$

Substituting (1.38) in (1.36) we obtain Laplace's equation for the so-called velocity potential ϕ:

$$\nabla^2\phi = 0. \tag{1.39}$$

It takes a bit more work to demonstrate the consistency of (1.37) for a time-dependent problem. We wish to show that if (1.37) is satisfied in some region of space at some initial time, then it will be satisfied at later times. To establish this, we manipulate Newton's second law (1.25), which for the case $\rho = $ const. can be written in the form

$$\frac{\partial \mathbf{v}}{\partial t} + (\mathbf{v} \cdot \nabla)\mathbf{v} = -\nabla\frac{p}{\rho}. \tag{1.40}$$

The nonlinear term can be transformed using the identity

$$(\mathbf{v} \cdot \nabla)\mathbf{v} = \nabla\frac{\mathbf{v}^2}{2} - \mathbf{v} \times (\nabla \times \mathbf{v}), \tag{1.41}$$

which is easily checked by writing it out in components. Thus we find

$$\frac{\partial \mathbf{v}}{\partial t} - \mathbf{v} \times (\nabla \times \mathbf{v}) = -\nabla\left[\frac{\mathbf{v}^2}{2} + \frac{p}{\rho}\right]. \tag{1.42}$$

Take the curl of this equation and set

$$\boldsymbol{\omega} \equiv \nabla \times \mathbf{v}. \tag{1.43}$$

Since $\nabla \times \nabla \equiv 0$, we find

$$\frac{\partial \boldsymbol{\omega}}{\partial t} - \nabla \times (\mathbf{v} \times \boldsymbol{\omega}) = 0. \tag{1.44}$$

We now use the vector identity

$$\nabla \times (\mathbf{v} \times \boldsymbol{\omega}) = (\boldsymbol{\omega} \cdot \nabla)\mathbf{v} - (\mathbf{v} \cdot \nabla)\boldsymbol{\omega} + \mathbf{v}(\nabla \cdot \boldsymbol{\omega}) - \boldsymbol{\omega}(\nabla \cdot \mathbf{v}), \tag{1.45}$$

which simplifies considerably for the case in hand since $\nabla \cdot \mathbf{v} = 0$ and $\nabla \cdot \boldsymbol{\omega} = \nabla \cdot (\nabla \times \mathbf{v}) = 0$. Thus we find that (1.44) can be rewritten as

$$\frac{\partial \boldsymbol{\omega}}{\partial t} + (\mathbf{v} \cdot \nabla)\boldsymbol{\omega} = (\boldsymbol{\omega} \cdot \nabla)\mathbf{v} \tag{1.46}$$

or

$$\frac{d\boldsymbol{\omega}}{dt} = (\boldsymbol{\omega} \cdot \nabla)\mathbf{v}, \tag{1.47}$$

where d/dt is the total time derivative following the matter (see equation (1.23)).

From equation (1.47) follows a theorem.

$$\omega = 0 \qquad \qquad \omega = 0$$

$$t \qquad \qquad \qquad t' > t$$

Figure 1.5: Irrotational flow remains irrotational over time

Theorem 1.7.1. *If $\omega = 0$ for an element of the fluid at some instant of time, $\omega = 0$ for that fluid element always. Similarly, if $\omega = 0$ in a connected region of fluid at some instant of time, ω remains 0 throughout that region as it moves along and distorts.*

This establishes the possibility of irrotational flow with $\omega = \nabla \times \mathbf{v} = 0$. If the flow is set up in such a way that it is irrotational to start with, it will remain irrotational, as illustrated in Figure 1.5.

For this irrotational flow the equation of motion of (1.42) reduces to

$$\frac{\partial \mathbf{v}}{\partial t} = -\nabla \left[\frac{\mathbf{v}^2}{2} + \frac{p}{\rho} \right]. \tag{1.48}$$

For steady-state flow, $\partial \mathbf{v}/\partial t = 0$ and we derive

$$\frac{1}{2}\rho \mathbf{v}^2 + p = \text{const.}. \tag{1.49}$$

With a gravitational potential there would be another term in the force:

$$\mathbf{F}/\text{unit volume} = -\nabla U, \quad U = \rho g z,$$

and we would obtain in place of (1.49)

$$\boxed{\frac{1}{2}\rho \mathbf{v}^2 + p + \rho g z = \text{const.}.} \tag{1.50}$$

This is the famous result known as Bernoulli's equation.

1.8 ELECTRODYNAMICS

We shall assume the reader is familiar with Maxwell's equations. These describe the electric and magnetic fields $\mathbf{E}$ and $\mathbf{B}$, defined by the force $\mathbf{F}$ on a point charge e moving with velocity $\mathbf{v}$:

$$\mathbf{F} = e \left(\mathbf{E} + \frac{\mathbf{v}}{c} \times \mathbf{B} \right). \tag{1.51}$$

The sources of the electromagnetic fields are the charge and current densities ρ and $\mathbf{j}$. In terms of these quantities, using Gaussian units, Maxwell's equations assume the form

$$\nabla \cdot \mathbf{E} = 4\pi \rho, \tag{1.52}$$

$$\mathbf{\nabla} \times \mathbf{B} = \frac{4\pi}{c}\mathbf{j} + \frac{1}{c}\frac{\partial \mathbf{E}}{\partial t}, \tag{1.53}$$

$$\mathbf{\nabla} \times \mathbf{E} = -\frac{1}{c}\frac{\partial \mathbf{B}}{\partial t}, \tag{1.54}$$

$$\mathbf{\nabla} \cdot \mathbf{B} = 0. \tag{1.55}$$

Here c is a constant which results from our choice of Gaussian units; as we see below it turns out to be the speed of light.

Maxwell's equations have a complicated form with two vector fields $\mathbf{E}$ and $\mathbf{B}$ coupled to each other. Fortunately they are linear equations. By making linear transformations we can usually reduce Maxwell's equations to the wave equation or Laplace's equation. We consider various cases of increasing complexity:

1.8.1 Time Independent Phenomena

In this case the electric and magnetic fields decouple from each other. For the electric field we find

$$\mathbf{\nabla} \cdot \mathbf{E} = 4\pi\rho \tag{1.56}$$
$$\mathbf{\nabla} \times \mathbf{E} = 0. \tag{1.57}$$

As discussed in connection with (1.37), (1.38) a standard theorem of vector analysis tells us that (1.57) implies that $\mathbf{E}$ is the gradient of a scalar field:

$$\mathbf{E} = -\mathbf{\nabla}\phi. \tag{1.58}$$

Substituting in (1.56) we obtain Poisson's equation for the electrostatic potential ϕ:

$$\mathbf{\nabla}^2\phi = -4\pi\rho. \tag{1.59}$$

This is of course Laplace's equation with an added inhomogeneous term.

Similarly for the time independent magnetic field we find

$$\mathbf{\nabla} \cdot \mathbf{B} = 0 \tag{1.60}$$

$$\mathbf{\nabla} \times \mathbf{B} = \frac{4\pi}{c}\mathbf{j}. \tag{1.61}$$

Another standard theorem of vector analysis teaches us that (1.60) implies that $\mathbf{B}$ is the curl of a vector field $\mathbf{A}$, called the vector potential:

$$\mathbf{B} = \mathbf{\nabla} \times \mathbf{A}. \tag{1.62}$$

Substituting (1.62) in (1.61) and using the vector identity

$$\mathbf{\nabla} \times (\mathbf{\nabla} \times \mathbf{A}) = \mathbf{\nabla}(\mathbf{\nabla} \cdot \mathbf{A}) - \mathbf{\nabla}^2\mathbf{A}, \tag{1.63}$$

we find

$$\nabla^2 \mathbf{A} - \nabla(\nabla \cdot \mathbf{A}) = -\frac{4\pi}{c}\mathbf{j}. \tag{1.64}$$

This would have the form of a vector Poisson equation if we could get rid of the term $\nabla(\nabla \cdot \mathbf{A})$.

To accomplish this we make a so-called gauge transformation. We introduce a new vector potential $\mathbf{A}'$ related to the original vector potential $\mathbf{A}$ by

$$\mathbf{A}' = \mathbf{A} + \nabla\Lambda. \tag{1.65}$$

Here Λ is so far an arbitrary scalar field. We see that $\mathbf{A}'$ deserves the name vector potential since using curl of grad is zero (in operator notation: $\nabla \times \nabla \equiv 0$), we find

$$\mathbf{B} = \nabla \times \mathbf{A}' = \nabla \times \mathbf{A}. \tag{1.66}$$

There is thus an arbitrariness in the definition of $\mathbf{A}$ corresponding to the scalar field Λ. We now choose Λ to eliminate the unwanted term in (1.64). Because of (1.66) the equation (1.64) holds for $\mathbf{A}'$ as well as $\mathbf{A}$. Choose Λ so that

$$\nabla \cdot \mathbf{A}' = 0. \tag{1.67}$$

To determine $\mathbf{A}'$ we then have the vector Poisson equation

$$\nabla^2 \mathbf{A}' = -\frac{4\pi}{c}\mathbf{j}. \tag{1.68}$$

All that remains is to check that it is possible to satisfy (1.67). Substituting (1.65) we find

$$\nabla^2\Lambda = -\nabla \cdot \mathbf{A} \tag{1.69}$$

as the equation for the appropriate gauge function Λ to achieve the desired goal. This equation has the form of Poisson's equation for a scalar function. Such an equation does in fact have a solution. We spend a large part of the present book showing how to solve such an equation. Thus it is possible to find the gauge function leading to the appropriate simplified equations (1.66)-(1.68).

1.8.2 Vacuum Equations

Maxwell's equations (1.52)-(1.55) simplify considerably when $\rho = 0$, $\mathbf{j} = 0$. Taking the curl of (1.54), using the identity (1.63) and (1.52), (1.53) with $\rho = 0$, $\mathbf{j} = 0$ we find

$$\boxed{\nabla^2\mathbf{E} - \frac{1}{c^2}\frac{\partial^2\mathbf{E}}{\partial t^2} = 0.} \tag{1.70}$$

A similar argument leads to

$$\boxed{\nabla^2\mathbf{B} - \frac{1}{c^2}\frac{\partial^2\mathbf{B}}{\partial t^2} = 0.} \tag{1.71}$$

Thus $\mathbf{E}$ and $\mathbf{B}$ separately satisfy the wave equation for this case. We must still go back to Maxwell's equations, however, to find the relation between $\mathbf{E}$ and $\mathbf{B}$.

1.8.3 General Case

We can use the same theorems from vector analysis and the same type of gauge transformation discussed in Section 1.8.1. The Maxwell equation (1.55) implies that **B** can be written as the curl of a vector potential **A**:

$$\mathbf{B} = \mathbf{\nabla} \times \mathbf{A} \tag{1.72}$$

Substituting this in (1.54) we obtain

$$\mathbf{\nabla} \times \left[\mathbf{E} + \frac{1}{c} \frac{\partial \mathbf{A}}{\partial t} \right] = 0, \tag{1.73}$$

which implies the existence of a scalar potential ϕ such that

$$\mathbf{E} = -\mathbf{\nabla}\phi - \frac{1}{c} \frac{\partial \mathbf{A}}{\partial t}. \tag{1.74}$$

Substituting (1.72) and (1.74) in the two Maxwell equations with sources, (1.52), (1.53), and using the vector identity (1.63), we derive the equations for ϕ and **A**:

$$\mathbf{\nabla}^2\phi - \frac{1}{c^2} \frac{\partial^2 \phi}{\partial t^2} = -4\pi\rho - \frac{1}{c} \frac{\partial}{\partial t} \left(\mathbf{\nabla} \cdot \mathbf{A} + \frac{1}{c} \frac{\partial \phi}{\partial t} \right), \tag{1.75}$$

$$\mathbf{\nabla}^2\mathbf{A} - \frac{1}{c^2} \frac{\partial^2 \mathbf{A}}{\partial t^2} = -\frac{4\pi}{c}\mathbf{j} + \mathbf{\nabla} \left(\mathbf{\nabla} \cdot \mathbf{A} + \frac{1}{c} \frac{\partial \phi}{\partial t} \right). \tag{1.76}$$

We can eliminate the last term from each of these equations by making a suitable gauge transformation. Introducing new scalar and vector potentials

$$\phi' = \phi - \frac{1}{c} \frac{\partial \Lambda}{\partial t}, \tag{1.77}$$

$$\mathbf{A}' = \mathbf{A} + \mathbf{\nabla}\Lambda, \tag{1.78}$$

with Λ a so far arbitrary scalar field, we see that (1.72) and (1.74) remain invariant, i.e.,

$$\mathbf{B} = \mathbf{\nabla} \times \mathbf{A}' \tag{1.79}$$

$$\mathbf{E} = -\mathbf{\nabla}\phi' - \frac{1}{c} \frac{\partial \mathbf{A}'}{\partial t}. \tag{1.80}$$

Consequently, (1.75) and (1.76) hold with ϕ and **A** replaced by ϕ' and $\mathbf{A}'$. We now choose Λ so that

$$\mathbf{\nabla} \cdot \mathbf{A}' + \frac{1}{c} \frac{\partial \phi'}{\partial t} = 0 \tag{1.81}$$

and obtain the simplified equations for ϕ', **A**:

$$\boxed{\mathbf{\nabla}^2\phi' - \frac{1}{c^2} \frac{\partial^2 \phi'}{\partial t^2} = -4\pi\rho,} \tag{1.82}$$

$$\boxed{\nabla^2 \mathbf{A}' - \frac{1}{c^2}\frac{\partial^2 \mathbf{A}'}{\partial t^2} = -\frac{4\pi}{c}\mathbf{j}.}$$ (1.83)

In (1.79)-(1.83) we have succeeded in writing Maxwell's equations in a largely decoupled form.

Substituting (1.77) and (1.78) in (1.81) we obtain the equation for that Λ which achieves the desirable result (1.81):

$$\nabla^2 \Lambda - \frac{1}{c^2}\frac{\partial^2 \Lambda}{\partial t^2} = -\left(\boldsymbol{\nabla}\cdot\mathbf{A} + \frac{1}{c}\frac{\partial \phi}{\partial t}\right)$$ (1.84)

This inhomogeneous wave equation is of a type which does have a solution. That is all we need to know; we do not need to actually solve the equation in practice. It is perhaps worth noting that (1.79)-(1.83) are invariant with respect to a further gauge transformation of the form

$$\phi' \longrightarrow \phi'' = \phi' - \frac{1}{c}\frac{\partial \Lambda'}{\partial t},$$ (1.85)

$$\mathbf{A}' \longrightarrow \mathbf{A}'' = \mathbf{A}' + \boldsymbol{\nabla}\Lambda',$$ (1.86)

where Λ' satisfies the wave equation

$$\nabla^2 \Lambda' - \frac{1}{c^2}\frac{\partial^2 \Lambda'}{\partial t^2} = 0,$$ (1.87)

since such a transformation leaves invariant

$$\boldsymbol{\nabla}\cdot\mathbf{A}'' + \frac{1}{c}\frac{\partial \phi''}{\partial t} = \boldsymbol{\nabla}\cdot\mathbf{A}' + \frac{1}{c}\frac{\partial \phi'}{\partial t} = 0.$$ (1.88)

1.9 SUMMARY

The preceding sections are of course no substitute for complete treatments of the various subjects covered. Our purpose has been merely to demonstrate that the same partial differential equations come up over and over again in different fields. We have seen that when suitable approximations and simplifications are made, we come back to three types of equations:

$$\boxed{\nabla^2 \psi = -4\pi\rho,}$$ (1.89)

$$\boxed{\nabla^2 \psi - \frac{1}{\kappa}\frac{\partial \psi}{\partial t} = -4\pi\rho,}$$ (1.90)

$$\boxed{\nabla^2 \psi - \frac{1}{c^2}\frac{\partial^2 \psi}{\partial t^2} = -4\pi\rho.}$$ (1.91)

For many applications we are interested in the homogeneous equations with $\rho = 0$ on the right.

If all subjects lead to the same equations, what are the differences between different subjects? The answer is that the boundary conditions are different. Laplace's equation in electrostatics uses different boundary conditions from Laplace's equation for the flow of an incompressible fluid, etc. We have so far completely neglected this question of boundary conditions and initial conditions. Eventually we must return to it since we need both the partial differential equation, (1.89) or (1.90) or (1.91), and the appropriate set of initial values and/or boundary values to determine a unique solution.

PROBLEMS

1.1   a) Consider the wave equation in one dimension:

$$\frac{\partial^2 u}{\partial x^2} - \frac{1}{c^2}\frac{\partial^2 u}{\partial t^2} = 0.$$

Change to variables

$$\xi = x - ct, \quad \eta = x + ct,$$

and show that the wave equation assumes the form

$$\frac{\partial^2 u}{\partial \xi \partial \eta} = 0$$

and that this has the general solution

$$u = f(\xi) + g(\eta) = f(x - ct) + g(x + ct),$$

where f and g are arbitrary functions.

b) Interpret the two contributions $f(x - ct)$ and $g(x + ct)$ and justify the statement often made in the text that c is the velocity of the waves. Draw some pictures to explain your considerations.

c) Suppose that u and $\partial u/\partial t$ are given at $t = 0$:

$$u(x, 0) = U(x),$$

$$\left.\frac{\partial u(x,t)}{\partial t}\right|_{t=0} = V(x).$$

Show that these conditions determine the functions f and g:

$$f(x) = \frac{1}{2}U(x) - \frac{1}{2c}\int_a^x dx'\, V(x'),$$

$$g(x) = \frac{1}{2}U(x) + \frac{1}{2c}\int_a^x dx'\, V(x'),$$

so that

$$u(x, t) = f(\xi) + g(\eta) = f(x - ct) + g(x + ct)$$

$$= \frac{1}{2}U(x - ct) - \frac{1}{2c}\int_a^{x-ct} dx'\, V(x')$$

$$+ \frac{1}{2}U(x + ct) + \frac{1}{2c}\int_a^{x+ct} dx'\, V(x').$$

Note that the value of a is immaterial. Draw some pictures to illustrate the special cases $V = 0$ and $U = 0$.

d) Consider now waves in a finite region, e.g., waves on a string of finite length $0 \le x \le l$. Show that the initial conditions of section (c) for the region $0 \le x \le l$ determine $f(\xi)$, $0 \le \xi \le l$, and $g(\eta)$, $0 \le \eta \le l$, and hence $u(x, t)$ inside the triangle ABC in the (x, ct) plane as shown in Figure 1.6. Show how boundary conditions at the ends of the string,

$$u(0, t) = u(l, t) = 0,$$

lead to recursion relations

$$f(\xi) + g(-\xi) = 0$$
$$f(2l - \eta) + g(\eta) = 0,$$

which, in addition to the initial conditions determine the solution in the whole region $0 \le x \le l$, $t > 0$.

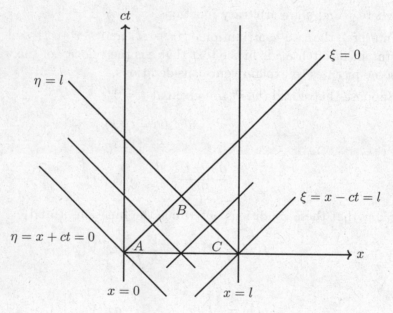

Figure 1.6: Boundary condition of waves in finite region for Problem 1.1

1.2 A cowboy twirls a lariat so that the rope forms a perfect circle in a horizontal plane, rotating with angular frequency ω. The cowboy pushes the rope in with his boot, creating a wave pulse in the circle, illustrated in Figure 1.7.

Assume that immediately after the pulse is created the <u>radial</u> velocity of the rope (i.e., transverse to the rope) in the region of the pulse vanishes. Describe the propagation of the pulse, including sketches of it at appropriately chosen times:

i) in a frame rotating with the lariat,

ii) in a fixed frame.

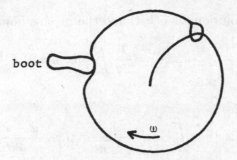

Figure 1.7: Deformation of swirling lariat for Problem 1.2

1.3 Using the method of Problem 1(a) show that the general solution of Laplace's equation in two dimensions,

$$\frac{\partial^2 u}{\partial x^2} + \frac{\partial^2 u}{\partial y^2} = 0,$$

is

$$u(x, y) = f(x + iy) + g(x - iy), \quad i = \sqrt{-1},$$

with f and g arbitrary functions, and that if u is to be a real function this reduces to

$$u(x, y) = 2 \operatorname{Re} f(x + iy),$$

where Re signifies "real part of".

1.4 According to the Fourier integral theorem an arbitrary function $u(x)$ can be expanded in the form

$$u(x) = \int_{-\infty}^{\infty} dk \, \widetilde{u}(k) e^{ikx},$$

where $\widetilde{u}(k)$ is given by

$$\widetilde{u}(k) = \frac{1}{2\pi} \int_{-\infty}^{\infty} dx \, u(x) e^{-ikx}.$$

Use this result to find the general solution of the one-dimensional heat conduction equation

$$\frac{\partial^2 u}{\partial x^2} = \frac{1}{\kappa} \frac{\partial u}{\partial t}.$$

Proceed via the following steps:

a) Expand:

$$u(x, t) = \int_{-\infty}^{\infty} dk \, \widetilde{u}(k, t) e^{ikx}.$$

Find and solve the ordinary differential equation in t for $\widetilde{u}(k, t)$.

b) Using your solution for $\tilde{u}(k,t)$ and the inverse Fourier transform at $t = 0$,

$$\tilde{u}(k,0) = \frac{1}{2\pi} \int_{-\infty}^{\infty} dx \; u(x,0)e^{-ikx},$$

show that

$$u(x,t) = \int_{-\infty}^{\infty} dx' \; \frac{1}{\sqrt{4\pi\kappa t}} \exp\left[-\frac{(x-x')^2}{4\kappa t}\right] u(x',0).$$

c) Suppose $u(x',0) = \delta(x'-a)$ as in Figure 1.8. Find $u(x,t)$ at later times $t > 0$. Draw some pictures of u at successively later times.

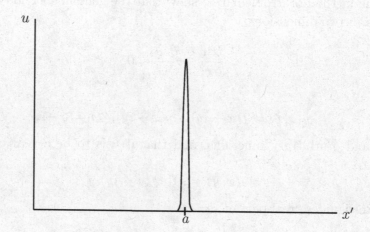

Figure 1.8: Dirac delta function centered at a for Problem (1.4c)

CHAPTER 2

Separation of Variables and Ordinary Differential Equations

2.1 INTRODUCTION

Various methods are available for the solution of partial differential equations. We consider in this chapter the elementary and generally useful method known as separation of variables. The separation procedure reduces the partial differential equation to several ordinary differential equations. In some cases these can be solved trivially. In several important cases we must resort to solution by expansion in series, so we discuss in a rather general way the series expansion method for second-order ordinary differential equations. Finally we begin our discussion of the eigenvalue problem by a study of Sturm-Liouville theory for second-order ordinary differential equations.

2.2 SEPARATION OF VARIABLES

The three partial differential equations under discussion are the homogeneous versions of (1.89) - (1.91):

$$\nabla^2 \psi = 0 \tag{2.1}$$

$$\nabla^2 \psi - \frac{1}{\kappa}\frac{\partial \psi}{\partial t} = 0 \tag{2.2}$$

$$\nabla^2 \psi - \frac{1}{c^2}\frac{\partial^2 \psi}{\partial t^2} = 0. \tag{2.3}$$

First we can separate off the time dependence. We assume a solution of the form

$$\psi(\mathbf{r}, t) = u(\mathbf{r})T(t). \tag{2.4}$$

Substituting this in the diffusion equation (2.2) and dividing by $\psi = uT$ leads to

$$\frac{1}{u(\mathbf{r})}\boldsymbol{\nabla}^2 u(\mathbf{r}) = -k^2 = \frac{1}{\kappa}\frac{1}{T(t)}\frac{\partial T(t)}{\partial t}, \tag{2.5}$$

where k^2 must be a constant since the first equality implies it is independent of t and the second equality implies it is independent of $\mathbf{r}$. The second part of (2.5) is a simple differential equation for $T(t)$ with solution

$$T = Ae^{-k^2\kappa t}, \tag{2.6}$$

and the first part of (2.5) is Helmholtz's equation for $u(\mathbf{r})$:

$$\boxed{\boldsymbol{\nabla}^2 u(\mathbf{r}) + k^2 u(\mathbf{r}) = 0.} \tag{2.7}$$

We can separate the time dependence in a similar way for the wave equation (2.3). Assuming a solution of the form (2.4), substituting in (2.3), and dividing by $\psi = uT$, we find

$$\frac{1}{u}\boldsymbol{\nabla}^2 u = -k^2 = \frac{1}{c^2}\frac{1}{T}\frac{\partial^2 T}{\partial t^2}, \tag{2.8}$$

with k^2 a constant independent of $\mathbf{r}$ and t. The second part of (2.8) is again a simple differential equation for the time dependence, with solution

$$T = Ae^{ikct} + Be^{-ikct}, \tag{2.9}$$

and for the space dependence we again obtain Helmholtz equation (2.7).

We note that Laplace's equation (2.1), which has no time dependence to begin with, is the special case of the Helmholtz equation (2.7) with $k^2 = 0$.

To proceed further, we must work in one or the other of the special coordinate systems.

2.3   RECTANGULAR COORDINATES (x, y, z)

In rectangular coordinates Helmholtz's equation has the form

$$\frac{\partial^2 u}{\partial x^2} + \frac{\partial^2 u}{\partial y^2} + \frac{\partial^2 u}{\partial z^2} + k^2 u = 0. \tag{2.10}$$

We assume a separated solution of the form

$$u = X(x)Y(y)Z(z), \tag{2.11}$$

substitute in (2.10), divide by $u = XYZ$, and find

$$\frac{1}{X}\frac{d^2 X}{dx^2} + \frac{1}{Y}\frac{d^2 Y}{dy^2} + \frac{1}{Z}\frac{d^2 Z}{dz^2} + k^2 = 0. \tag{2.12}$$

Since $X(x)$ is a function only of x, it is clear that $(1/X)d^2 X/dx^2$ is equal to a quantity

$$-k^2 - \frac{1}{Y}\frac{d^2 Y}{dy^2} - \frac{1}{Z}\frac{d^2 Z}{dz^2},$$

which is a function only of y and z. From these two statements we conclude that

$$\frac{1}{X}\frac{d^2X}{dx^2} = -k_1^2 \tag{2.13}$$

is a constant. Similarly we find

$$\frac{1}{Y}\frac{d^2Y}{dy^2} = -k_2^2, \tag{2.14}$$

$$\frac{1}{Z}\frac{d^2Z}{dz^2} = -k_3^2. \tag{2.15}$$

Substituting (2.13)-(2.15) in (2.12) leads to the relation

$$k_1^2 + k_2^2 + k_3^2 = k^2. \tag{2.16}$$

The differential equations (2.13)-(2.15) are easy to solve, and we find

$$X = A_1 e^{ik_1 x} + B_1 e^{-ik_1 x}$$

$$Y = A_2 e^{ik_2 y} + B_2 e^{-ik_2 y}$$

$$Z = A_3 e^{ik_3 x} + B_3 e^{-ik_3 z}. \tag{2.17}$$

Multiplying these together we obtain solutions for $u(\mathbf{r})$ of the form

$$\boxed{u(\mathbf{r}) = A e^{i\mathbf{k}\cdot\mathbf{r}}} \tag{2.18}$$

with $\mathbf{k}$ a three-dimensional vector satisfying (2.16),

$$\mathbf{k}^2 = k^2. \tag{2.19}$$

Note that all the various possibilities $\exp(\pm ik_1 x \pm ik_2 y \pm ik_3 z)$ are obtained from (2.18) by reversing the signs of the components of $\mathbf{k}$, an operation which does not affect (2.19).

2.4 CYLINDRICAL COORDINATES (r, θ, z)

In cylindrical coordinates, as shown in Figure 2.1, Helmholtz's equation assumes the form

$$\frac{\partial^2 u}{\partial r^2} + \frac{1}{r}\frac{\partial u}{\partial r} + \frac{1}{r^2}\frac{\partial^2 u}{\partial \theta^2} + \frac{\partial^2 u}{\partial z^2} + k^2 u = 0. \tag{2.20}$$

We assume a separated solution of the form

$$u = R(r)\Theta(\theta)Z(z), \tag{2.21}$$

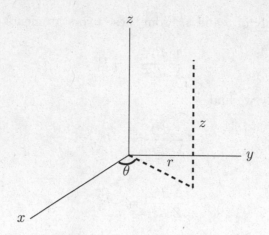

Figure 2.1: System of cylindrical coordinates

substitute in (2.20), divide by $u = R\Theta Z$, and we find

$$\frac{1}{R}\left[\frac{d^2R}{dr^2} + \frac{1}{r}\frac{dR}{dr}\right] + \frac{1}{r^2}\frac{1}{\Theta}\frac{d^2\Theta}{d\theta^2} + \frac{1}{Z}\frac{d^2Z}{dz^2} + k^2 = 0. \qquad (2.22)$$

The form of this equation implies

$$\frac{1}{\Theta}\frac{d^2\Theta}{d\theta^2} = -m^2, \qquad (2.23)$$

$$\frac{1}{Z}\frac{d^2Z}{dz^2} = \alpha^2 - k^2, \qquad (2.24)$$

$$\frac{1}{R}\left[\frac{d^2R}{dr^2} + \frac{1}{r}\frac{dR}{dr}\right] - \frac{m^2}{r^2} + \alpha^2 = 0, \qquad (2.25)$$

where α and m are constants.

The differential equations (2.23) and (2.24) are simple to solve and we find

$$\Theta = Ae^{im\theta} + Be^{-im\theta} = A'\cos m\theta + B'\sin m\theta \qquad (2.26)$$

$$Z = Ce^{\sqrt{\alpha^2-k^2}\,z} + De^{-\sqrt{\alpha^2-k^2}\,z} = C'\cosh\sqrt{\alpha^2-k^2}\,z + D'\sinh\sqrt{\alpha^2-k^2}\,z. \qquad (2.27)$$

For many (but not all) physical applications m must be an integer in order to ensure that Θ is a single valued function, i.e., $\Theta(\theta + 2\pi) = \Theta(\theta)$.

In the equation (2.25) for $R(r)$, if $\alpha \neq 0$, let $r = \rho/\alpha$ to obtain

$$\boxed{\frac{d^2R}{d\rho^2} + \frac{1}{\rho}\frac{dR}{d\rho} + \left(1 - \frac{m^2}{\rho^2}\right)R = 0.} \qquad (2.28)$$

This famous differential equation is called Bessel's equation. It has two linearly independent solutions, which we will study in some detail later:

$$R = GJ_m(\rho) + HN_m(\rho) = GJ_m(\alpha r) + HN_m(\alpha r). \tag{2.29}$$

The special case $\alpha = 0$ deserves special mention. This arises in the case of z independent solutions of Laplace's equation, for which $\alpha = k = 0$. The equation for $R(r)$ reduces to

$$\frac{d^2R}{dr^2} + \frac{1}{r}\frac{dR}{dr} - \frac{m^2}{r^2}R = 0. \tag{2.30}$$

One can easily find the elementary solutions of this equation:

$$R(r) = \begin{cases} Gr^m + Hr^{-m}, & m \neq 0, \\ G + H\ln r, & m = 0. \end{cases} \tag{2.31}$$

2.5 SPHERICAL COORDINATES (r, θ, ϕ)

In spherical coordinates, as shown in Figure 2.2, Helmholtz's equation assumes the form

$$\frac{1}{r}\frac{\partial^2(ru)}{\partial r^2} + \frac{1}{r^2\sin\theta}\left[\frac{\partial}{\partial\theta}\left(\sin\theta\frac{\partial u}{\partial\theta}\right) + \frac{1}{\sin\theta}\frac{\partial^2 u}{\partial\varphi^2}\right] + k^2u = 0. \tag{2.32}$$

We assume a separated solution of the form

$$u = R(r)Y(\theta, \varphi), \tag{2.33}$$

substitute in (2.32), divide by $u = RY$, and find

$$\frac{1}{R}\frac{1}{r}\frac{d^2}{dr^2}(rR) + \frac{1}{r^2}\frac{1}{Y\sin\theta}\left[\frac{\partial}{\partial\theta}\left(\sin\theta\frac{\partial Y}{\partial\theta}\right) + \frac{1}{\sin\theta}\frac{\partial^2 Y}{\partial\varphi^2}\right] + k^2 = 0. \tag{2.34}$$

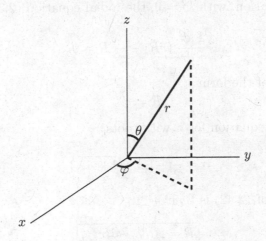

Figure 2.2: System of spherical coordinates

The form of this equation implies

$$\frac{1}{Y \sin\theta} \left[\frac{\partial}{\partial\theta}\left(\sin\theta \frac{\partial Y}{\partial\theta} \right) + \frac{1}{\sin\theta} \frac{\partial^2 Y}{\partial\varphi^2} \right] = -\lambda \tag{2.35}$$

and

$$\frac{1}{R} \frac{1}{r} \frac{d^2}{dr^2}(rR) + k^2 - \frac{\lambda}{r^2} = 0, \tag{2.36}$$

where λ is a constant.

Consider first the radial equation (2.36). If $k^2 \neq 0$, let $r = \rho/k$ and rewrite the equation in the form

$$\frac{d^2R}{d\rho^2} + \frac{2}{\rho}\frac{dR}{d\rho} + \left(1 - \frac{\lambda}{\rho^2} \right) R = 0. \tag{2.37}$$

This is almost Bessel's equation (2.28). To reduce it to the form of Bessel's equation, make the substitution

$$R = \frac{1}{\sqrt{\rho}} S. \tag{2.38}$$

After a short calculation one finds Bessel's equation for S:

$$\boxed{\frac{d^2S}{d\rho^2} + \frac{1}{\rho}\frac{dS}{d\rho} + \left(1 - \frac{\lambda + \frac{1}{4}}{\rho^2} \right) S = 0.} \tag{2.39}$$

Comparing with (2.28), (2.29) we see that this has the solution

$$S = A J_\beta(\rho) + B N_\beta(\rho), \quad \beta = \sqrt{\lambda + \frac{1}{4}}, \tag{2.40}$$

or equivalently

$$\boxed{R = A\frac{1}{\sqrt{kr}} J_\beta(kr) + B\frac{1}{\sqrt{kr}} N_\beta(kr).} \tag{2.41}$$

For Laplace's equation, with $k^2 = 0$, the radial equation (2.36) is much simpler:

$$\frac{1}{r}\frac{d^2}{dr^2}(rR) - \frac{\lambda}{r^2}R = 0. \tag{2.42}$$

Assuming a solution of the form

$$R = r^\alpha, \tag{2.43}$$

one finds a quadratic equation for α with roots

$$\alpha = \frac{1}{2}(-1 \pm \sqrt{1 + 4\lambda}). \tag{2.44}$$

The general solution of (2.42) is then

$$\boxed{R = A r^{\alpha_1} + B r^{\alpha_2},} \tag{2.45}$$

where α_1, α_2 are the two roots (2.44).

Returning now to the angular equation (2.35), we separate again,

$$Y = P(\theta)\Phi(\varphi), \tag{2.46}$$

and obtain

$$\frac{1}{P}\frac{1}{\sin\theta}\frac{d}{d\theta}\left(\sin\theta\frac{dP}{d\theta}\right) + \frac{1}{\sin^2\theta}\frac{1}{\Phi}\frac{d^2\Phi}{d\varphi^2} + \lambda = 0, \tag{2.47}$$

which implies

$$\frac{1}{\Phi}\frac{d^2\Phi}{d\varphi^2} = -m^2 \tag{2.48}$$

and

$$\frac{1}{\sin\theta}\frac{d}{d\theta}\left(\sin\theta\frac{dP}{d\theta}\right) + \left(\lambda - \frac{m^2}{\sin^2\theta}\right)P = 0. \tag{2.49}$$

The ϕ equation (2.48) has a simple solution

$$\boxed{\Phi = Ae^{im\varphi} + Be^{-im\varphi}.} \tag{2.50}$$

The θ equation (2.49) is Legendre's equation. To eliminate the trigonometric functions let $x = \cos\theta$. Equation (2.49) then assumes the form

$$\frac{d}{dx}\left[(1-x^2)\frac{dP}{dx}\right] + \left[\lambda - \frac{m^2}{1-x^2}\right]P = 0 \tag{2.51}$$

or

$$\boxed{\frac{d^2P}{dx^2} - \frac{2x}{1-x^2}\frac{dP}{dx} + \frac{1}{1-x^2}\left[\lambda - \frac{m^2}{1-x^2}\right]P = 0,} \tag{2.52}$$

which has two linearly independent solutions to be studied in detail later:

$$\boxed{P = CP_l^m(x) + DQ_l^m(x), \quad l(l+1) = \lambda.} \tag{2.53}$$

We have considered here only the three most common coordinate systems. Evidently the method can be employed in the case of the more exotic coordinate systems such as elliptic coordinates.

2.6 SERIES SOLUTIONS OF ORDINARY DIFFERENTIAL EQUATIONS: PRELIMINARIES

We have seen that the separation of variables procedure reduces a partial differential equation to ordinary differential equations. In some cases the ordinary differential equations were trivial. In Sections 2.2-2.5 we came across two nontrivial cases: Bessel's equation (2.28) and Legendre's equation (2.52). For these and other cases we need a more general method of obtaining a solution.

We shall study the general class of second-order linear differential equations of the form

$$\frac{d^2u}{dx^2} + p(x)\frac{du}{dx} + q(x)u = 0. \tag{2.54}$$

We restrict ourselves to the case of a real variable x and assume that in some region in x of interest $p(x)$ and $q(x)$ are analytic functions except for isolated poles. When we say that a function $f(x)$ is analytic at a point x_0 we shall mean that it can be expanded in a Taylor series

$$f(x) = \sum_{n=0}^{\infty} \frac{1}{n!} f^{(n)}(x_0)(x - x_0)^n \tag{2.55}$$

which converges in some neighborhood of the point x_0. A function is analytic in a region if it is analytic at every point in the region. Thus we assume that $p(x)$ and $q(x)$ can be expanded in convergent Taylor series, except that they cannot be expanded about certain special points, say $x = a$, near which they behave like

$$\frac{1}{x - a} \quad \text{first-order pole}$$

or

$$\frac{1}{(x - a)^2} \quad \text{second-order pole},$$

etc.[1]

If $p(x)$ and $q(x)$ are analytic near x_0, this is called an <u>ordinary point</u> of the differential equation. If $p(x)$ and/or $q(x)$ have poles at x_0 but

$$(x - x_0)p(x) \quad \text{and} \quad (x - x_0)^2 q(x)$$

are analytic near x_0, x_0 is called a <u>regular singular point</u> of the differential equation. If the singularities are worse than this, we have an <u>irregular singular point</u>. As we shall see shortly, near ordinary points and regular singular points (except when the roots of the indicial equation differ by an integer) we can find two linearly independent power series solutions of the form

$$(x - x_0)^\alpha \sum_{n=0}^{\infty} a_n (x - x_0)^n.$$

Before actually finding these solutions, we dispose of a few preliminaries. A quantity which is sometimes useful is the Wronskian. Suppose we have two solutions $u_1(x)$ and $u_2(x)$ of the differential equation (2.54). Then the Wronskian is defined by

$$W(x) = u_1(x)\frac{du_2(x)}{dx} - u_2(x)\frac{du_1(x)}{dx} = \begin{vmatrix} u_1(x) & u_2(x) \\ u_1'(x) & u_2'(x) \end{vmatrix}. \tag{2.56}$$

Two solutions of the differential equation are linearly dependent if and only if their Wronskian vanishes identically. For if they are linearly dependent there exist two constants A and B, not both zero, such that

$$Au_1(x) + Bu_2(x) = 0. \tag{2.57}$$

---

[1] We note that the developments which follow can be carried through in the complex plane with p and q being analytic functions except for poles, the words now having their usual significance in complex variable theory.

Differentiating, we find

$$Au_1'(x) + Bu_2'(x) = 0. \tag{2.58}$$

The condition that (2.57) and (2.58) have a solution for A and B, with not both A and B zero, is

$$W = \begin{vmatrix} u_1 & u_2 \\ u_1' & u_2' \end{vmatrix} = 0. \tag{2.59}$$

On the other hand, if $W = 0$, we have

$$u_1 u_2' = u_2 u_1', \tag{2.60}$$

which implies

$$\frac{d}{dx}\left[\frac{u_2}{u_1}\right] = 0, \quad \frac{u_2}{u_1} = C. \tag{2.61}$$

We can find a differential equation for the Wronskian. The functions $u_1(x)$ and $u_2(x)$ are assumed to be two solutions of (2.54):

$$\frac{d^2 u_1}{dx^2} + p\frac{du_1}{dx} + qu_1 = 0, \tag{2.62}$$

$$\frac{d^2 u_2}{dx^2} + p\frac{du_2}{dx} + qu_2 = 0 \tag{2.63}$$

Multiple (2.62) by u_2, (2.63) by u_1 and subtract to obtain

$$\frac{dW}{dx} + pW = 0. \tag{2.64}$$

This has the solution

$$W(x) = W(x_1)\exp\left(-\int_{x_1}^{x} p(x')dx'\right). \tag{2.65}$$

We can use this result to find a second solution of our differential equation if we already know one solution:

$$u_1^2\frac{d}{dx}\left[\frac{u_2}{u_1}\right] = W, \tag{2.66}$$

$$u_2(x) = u_1(x)\int_{x_1}^{x} dx'\frac{W(x')}{u_1^2(x')}, \tag{2.67}$$

where W is given by (2.65).

If we have two linearly independent solutions $u_1(x)$ and $u_2(x)$ of (2.54), we can construct from them a solution

$$u_3(x) = Au_1(x) + Bu_2(x) \tag{2.68}$$

which has an arbitrarily assigned value and derivative at some point x_0. We want to choose A and B such that

$$u_3(x_0) = Au_1(x_0) + Bu_2(x_0),$$
$$u_3'(x_0) = Au_1'(x_0) + Bu_2'(x_0). \tag{2.69}$$

Since the Wronskian does not vanish for linearly independent solutions, we can solve by Cramer's rule for A and B. We can now prove the following theorem.

Theorem 2.6.1. *There can be no more than two linearly independent analytic solutions of the differential equation* (2.54) *in the vicinity of a point x_0.*

Proof. Suppose there were three linearly independent solutions $u_1(x)$, $u_2(x)$ and $u_3(x)$. Then $u_3(x)$ has a certain value and derivative at x_0 and we know that we can field A and B such that (2.69) is satisfied. From (2.69) and the differential equation (2.54) for $u_1(x)$, $u_2(x)$ and $u_3(x)$ we then find

$$u_3''(x_0) = -p(x_0)u_3'(x_0) - q(x_0)u_3(x_0)$$
$$= Au_1''(x_0) + Bu_2''(x_0). \tag{2.70}$$

Similarly, from derivatives of the differential equation we have

$$u_3'''(x_0) = Au_1'''(x_0) + Bu_2'''(x_0),$$
$$\vdots$$
$$u_3^{(n)}(x_0) = Au_1^{(n)}(x_0) + Bu_2^{(n)}(x_0). \tag{2.71}$$

Since the functions are all assumed to be analytic in the neighborhood of x_0, we can expand in a Taylor series:

$$u_3(x) = u_3(x_0) + u_3'(x_0)(x - x_0) + \frac{1}{2}u_3''(x_0)(x - x_0)^2 + \dots$$
$$= Au_1(x) + Bu_2(x).$$

Thus as soon as we find two linearly independent solutions of our differential equation (2.54) we are through because we can express any other solution in terms of them. □

2.7 EXPANSION ABOUT A REGULAR SINGULAR POINT

We can now carry through the power series solution of (2.54) in the neighborhood of a regular singular point x_0. Near such a point $p(x)$ and/or $q(x)$ may be singular, but $(x - x_0)p(x)$ and $(x - x_0)^2 q(x)$ are analytic. Thus we multiply (2.54) by $(x - x_0)^2$ to obtain

$$(x - x_0)^2 \frac{d^2u}{dx^2} + (x - x_0)P(x)\frac{du}{dx} + Q(x)u(x), \tag{2.72}$$

where

$$P(x) = (x - x_0)p(x) = \sum_{n=0}^{\infty} p_n(x - x_0)^n = p_0 + p_1(x - x_0) + p_2(x - x_0)^2 + \dots \tag{2.73}$$

and

$$Q(x) = (x - x_0)^2 q(x) = \sum_{n=0}^{\infty} q_n (x - x_0)^n = q_0 + q_1(x - x_0) + q_2(x - x_0)^2 + \dots \quad (2.74)$$

are analytic in the neighborhood of x_0 and so have Taylor series expansions. Note that an ordinary point is the special case with

$$p_0 = q_0 = q_1 = 0. \quad (2.75)$$

We try to find a series solution of (2.72) of the form

$$u(x) = (x - x_0)^\alpha \left[1 + \sum_{n=1}^{\infty} u_n (x - x_0)^n \right]. \quad (2.76)$$

Here the leading term in the series inside the bracket has been normalized to unity; this represents no loss in generality for a linear equation like (2.72). The exponent α is to be determined and sometimes will be nonintegral. Substituting (2.73), (2.74) and (2.76) in (2.72), we find

$$(x - x_0)^\alpha \left[\alpha(\alpha - 1) + \sum_{n=1}^{\infty} (n + \alpha)(n + \alpha - 1) u_n (x - x_0)^n \right]$$

$$+ (x - x_0)^\alpha \left[\alpha + \sum_{n=1}^{\infty} (n + \alpha) u_n (x - x_0)^n \right] \times \sum_{n=0}^{\infty} p_n (x - x_0)^n$$

$$+ (x - x_0)^\alpha \left[1 + \sum_{n=1}^{\infty} u_n (x - x_0)^n \right] \times \sum_{n=0}^{\infty} q_n (x - x_0)^n$$

$$= 0. \quad (2.77)$$

Equating coefficients of successive powers of $(x - x_0)$ to zero, with an intent toward solving for α, we obtain

$$\alpha(\alpha - 1) + p_0 \alpha + q_0 = 0 \quad (2.78)$$

$$u_1 \left[(\alpha + 1)\alpha + p_0(\alpha + 1) + q_0 \right] + \alpha p_1 + q_1 = 0,$$

$$u_2 \left[(\alpha + 2)(\alpha + 1) + p_0(\alpha + 2) + q_0 \right] + u_1 \left[(\alpha + 1)p_1 + q_1 \right] + \alpha p_2 + q_2 = 0,$$

$$\vdots$$

$$u_n \left[(\alpha + n)(\alpha + n - 1) + p_0(\alpha + n) + q_0 \right] + \dots = 0. \quad (2.79)$$

$$\vdots$$

The equation (2.78) is called the indicial equation. Since it is quadratic, it will in general have two roots:

$$\alpha_1, \ \alpha_2.$$

For each root we can then solve the remaining equations (2.79) successively for $u_1, u_2, \ldots u_n, \ldots$. Thus we will obtain two solutions of the form (2.76) for the differential equation. We note the following points:

1) If x_0 is an ordinary point, (2.75) holds and the two solutions of the indicial equation are $\alpha = 0, 1$. Thus we have two solutions of the form

$$u = \sum_{n=0}^{\infty} C_n(x - x_0)^n, \tag{2.80}$$

with only integral powers of n. These functions are analytic near $x = x_0$. (See point (4) below.)

2) It is easy to verify that for $p(x)$, $q(x)$ more singular than assumed above, i.e., for expansion about an irregular singular point, the term $\alpha(\alpha - 1)$ would not appear in the indicial equation. Thus the indicial equation would have at most one root, so we would be able to find at most one, and in some cases no, solution of the assumed form. Evidently the treatment of the irregular singular point is more difficult.

3) Even in the case of the regular singular point there is a difficulty if the roots of the indicial equation differ by zero or an integer. If the two roots are the same, we obtain only one solution. If

$$\alpha_1 - \alpha_2 = n = \text{positive integer}, \tag{2.81}$$

then since α_1 satisfies the indicial equation (2.78) we find

$$(\alpha_2 + n)(\alpha_2 + n - 1) + p_0(\alpha_2 + n) + q_0 = 0. \tag{2.82}$$

Since this quantity is the coefficient of u_n in the nth of the equations (2.79), we see that it will not be possible to solve this equation for u_n. If the other terms in the equation are nonzero, the equation yields $u_n = \infty$ and the procedure fails; in this case only α_1, the larger of the two roots of the indicial equation, will yield a solution of the differential equation. If the other terms in the equation vanish, u_n is not determined; in this case we can find two solutions of the differential equation. The series corresponding to α_2 contains an arbitrary constant u_n which just multiplies an admixture of the solution corresponding to α_1. The ordinary point provides an example of this second case.

4) It is shown in mathematical works on differential equations[2] that the series in the square bracket in (2.76) actually converges and defines an analytic function in some neighborhood of the point x_0. The factor $(x - x_0)^\alpha$ has a branch point singularity for α nonintegral.

---

[2] See for example E. T. Whittaker and G. N. Watson, A Course of Modern Analysis (Cambridge University Press, London, 1952), page 199.

Consider now how to obtain a second solution near a regular singular point when the roots of the indicial equation differ by zero or an integer. We can use the formulas (2.65), (2.67) which give a second solution in terms of a first solution and the Wronskian. Using (2.73), which depends on the assumption that x_0 is a regular singular point, we find

$$p(x') = \frac{p_0}{x' - x_0} + p_1 + p_2(x' - x_0) + \cdots,$$

$$\int_{x_1}^{x} p(x') dx' = p_0 \ln(x - x_0) + p_1(x - x_0) + \frac{p_2(x - x_0)^2}{2} + \cdots + \text{const.},$$

$$W(x) = \frac{\text{const.}}{(x - x_0)^{p_0}} \exp\left[-p_1(x - x_0) - \frac{p_2(x - x_0)^2}{2} + \cdots\right]. \tag{2.83}$$

Also we have

$$u_1(x) = (x - x_0)^{\alpha_1} \left[1 + \sum_n u_n(x - x_0)^n\right], \tag{2.84}$$

where α_1 is the larger root of the indicial equation. Substituting (2.83) and (2.84) in (2.67), we find

$$u_2(x) = \text{const.} \, u_1(x) \int_{x_1}^{x} dx' \frac{1}{(x' - x_0)^{p_0 + 2\alpha_1}} f(x'), \tag{2.85}$$

where $f(x')$ is an analytic function which can be expanded as

$$f(x') = 1 + \sum_n f_n(x' - x_0)^n. \tag{2.86}$$

From the indicial equation $\alpha^2 + (p_0 - 1)\alpha + q_0 = 0$ we can see that the sum of the two roots of the indicial equation is given by $\alpha_1 + \alpha_2 = 1 - p_0$ so that $p_0 + 2\alpha_1 = 1 + s$ with $s = \alpha_1 - \alpha_2$. Substituting (2.86) in (2.85) and integrating term-by-term, we find

$$u_2(x) = \text{const.} \, u_1(x) \left[-\frac{1}{s}\frac{1}{(x - x_0)^2} - \frac{f_1}{s - 1}\frac{1}{(x - x_0)^{s-1}} + \right.$$

$$\left. \cdots + f_s \ln(x - x_0) + f_{s+1}(x - x_0) + \cdots + \text{const.}\right]$$

$$= \text{const.} \, f_s u_1(x) \ln(x - x_0) + (x - x_0)^{\alpha_2} \sum_{n=0}^{\infty} g_n(x - x_0)^n. \tag{2.87}$$

For the case $\alpha_1 = \alpha_2$ the last term in (2.87) can be replaced by $(x - x_0)^{\alpha_1 + 1}$ $\sum_{n=0}^{\infty} g_n(x - x_0)^n$; this amounts to the addition of some multiple of the other solution $u_1(x)$. The significance of the formula (2.87) is that it tells us the form of the solution. The assumption (2.76) did not work because we did not include the logarithmic term $u_1(x)\ln(x - x_0)$. To actually proceed, first find the solution $u_1(x)$. To then find $u_2(x)$, substitute a series of the form (2.87) into the differential equation and determine the coefficients g_n so as to satisfy the equation.

2.8   STURM-LIOUVILLE EIGENVALUE PROBLEM

In the previous sections we have developed a technique (expansion in series) for solving linear second-order differential equations. Frequently one uses this technique to solve a Sturm-Liouville eigenvalue problem. This is the problem of finding solutions of a differential equation of the form

$$\mathcal{L}u_n(x) \equiv \frac{d}{dx}\left[p(x)\frac{du_n(x)}{dx}\right] - q(x)u_n(x) = -\lambda_n\rho(x)u_n(x) \qquad (2.88)$$

subject to specified boundary conditions at the beginning and end of some interval (a, b) on the real line, e.g.,

$$Au(a) + Bu'(a) = 0, \qquad (2.89)$$

$$Cu(b) + Du'(b) = 0, \qquad (2.90)$$

where A, B, C, D are given constants. The form assumed for the differential operator $\mathcal{L}$ in (2.88) is rather general since after multiplication by a suitable factor any second-order linear differential operator can be put in this form. The form (2.89), (2.90) for the boundary conditions is also rather general, although other possibilities such as periodic boundary conditions

$$u(x) = u(x + b - a) \qquad (2.91)$$

are sometimes of interest.

The eigenvalue problem (2.88)-(2.90) is an infinite dimensional generalization of the finite dimensional matrix eigenvalue problem

$$\mathbb{M}\mathbf{u} = \lambda\mathbf{u} \qquad (2.92)$$

with $\mathbb{M}$ an $n \times n$ matrix and $\mathbf{u}$ an n-dimensional column vector. As in the matrix case (2.88) - (2.90) will have solutions only for certain values of the eigenvalue λ_n. The solutions u_n corresponding to these λ_n are the eigenfunctions. For the finite dimensional case with an $n \times n$ matrix $\mathbb{M}$, there can be at most n linearly independent eigenfunctions. For the Sturm- Liouville case (2.88)-(2.90) there will in general be an infinite set of eigenvalues λ_n with corresponding eigenfunctions $u_n(x)$.

The differential equations derived by separating variables in Sections 2.3-2.5 are all of the form (2.88), the separation constants being the eigenvalue parameters λ. The boundary conditions (2.89), (2.90) are determined by the physical application under study.

We can illustrate all these remarks with a simple example which is a special case of (2.88)-(2.90):

$$\mathcal{L}u_n(x) \equiv \frac{d^2}{dx^2}u_n(x) = -k_n^2 u_n(x), \quad \text{for} \tag{2.93}$$

$$u(0) = 0, \tag{2.94}$$
$$u(b) = 0. \tag{2.95}$$

The solutions of (2.93) are

$$u_n(x) = A_n \sin k_n x + B_n \cos k_n x. \tag{2.96}$$

To satisfy (2.94) we must have $B_n = 0$. The eigenfunctions are then of the form

$$u_n(x) = A_n \sin k_n x, \tag{2.97}$$

and the boundary condition (2.95) determines the eigenvalues:

$$k_n = \frac{n\pi}{b}, \quad n = 0, 1, 2 \dots. \tag{2.98}$$

We note that there are an infinite number of eigenvalues and eigenfunctions. The multiplicative constant A_n in (2.97) is not determined by the linear equations (2.88)-(2.90). Usually it is fixed by some nonlinear normalization condition such as

$$\int_0^b dx\, [u_n(x)]^2 = 1. \tag{2.99}$$

The differential equation (2.93) also describes the variation with azimuthal angle (θ in Section 2.4, φ in Section 2.5) when the variables are separated in the Helmholtz equation:

$$\frac{d^2}{d\varphi^2}u_m(\varphi) = -m^2 u_m(\varphi). \tag{2.100}$$

In physical problems where the complete range of φ is available to the system, i.e., not a wedge, the boundary condition is usually that $u_m(\varphi)$ be single valued:

$$u_m(\varphi) = u_m(\varphi + 2\pi). \tag{2.101}$$

In this case the eigenvalue m must be an integer

$$m = 0, \pm 1, \pm 2, \dots \tag{2.102}$$

to ensure that the eigenfunctions

$$u_m = \begin{cases} \sin m\varphi \\ \cos m\varphi \end{cases} \tag{2.103}$$

be single valued.

The solutions $u_n(x)$, λ_n of a Sturm-Liouville eigenvalue problem have some general properties of basic importance. First consider the differential operator $\mathcal{L}$ of (2.88). We note that if $u(x)$ and $v(x)$ are arbitrary twice differentiable functions and

$$\mathcal{L}u \equiv \frac{d}{dx}\left[p(x)\frac{du(x)}{dx}\right] - q(x)u(x),$$

$$\mathcal{L}v \equiv \frac{d}{dx}\left[p(x)\frac{dv(x)}{dx}\right] - q(x)v(x), \tag{2.104}$$

we find by integrating by parts

$$\int_a^b dx\, [v\mathcal{L}u - u\mathcal{L}v] = \left[p\left(v\frac{du}{dx} - u\frac{dv}{dx}\right)\right]_a^b. \tag{2.105}$$

An operator $\mathcal{L}$ which satisfies (2.105) is said to be self-adjoint. We have noted above that any second-order linear differential operator can be put in this self-adjoint form by multiplication by a suitable factor.

It is easy to show that for functions $u(x)$, $v(x)$ satisfying boundary conditions of the form (2.89), (2.90) or for functions $u(x)$, $v(x)$ satisfying the periodic boundary condition (2.91) the right-hand side of (2.105) vanishes. For both of these cases we then have

$$\int_a^b dx\, v\mathcal{L}u = \int_a^b dx\, u\mathcal{L}v. \tag{2.106}$$

Consider now two different eigenfunctions $u_n(x)$, $u_m(x)$ belonging to different eigenvalues $\lambda_n \neq \lambda_m$:

$$\mathcal{L}u_n(x) = -\lambda_n\rho(x)u_n(x), \tag{2.107}$$
$$\mathcal{L}u_m(x) = -\lambda_m\rho(x)u_m(x). \tag{2.108}$$

Multiplying (2.107) by u_m, (2.108) by u_n, integrating and subtracting, we find

$$\int_a^b dx\, [u_m\mathcal{L}u_n - u_n\mathcal{L}u_m] = -(\lambda_n - \lambda_m)\int_a^b dx\, \rho(x)u_n(x)u_m(x). \tag{2.109}$$

As we have shown in (2.106) the left-hand side of (2.109) will vanish for either set of boundary conditions, (2.89), (2.90) or (2.91), so for either of these cases we find

$$\boxed{\int_a^b dx\, \rho(x)u_n(x)u_m(x) = 0, \quad \lambda_n \neq \lambda_m.} \tag{2.110}$$

Two functions $u_n(x)$, $u_m(x)$ satisfying this condition are said to be orthogonal with weight function $\rho(x)$. If $\rho(x)$ is nonnegative we can normalize the u_n:

$$\boxed{\int_a^b dx\, \rho(x)[u_n(x)]^2 = 1.} \tag{2.111}$$

The most important property of the eigenfunctions of a Sturm-Liouville problem is that they form a <u>complete set</u>. This means that an arbitrary function $\psi(x)$ can be expanded in an infinite series of the form

$$\psi(x) = \sum_n a_n u_n(x). \tag{2.112}$$

The expansion coefficients a_n are determined by multiplying (2.112) by $\rho(x)u_n(x)$, integrating term-by-term, and using the orthogonality relation (2.110):

$$a_n = \frac{\int_a^b dx\ \rho(x)u_n(x)\psi(x)}{\int_a^b dx\ \rho(x)[u_n(x)]^2}. \tag{2.113}$$

A precise mathematical discussion of the completeness of the eigenfunctions $u_n(x)$ and the validity of the expansion (2.112) is quite involved and certainly well beyond the scope of the present book. Detailed discussions are given by Courant and Hilbert,[3] Titchmarsh,[4] and Stakgold.[5] The latter authors discuss the interesting singular cases in which the integration region is infinite or one of the functions in the differential equation contains singularities. We content ourselves here with quoting a few results from Courant and Hilbert. Normalize the eigenfunctions,

$$\boxed{\int_a^b dx\ \rho(x)[u_n(x)]^2 = 1,} \tag{2.114}$$

so that the expansion (2.112), (2.113) assumes the form

$$\boxed{\psi(x) = \sum_n c_n u_n(x),} \tag{2.115}$$

$$\boxed{c_n = \int_a^b dx\ \rho(x)u_n(x)\psi(x).} \tag{2.116}$$

According to Courant and Hilbert[6] we have the following theorem.

Theorem 2.8.1. *The system of eigenfunctions $u_n(x)$ is complete in the following sense: For any continuous function $\psi(x)$ and any positive ϵ, arbitrarily small, we can find a finite linear combination*

$$\alpha_1 u_1(x) + \alpha_2 u_2(x) + \ldots + \alpha_N u_N(x) = S_N \tag{2.117}$$

---

[3]R. Courant and D. Hilbert, <u>Methods of Mathematical Physics</u> (Interscience Publishers, New York, 1953), Vol. 1.

[4]E. C. Titchmarsh, <u>Eigenfunction Expansions Associated with Second Order Differential Equations</u> (Oxford University Press, New York, 1962), 2 volumes.

[5]I. Stakgold, <u>Boundary Value Problems of Mathematical Physics</u> (Macmillan, New York, 1967), Vol. 1.

[6]R. Courant and D. Hilbert, <u>op. cit.</u>, p. 424.

of the eigenfunctions such that

$$\int_a^b dx \, \rho(x)[\psi(x) - S_N]^2 < \epsilon. \tag{2.118}$$

We obtain the best approximation for a given N, i.e., the smallest value for the left-hand side of (2.118) with the coefficients

$$\alpha_n = c_n = \int_a^b dx \, \rho(x) u_n(x) \psi(x). \tag{2.119}$$

For these coefficients we have

$$\lim_{N \to \infty} \int_a^b dx \, \rho(x) \, [\psi(x) - S_N]^2 = \int_a^b dx \, \rho(x)[\psi(x)]^2 - \sum_{n=1}^{\infty} c_n^2 = 0. \tag{2.120}$$

The result (2.120) is described by saying that $\sum_n c_n u_n(x)$ converges in the mean to $\psi(x)$. For the one-dimensional cases under discussion here, this theorem can be extended to an actual expansion theorem. According to Courant and Hilbert[7] we have the following theorem.

Theorem 2.8.2. *Every piecewise continuous function defined in the fundamental domain with a square-integrable first derivative may be expanded in an eigenfunction series which converges absolutely and uniformly in all subdomains free of points of discontinuity; at the points of discontinuity it represents (like the Fourier series) the arithmetic mean of the right- and left-hand limits. (It should be remarked that this theorem does not require that the functions expanded satisfy the boundary conditions.)*

Finally we mention the general result which in the physics literature is called the completeness relation. Assuming normalized eigenfunctions, (2.114), substitute (2.116) in (2.115) to obtain

$$\psi(x) = \int_a^b dx' \left\{ \rho(x') \sum_n u_n(x) u_n(x') \right\} \psi(x'). \tag{2.121}$$

Since $\psi(x)$ is an arbitrary function, this implies

$$\boxed{\rho(x') \sum_n u_n(x) u_n(x') = \delta(x - x'),} \tag{2.122}$$

where δ is the so-called Dirac δ function, illustrated in Figure 2.3:

$$\delta(x - a) = \begin{cases} 0, & x \neq a \\ \infty, & x = a \end{cases}$$

in such a way that

---

[7]Ibid., p. 426, p. 359.

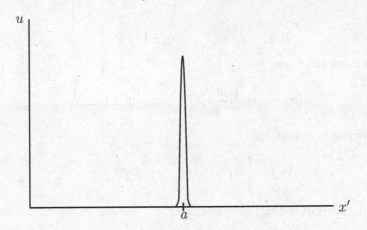

Figure 2.3: Dirac delta function centered at a

$$\int dx \; \delta(x - a) = 1. \tag{2.123}$$

An alternate way to obtain the completeness relation (2.122) is to apply the expansion theorem (2.115), (2.116) to the special function $\delta(x - x')$ and use the general formula

$$\int dx \; \delta(x - x')f(x) = f(x') \tag{2.124}$$

to perform the integral in (2.116).

2.9 FOURIER SERIES AND INTEGRALS

The most familiar examples of complete sets of functions are the Fourier series and Fourier integrals. Consider the Sturm-Liouville eigenvalue problem

$$\frac{d^2 u_n(x)}{dx^2} = -k_n^2 u_n(x) \tag{2.125}$$

on an interval

$$a < x < a + T \tag{2.126}$$

with a periodic boundary condition

$$u_n(x) = u_n(x + T). \tag{2.127}$$

The solutions of (2.125) are

$$u_n(x) = \begin{cases} e^{ik_n x} \\ \\ e^{-ik_n x} \end{cases}. \tag{2.128}$$

To satisfy the periodic boundary condition, we must have

$$e^{ik_n x} = e^{ik_n(x+T)}, \tag{2.129}$$

which implies

$$k_n T = 2\pi n, \quad n = 0, \pm 1, \pm 2, \ldots . \tag{2.130}$$

The eigenfunctions are then

$$u_n(x) = \exp\left[i\frac{2\pi n}{T}x\right], \quad n = 0, \pm 1, \pm 2, \ldots \tag{2.131}$$

with the orthogonality relation

$$\int_a^{a+T} dx\ u_n^*(x)u_m(x) = T\delta_{nm}. \tag{2.132}$$

The expansion theorem is the Fourier series

$$\psi(x) = \sum_{n=-\infty}^{\infty} c_n \exp\left[i\frac{2\pi n}{T}x\right]$$

$$= a_0 + \sum_{n=1}^{\infty}\left[a_n \cos\left(\frac{2\pi n}{T}x\right) + b_n \sin\left(\frac{2\pi n}{T}x\right)\right]. \tag{2.133}$$

Using the orthogonality relation (2.132) we can find the expansion coefficients:

$$c_n = \frac{1}{T}\int_a^{a+T} dx\ \exp\left[-i\frac{2\pi n}{T}x\right]\psi(x). \tag{2.134}$$

At the heuristic level the Fourier integral may be obtained from the Fourier series by passing to the limit of an infinite interval. Take $a = -T/2$ so that the interval is $-T/2 < x < T/2$, and let $T \to \infty$. Let

$$k = \frac{2\pi n}{T}, \tag{2.135}$$

and note that the interval

$$dk = \frac{2\pi(n+1)}{T} - \frac{2\pi n}{T} = \frac{2\pi}{T}dn \tag{2.136}$$

between allowed k values becomes smaller and smaller so that the series (2.133) may be replaced by an integral,

$$\sum_n \to \int dn = \frac{T}{2\pi}\int dk. \tag{2.137}$$

Equations (2.133) and (2.134) thus assume the form

$$\psi(x) = \frac{T}{2\pi}\int_{-\infty}^{\infty} dk\ c(k)e^{ikx}, \tag{2.138}$$

$$c(k) = \frac{1}{T}\int_{-\infty}^{\infty} dx\ e^{-ikx}\psi(x), \tag{2.139}$$

or, if we let $\widetilde{\psi}(k) = c(k)\,T/\sqrt{2\pi}$,

$$\psi(x) = \frac{1}{\sqrt{2\pi}} \int_{-\infty}^{\infty} dk\ \widetilde{\psi}(x)e^{ikx}, \tag{2.140}$$

$$\widetilde{\psi}(k) = \frac{1}{\sqrt{2\pi}} \int_{-\infty}^{\infty} dx\ e^{-ikx}\psi(x). \tag{2.141}$$

This is the Fourier integral theorem for expansion of an arbitrary function $\psi(x)$ on the infinite interval $-\infty < x < \infty$ in the eigenfunctions e^{ikx} of the operator d^2/dx^2:

$$\frac{d^2}{dx^2}e^{ikx} = -k^2 e^{ikx}. \tag{2.142}$$

For this case of an infinite interval, we have a continuous distribution of the eigenvalues k^2 rather than a discrete set.

The completeness relation for the Fourier integral (or Fourier transform of the δ function) is easily obtained by the method outlined at the end of Section 2.8 and one finds

$$\delta(x - x') = \frac{1}{2\pi} \int_{-\infty}^{\infty} dk\ e^{ik(x-x')}. \tag{2.143}$$

2.10 NUMERICAL SOLUTION OF ORDINARY DIFFERENTIAL EQUATIONS

Numerical methods provide a useful and often unavoidable alternative to the analytic methods discussed above for the solution of ordinary differential equations and eigenvalue problems. This is a big subject, and we can only scratch the surface here. Fortunately, libraries exist for programming languages for the solution of differential equations, which will work in all but the most pathological cases, so that all the user need supply is a subroutine to calculate the particular functions in his differential equation.

A second-order differential equation (linear or nonlinear) can be solved for the second derivative:

$$u''(x) = F(x, u, u'). \tag{2.144}$$

This can be further reduced to a pair of first-order equations

$$\begin{aligned} u' &= v, \\ v' &= F(x, u, v). \end{aligned} \tag{2.145}$$

Since the generalization to a pair of equations is straightforward, we shall consider in what follows only a single first-order equation

$$y' = G(x, y) \tag{2.146}$$

with the given initial value

$$y(x_0) = y_0. \tag{2.147}$$

The most direct method for numerically solving (2.146), (2.147) would be to approximate

$$y' = \frac{dy}{dx} \approx \frac{\Delta y}{\Delta x} \tag{2.148}$$

so that (2.146) becomes

$$\Delta y \approx G(x, y)\Delta x. \tag{2.149}$$

The solution $y_n = y(x_n)$ at successive grid points

$$x_n = x_0 + nh, \tag{2.150}$$

separated by an interval h, is then given by

$$y_{n+1} \approx y_n + G(x_n, y_n)h. \tag{2.151}$$

For a given size interval h, more accurate results can be obtained by using an interpolation formula for the derivative more accurate than merely its value $G(x_n, y_n)$ at the beginning of the interval. One of many possibilities is the Runge-Kutta scheme:

$$y_{n+1} \approx y_n + \frac{1}{6}\left(k_0 + 2k_1 + 2k_2 + k_3\right), \tag{2.152}$$

where

$$k_0 = hG(x_n, y_n),$$

$$k_1 = hG\left(x_n + \frac{1}{2}h, y_n + \frac{1}{2}k_0\right),$$

$$k_2 = hG\left(x_n + \frac{1}{2}h, y_n + \frac{1}{2}k_1\right),$$

$$k_3 = hG\left(x_n + h, y_n + k_2\right). \tag{2.153}$$

One can show by expanding all quantities in Taylor series in h that these formulas are accurate through terms of order h^4. Also, the formula (2.152) reduces to Simpson's rule for G independent of y.

The Runge-Kutta and other more elaborate schemes are described in detail in works on numerical methods.[8] An important question discussed in these works is that of stability. The differential equation is approximated by a difference equation. If the order of the difference equation is greater than that of the differential equation, it will have extraneous solutions. If some of these extraneous solutions grow in the direction of integration, the numerical integration scheme is unstable. Fortunately, with modern computers it is easy to check on questions of accuracy and stability by

---

[8]See, for example, F. B. Hildebrand, Introduction to Numerical Analysis (McGraw-Hill, New York, 1956), Chap. 6; C. W. Gear, Initial Value Problems in Ordinary Differential Equations (Prentice-Hall, Englewood Cliffs, N.Y., 1971).

reducing the step size h and repeating the calculation. Some of the more elaborate programs for solving differential equations perform this sort of check during the course of the calculation and automatically reduce the step size h as much as needed in regions where the solution of the differential equation tends to be inaccurate and/or in regions where the functions in the differential equation vary rapidly.

Given numerical integration scheme such as the Runge-Kutta method, we can use it to solve a Sturm-Liouville eigenvalue problem, (2.88)-(2.90). To do this, we proceed as follows: Assume some value λ for the eigenvalue parameter and integrate the differential equation

$$\frac{d}{dx}\left[p(x)\frac{du}{dx}\right] - q(x)u(x) = -\lambda\rho(x)u(x), \tag{2.154}$$

starting at the point $x = a$ with the boundary condition (2.89):

$$Au(a) + Bu'(a) = 0. \tag{2.155}$$

The normalization of $u(x)$ is not fixed by the linear homogeneous equations, so choose for purposes of the numerical integration

$$u(a) = 1,$$
$$u'(a) = -\frac{A}{B}, \tag{2.156}$$

or, if $B = 0$, $u(a) = 0$, $u'(a) = 1$. Then the solution of the differential equation and the boundary condition at $x = 1$ is

$$Ku(x), \tag{2.157}$$

where K is an arbitrary constant.

On the other hand we can start at the point $x = b$ with the boundary condition (2.90) and integrate backward with the same value λ for the eigenvalue parameter. Call this solution $v(x)$; it satisfies the differential equation (2.154) and the boundary condition

$$Cv(b) + Dv'(b) = 0. \tag{2.158}$$

Again we can normalize in an arbitrary way, so for purposes of the numerical integration take

$$v(b) = 1,$$
$$v'(b) = -\frac{C}{D}, \tag{2.159}$$

or, if $D = 0$, $v(b) = 0$, $v'(b) = 1$. Then the solution of the differential equation and the boundary condition at $x = b$ is

$$Lv(x), \tag{2.160}$$

where L is an arbitrary constant.

We can now try to choose the constants K and L so that our two solutions match in value and derivative at some point $x = X$ between a and b. Thus we require

$$Ku(X) - Lv(X) = 0,$$
$$Ku'(X) - Lv'(X) = 0. \tag{2.161}$$

The condition that these two equations have a nonzero solution for K and L is

$$D(\lambda) \equiv \begin{vmatrix} u(X) & -v(X) \\ u'(X) & -v'(X) \end{vmatrix} \equiv v(X)u'(X) - u(X)v'(X) = 0. \tag{2.162}$$

The determinant $D(\lambda)$ here depends on the assumed value λ used in solving the differential equation (2.154). For most values of λ, $D(\lambda) \neq 0$, i.e., it is not possible to match the two solutions at $x = X$. The values of λ for which $D(\lambda) = 0$, when it is possible to match the solutions, are the eigenvalues λ_n, and the corresponding solutions $Ku(x) = Lv(x)$ are the eigenfunctions.

To find these special values of λ one can start at small $\lambda = \lambda_0$ and then successively larger values $\lambda = \lambda_0 + n\Delta\lambda$ solve the differential equations, evaluate $D(\lambda)$, and locate the intervals in which $D(\lambda)$ changes sign. Inside these intervals one can use Newton's method to find precise values of λ for which $D(\lambda) = 0$. In this way one can find numerically to any desired accuracy the eigenvalues and eigenfunctions of a Sturm-Liouville problem.

PROBLEMS

2.1 Show that, if m is not zero or an integer, the equation

$$u'' + \left[\frac{\frac{1}{4} - m^2}{x^2} - \frac{1}{4} \right] u = 0$$

is satisfied by two series about $x = 0$ with leading terms

$$x^{\frac{1}{2}+m} \left\{ 1 + \frac{x^2}{16(1+m)} + \ldots \right\},$$

$$x^{\frac{1}{2}-m} \left\{ 1 + \frac{x^2}{16(1-m)} + \ldots \right\}.$$

Determine the recursion relation for the coefficient of the general term in each series and show that the series converge for all values of x. (Whittaker and Watson)

2.2 Show that the two solutions of the hypergeometric equation

$$x(1 - x)y'' + [c - (a + b + 1)x]y' - aby = 0$$

are

$$y = \begin{cases} {}_2F_1(a, b; c; x) \\ \\ x^{1-c} \, {}_2F_1(1 - c + a, 1 - c + b; 2 - c; x), \end{cases}$$

where the hypergeometric function is given by the series

$${}_2F_1(a, b; c; x) = 1 + \frac{ab}{c} \frac{x}{1!} + \frac{a(a + 1)b(b + 1)}{c(c + 1)} \frac{x^2}{2!} + \ldots .$$

Show that the series converges for $|x| < 1$.

2.3 By expanding about the regular singular point $x = 1$, find the series solutions of Legendre's equation of order zero:

$$(1 - x^2)\frac{d^2y}{dx^2} - 2x\frac{dy}{dx} + \lambda y = 0.$$

Show that both roots of the indicial equation vanish: $\alpha = 0, 0$.

Find the first few terms in the regular solutions:

$$y = P(x) = 1 + \frac{\lambda}{2}(x - 1) - \frac{2 - \lambda}{2 \cdot 2^2}\frac{\lambda}{2}(x - 1)^2 + \ldots .$$

The other solution is supposed to be of the form

$$y = P(x) \ln(x - 1) + (x - 1)[b_0 + b_1(x - 1) + b_2(x - 1)^2 + \ldots].$$

Substitute in the differential equation and find b_0 and b_1.

2.4 Show that the solutions of the equation

$$u'' + \frac{1}{x}u' - m^2 u = 0$$

near $x = 0$ are

$$u_1(x) = 1 + \sum_{n=1}^{\infty} \frac{m^{2n}x^{2n}}{2^{2n}(n!)^2},$$

$$u_2(x) = u_1(x)\ln(x) - \sum_{n=1}^{\infty} \frac{m^{2n}x^{2n}}{2^{2n}(n!)^2}\left[1 + \frac{1}{2} + \ldots + \frac{1}{n}\right].$$

Show that these series converge for all values of x. (Whittaker and Watson)

2.5 Show that the equation

$$u'' \frac{1}{4x^2}(1 - x^2)u = 0$$

has two solutions of the form

$$u_1 = x^{\frac{1}{2}}\left\{1 + \frac{x^2}{16} + \frac{x^4}{1024} + \ldots\right\},$$

$$u_2 = u_1(x)\ln(x) - \frac{x^{\frac{5}{2}}}{16} + \ldots,$$

near $x = 0$. (Whittaker and Watson)

2.6 Find the eigenfunctions $u_n(x)$ and eigenvalues λ_n for the differential equation

$$\frac{d^2 u_n(x)}{dx^2} = -\lambda_n u_n(x)$$

in the interval $0 \leq x \leq a$ for the following sets of boundary conditions:

a) $u(0) = 0 \qquad u(a) = 0,$
b) $u(0) = 0 \qquad u'(a) = 0,$
c) $u'(0) = 0 \qquad u'(a) = 0,$
d) $u(0) + au'(0) = 0 \qquad u(a) - au'(a) = 0.$

Each is a separate case. For case d) find the equation which determines the eigenvalue and verify that there is an infinite set of eigenfunctions and eigenvalues.

2.7 Find the factor which, when multiplied into the general second-order linear differential equation

$$f(x)\frac{d^2 u}{dx^2} + g(x)\frac{du}{dx} - h(x)u(x) = -\lambda\sigma(x)u(x),$$

converts it to the self-adjoint form (2.88).

2.8 Consider approximating the given function $\psi(x)$ by a linear combination of the orthonormal set $u_n(x)$ on the interval $a \le x \le b$,

$$\int_a^b dx \, u_n(x)u_m(x) = \delta_{nm}.$$

a) Show that

$$I_N = \int_a^b dx \left[\psi(x) - \sum_{n=1}^N \alpha_n u_n(x) \right]^2$$

$$= \int_a^b dx \, \psi^2(x) - \sum_{n=1}^N c_n^2 + \sum_{n=1}^N (\alpha_n - c_n)^2,$$

where the c_n are the Fourier expansion coefficients

$$c_n = \int_a^b dx \, u_n(x)\psi(x).$$

b) Show that I_N is a minimum for a given N, and thus that we obtain the best approximation in the mean for a given N, when $\alpha_n = c_n$.

c) Derive Bessel's inequality

$$\sum_{n=1}^N c_n^2 \le \int_a^b dx \, \psi^2(x).$$

d) We say that $\psi(x)$ is approximated in the mean by $\sum_{n=1}^\infty c_n u_n(x)$ provided

$$\lim_{N \to \infty} \int_a^b dx \left[\psi(x) - \sum_{n=1}^N c_n u_n(x) \right]^2 = 0.$$

Show that the necessary and sufficient condition to approximate in mean a square integrable function $\psi(x)$ is the Parseval relation

$$\sum_{n=1}^\infty c_n^2 = \int_a^b dx \, \psi^2(x).$$

The following two computer problems will require a good deal of time and effort as well as the availability of a computer.

2.9 Separating the variables,

$$\psi(\mathbf{r}) = \frac{u_l(r)}{r} Y_l^m(\theta, \varphi),$$

in Schrödinger's equation,

$$-\frac{\hbar^2}{2m} \mathbf{\nabla}^2 \psi + V(r)\psi = E\psi,$$

leads to the radial equation

$$-\frac{\hbar^2}{2m}\left[\frac{d^2 u_l(r)}{dr^2} - \frac{l(l+1)}{r^2}u_l(r)\right] + V(r)u_l(r) = Eu_l(r),$$

. where $l = 0, 1, 2, \ldots$ is the angular momentum quantum number. Consider the following three attractive potentials with finite range β:

i) $V(r) = -V_0 \exp(-r^2/\beta^2)$ Gaussian well,

ii) $V(r) = -V_0 \exp(-r/\beta)$ Exponential well,

iii) $V(r) = -V_0 \frac{\exp(-r/\beta)}{r/\beta}$ Yukawa well.

The bound states are determined by the Sturm-Liouville eigenvalue problem · which consists of the radial equation above plus the boundary conditions

$$u_l(r) \xrightarrow[r \to 0]{} 0, \quad u_l(r) \xrightarrow[r \to \infty]{} 0.$$

a) Show that for $E < 0$ and $r \gg \beta$ the acceptable solution of the radial wave equation varies as

$$u_l(r) \xrightarrow[r \to \infty]{} Ce^{-\kappa r}, \quad \kappa = \sqrt{\frac{-2mE}{\hbar^2}}.$$

Find the first few terms in the power series solution consistent with the boundary condition for $u_l(r)$ near $r = 0$ for the three potentials above. These power series are necessary for numerical work to bridge over the singular behavior as $r \to 0$ of the Yukawa well and/or of the term $l(l+1)/r^2$ for $l > 0$.

b) Write a computer program to find the eigenfunctions and eigenvalues for the radial Schrödinger equation. Integrate out from $r = 0$, starting from a small nonzero $r = r_0 \ll \beta$, where the power series of part a) is valid. Integrate in from a large $r = r_\infty \gg \beta$ where the exponential form of part a) is valid. Match the solutions for an r near $r = \beta$. If you wish use a library program to integrate the differential equation. Subdivide your program into several subroutines so that the different cases i), ii), iii) can be handled easily by changing subroutines.

c) Make plots of the eigenvalues E for several bound states for $l = 0, 1, 2$ as functions of the depth of the potential V_0. For a fairly deep well with several bound states plot the eigenfunctions u_l as functions of r. Repeat for each potential i), ii), iii).

2.10 For positive energies the radial equation of Problem 2.9 determines the phase shift $\delta_l(k)$. The boundary conditions are now

$$u_l(r) \xrightarrow[r \to 0]{} 0,$$

$$u_l(r) \xrightarrow[r \to \infty]{} C \sin\left[kr - l\frac{\pi}{2} + \delta_l(k)\right],$$

with

$$k = \sqrt{\frac{2mE}{\hbar^2}}.$$

a) Consider the attractive potential wells of Problem 2.9. Show that for $E > 0$ and $r \gg \beta$, $u_l(r)$ approaches the form given above. In Chapter 6 [see equations (6.28), (6.32)] it is shown that $\delta_l(k) = 0$ when $V(r) = 0$; thus $\delta_l(k)$ is the shift in phase due to the presence of the potential.

b) Write a computer program to determine $\delta_l(k)$ for $l = 0, 1, 2$ as a function of k for the potentials i), ii), iii) of Problem 2.9. Integrate out from $r = 0$, starting from a small nonzero $r = r_0 \ll \beta$, where the power series solutions of Problem 2.9(a) are valid. Determine $\delta_l(k)$ from the variation with r of the numerically determined wave function $u_l(r)$ at $r = r_\infty \gg \beta$:

$$\delta_l(k) = \frac{l\pi}{2} - kr_\infty + \tan^{-1}\left[\frac{ku_l(r_\infty)}{u_l'(r_\infty)}\right].$$

For numerical purposes choose r_∞ as small as possible.

Spherical Harmonics and Applications

3.1 INTRODUCTION

Spherical harmonics enter our discussion as the eigenfunctions of the angular part of the Laplacian in spherical coordinates. These functions are of basic importance, not only for the applications in classical physics which are the main focus of the present work, but also in quantum physics where they form the wave functions describing orbital angular momentum. Nearly every calculation dealing with an atomic, molecular, nuclear, or elementary particle system employs spherical harmonics in one way or another. Because of this basic importance we give a fairly detailed treatment including many useful but sometimes tiresome algebraic details.

According to the results of Section 2.5, the separation procedure in spherical coordinates leads to (2.35) for the angular dependence:

$$\frac{1}{\sin\theta}\frac{\partial}{\partial\theta}\left[\sin\theta\frac{\partial Y(\theta,\varphi)}{\partial\theta}\right] + \frac{1}{\sin^2\theta}\frac{\partial^2 Y(\theta,\varphi)}{\partial\varphi^2} = -\lambda Y(\theta,\varphi). \tag{3.1}$$

According to (2.46) - (2.52) we can then separate the ϕ dependence rather trivially to obtain

$$Y(\theta,\varphi) = P(\theta)e^{im\varphi}, \tag{3.2}$$

where P satisfies Legendre's equation:

$$\frac{d^2P}{dx^2} - \frac{2x}{1-x^2}\frac{dP}{dx} + \frac{1}{1-x^2}\left[\lambda - \frac{m^2}{1-x^2}\right]P = 0. \tag{3.3}$$

In (3.3) we use as variable $x = \cos\theta$.

As discussed in Section 2.8 [see (2.101)], in the usual physical applications m in (3.2) must be an integer,

$$m = 0, \pm 1, \pm 2, \ldots, \tag{3.4}$$

to ensure that $Y(\theta,\varphi)$ be single valued:

$$e^{im\varphi} = e^{im(\varphi+2\pi)}. \tag{3.5}$$

We consider only this case in what follows.

3.2 SERIES SOLUTION OF LEGENDRE'S EQUATION - LEGENDRE POLYNOMIALS

The points $x = \pm 1$ are regular singular points of the differential equation (3.3). If we attempt a series solution about either of these points,

$$P = (x \mp 1)^\alpha \left[1 + \sum_{n=1}^{\infty} f_n (x \mp 1)^n \right], \qquad (3.6)$$

we find an indicial equation [see (2.78) and (2.72)-(2.74)]:

$$\alpha(\alpha - 1) + \alpha - \frac{m^2}{4} = 0$$

$$\alpha = \pm \frac{|m|}{2}. \qquad (3.7)$$

For m an integer we then have the troublesome case in which the roots of the indicial equation differ by an integer, namely $\alpha_1 - \alpha_2 = |m|$. If we expand about $x = 1$, one of our solutions will be finite near $x = 1$; in fact very near $x = 1$ it will vary as $(1 - x)^{\frac{|m|}{2}}$. The other solution will be infinite at $x = 1$, varying, according to (2.87) like $(1 - x)^{-\frac{|m|}{2}}$, $m \neq 0$, or $\ln(1 - x)$, $m = 0$, when we are very near $x = 1$. Expansion about $x = -1$ leads to similar results; one solution finite, the other singular. Even if we take a solution which is finite at $x = +1$ say, when we continue it to the point $x = -1$, we will in general have unavoidably included some of the solution which is infinite at $x = -1$. Thus in general the solutions of the differential equations will be infinite at one or the other or both of the points $x = \pm 1$. A solution finite at both $x + 1$ and $x = -1$ can occur only under special circumstances, in fact special values of the parameter λ in the differential equation (3.3).

Now it is just these special circumstances which are of interest in most physical applications. In physical applications there is usually nothing special about the points $x = \cos\theta = \pm 1$, i.e., $\theta = 0, \pi$. Usually we want a solution which is finite at these points, just as it is finite at other points. It is therefore appropriate to introduce as a boundary condition that the solutions $P(x)$ be finite for all $-1 \leq x = \cos\theta \leq +1$. The differential equation (3.3) together with this boundary condition forms a type of Sturm-Liouville eigenvalue problem, the separation constant λ which appears in (3.3) being the eigenvalues. Such problems were discussed in a rather general context in Section 2.8. The case under discussion in this section is slightly different from that envisioned in Section 2.8 due to the singularities at $x = \pm 1$ which appear in the differential equation. Thus the boundary conditions of the type (2.89), (2.90) or (2.91) have been replaced by the finiteness boundary condition discussed above.

We consider first the case $m = 0$. We write the eigenvalue parameter λ in the form

$$\lambda = l(l + 1) \qquad \text{[Below we find that } l = \text{integer.]}. \qquad (3.8)$$

The differential equation (3.3) assumes the form

$$(1 - x^2)\frac{d^2 P}{dx^2} - 2x\frac{dP}{dx} + l(l+1)P = 0. \tag{3.9}$$

Let us obtain the series solutions of this equation by expanding about the ordinary point $x = 0$,

$$P = \sum_{n=0}^{\infty} c_n x^n = c_0 + c_1 x + c_2 x^2 + \dots . \tag{3.10}$$

Substituting this series in the differential equation immediately leads to the recursion relation

$$\frac{c_{n+2}}{c_n} = \frac{n(n+1) - l(l+1)}{(n+2)(n+1)}. \tag{3.11}$$

Thus we find two solutions of the form

$$c_0 + c_2 x^2 + c_4 x^4 + \dots = c_0 \left[1 + \frac{c_2}{c_0}x^2 + \frac{c_4}{c_0}x^4 + \dots \right] \tag{3.12}$$

and

$$c_1 x + c_3 x^3 + c_5 x^5 + \dots = c_1 x \left[1 + \frac{c_3}{c_1}x^2 + \frac{c_5}{c_1}x^4 + \dots \right]. \tag{3.13}$$

According to the ratio test, these series converge for $|x| < 1$ since we find

$$\lim_{n \to \infty} \frac{c_{n+2}\, x^{n+2}}{c_n\, x^n} = x^2. \tag{3.14}$$

The ratio test does not tell us if the series converge or not for $x = \pm 1$. According to the discussion at the beginning of this section we expect the series in general to diverge at $x = \pm 1$, and this can be shown explicitly.[1]

The only possibility to obtain functions finite at $x = \pm 1$ is to make the series break off and so obtain a polynomial solution. From the recursion relation (3.11) we see that this will happen for

$$l = \text{nonnegative integer} = 0, 1, 2, \dots . \tag{3.15}$$

Thus the finiteness condition determines the eigenvalues $\lambda = l(l+1)$. For $l = $ integer one of the series (3.12), (3.13) breaks off to form a polynomial; the other series is discarded since it does not satisfy the boundary condition.

From the form of (3.12), (3.13) we see that our solutions are either even or odd polynomials of degree l. We shall call them Legendre polynomials $P_l(x)$ and rewrite the series in the form

$$P_l(x) = \sum_{r=0}^{r_{max}} a_r x^{l-2r}, \tag{3.16}$$

---

[1]Courant and Hilbert, op. cit., page 326.

$$r_{max} = \begin{cases} \dfrac{l}{2}, & l \quad \text{even} \\[4mm] \dfrac{l-1}{2}, & l \quad \text{odd} \end{cases} . \tag{3.17}$$

With this notation we can handle the l even and l odd cases together. From the recursion relation for the c's we can find the recursion relation for the a's:

$$\frac{a_r}{a_{r-1}} = \frac{c_{l-2r}}{c_{l-2r+2}}$$

$$= \frac{(l-2r+1)(l-2r+2)}{(l-2r)(l-2r+1) - l(l+1)}$$

$$= -\frac{(l-2r+1)(l-2r+2)}{2r(2l-2r+1)}. \tag{3.18}$$

Employing this repeatedly we can find a_r in terms of a_0:

$$a_r = -\frac{(l-2r+1)(l-2r+2)}{2r(2l-2r+1)} a_{r-1}$$

$$= (-1)^2 \frac{(l-2r+1)(l-2r+2)(l-2r+3)(l-2r+4)}{2r\,2(r-1)(2l-2r+1)(2l-2r+3)} a_{r-2}$$

$$= (-1)^r \frac{l!}{(l-2r)!} \frac{1}{2^r r!} \frac{1}{(2l-2r+1)(2l-2r+3)\dots(2l-1)} a_0$$

$$= (-1)^r \frac{l!}{(l-2r)!} \frac{1}{2^r r!} \frac{(2l-2r)!\,2^l\,l!}{(2l)!\,2^{l-r}(l-r)!} a_0. \tag{3.19}$$

The constant a_0 is a multiplicative constant which determines the normalization of $P_l(x)$. By convention we choose it to have the value $a_0 = (2l)!\,[2^l(l!)^2]^{-1}$ in defining the Legendre polynomial. Then we find

$$P_l(x) = \frac{1}{2^l} \sum_{r=0}^{r_{max}} \frac{(-1)^r}{r!} \frac{(2l-2r)!}{(l-r)!(l-2r)!} x^{l-2r}. \tag{3.20}$$

The explicit expression (3.20) for $P_l(x)$ can be rewritten in a more useful form:

$$P_l(x) = \frac{1}{2^l} \frac{d^l}{dx^l} \sum_{r=0}^{r_{max}} \frac{(-1)^r}{r!} \frac{1}{(l-r)!} x^{2l-2r}, \tag{3.21}$$

or using the expansion

$$(1 - x^2)^l = \sum_r (-1)^r \frac{l!}{r! \, (l-r)!} \, x^{2r}$$

$$= \sum_r (-1)^{l-r} \frac{l!}{(l-r)! \, r!} \, x^{2l-2r}, \tag{3.22}$$

$$\boxed{P_l(x) = \frac{1}{2^l \, l!} \frac{d^l}{dx^l} (x^2 - 1)^l.} \tag{3.23}$$

This convenient result is known as Rodrigues's formula.

3.3 PROPERTIES OF LEGENDRE POLYNOMIALS

In this section we review some important properties of the Legendre polynomials.

- From Rodrigues's formula (3.23) it is easy to evaluate the low order polynomials:

$$P_0(x) = 1,$$
$$P_1(x) = x,$$

$$P_2(x) = \frac{1}{2}(3x^2 - 1),$$

$$P_3(x) = \frac{1}{2}(5x^3 - 3x),$$

$$P_4(x) = \frac{1}{8}(35x^4 - 30x^2 + 3). \tag{3.24}$$

- We can also easily find $P_l(x)$ at $x = \pm 1$ from (3.23).

  The factors $(x + 1)^l (x - 1)^l$ will vanish at the end points unless they are completely differentiated out, and we find

$$\boxed{\begin{aligned} P_l(1) &= 1, \\ P_l(-1) &= (-1)^l. \end{aligned}}$$

$$(3.25)$$
$$(3.26)$$

The constant a_0 in (3.19) was chosen to achieve the supposedly desirable normalization (3.25), and this choice is by now frozen by convention.

We can establish a couple more results which tell us the general from of $P_l(x)$.

- From Rolle's theorem, that between every two zeros of a polynomial there is an odd number (hence at least one) of zeros of its derivative, we find from (3.23)

$$P_l(x) \text{ has } l \text{ zeros in the interval } -1 < x < l. \tag{3.27}$$

- Another result is

$$|P_l(x)| \le 1, \quad -1 \le x \le 1. \tag{3.28}$$

This can be proved from the identity (3.43) to be established below.

Proof. Writing this in the form

$$\sum_l P_l(\cos\theta)u^l = \frac{1}{\sqrt{1 - 2u\cos\theta + u^2}}$$

$$= \frac{1}{\sqrt{1 - ue^{i\theta}}}\frac{1}{\sqrt{1 - ue^{-i\theta}}}$$

$$= \left[1 + \frac{1}{2}ue^{i\theta} + \frac{1}{2}\cdot\frac{3}{2}\cdot\frac{1}{2!}e^{2i\theta} + \frac{1}{2}\cdot\frac{3}{2}\cdot\frac{5}{2}\cdot\frac{1}{3!}u^3 e^{3i\theta} + \ldots\right] \times$$

$$\left[1 + \frac{1}{2}ue^{-i\theta} + \frac{1}{2}\cdot\frac{3}{2}\cdot\frac{1}{2!}e^{-2i\theta} + \frac{1}{2}\cdot\frac{3}{2}\cdot\frac{5}{2}\cdot\frac{1}{3!}u^3 e^{-3i\theta} + \ldots\right], \tag{3.29}$$

we see that $P_l(\cos\theta)$, which is the coefficient of u^l when we multiply out the right-hand side of (3.29), is a polynomial in $e^{i\theta}$ and $e^{-i\theta}$ with positive coefficients. From the properties of complex numbers we then see that $|P_l(\cos\theta)| \le$ the same polynomial with $e^{\pm i\theta}$ replaced by $|e^{\pm i\theta}| = 1$, i.e., the same polynomial evaluated at $\theta = 0$, or $x = \cos\theta = 1$: $|P_l(\cos\theta)| \le P_l(1)$. Using (3.25) we then obtain (3.28). $\qquad\square$

- For large $|x| \gg 1$ we obtain easily from (3.23)

$$P_l(x) \to \frac{(2l)!}{2^l(l!)^2}x^l, \quad |x| \gg 1. \tag{3.30}$$

- The Legendre polynomials $P_l(x)$ are the solutions of the Sturm-Liouville eigenvalue problem

$$\frac{d}{dx}\left[(1 - x^2)\frac{dP_l}{dx}\right] = -l(l+1)P_l(x) \tag{3.31}$$

with the boundary condition

$$P_l(x) \text{ finite } -1 \le x \le +1. \tag{3.32}$$

It is easy to check that the differential operator on the left of (3.31) satisfies the relation (2.106) for functions satisfying this boundary condition, so that the standard manipulations of Section 2.8 lead to an orthogonality relation (2.110) in the usual way:

$$\int_{-l}^{l} dx \ P_l(x) P_m(x) = 0, \quad l \neq m. \tag{3.33}$$

For $l = m$ we can calculate the normalization integral from Rodrigues's formula by integration by parts:

$$\int_{-1}^{+1} dx \ [P_l(x)]^2 = \frac{1}{[2^l \ l!]^2} \int_{-1}^{+1} dx \ \frac{d^l}{dx^l} (x^2 - 1)^l \frac{d^l}{dx^l} (x^2 - 1)^l$$

$$= \frac{(-1)^l}{[2^l \ l!]^2} \int_{-1}^{+1} dx \ (x^2 - 1)^l \frac{d^{2l}}{dx^{2l}} (x^2 - 1)^l$$

$$= \frac{(-1)^l (2l)!}{[2^l \ l!]^2} \int_{-1}^{+1} dx \ (x - l)^l (x + 1)^l$$

$$= \frac{(-1)^l (2l)!}{[2^l \ l!]^2} (-1)^l \frac{l!^2}{(2l)!} \int_{-1}^{+1} dx \ (x + 1)^{2l}$$

$$= \frac{2}{2l + 1}. \tag{3.34}$$

In summary the Legendre polynomials $P_l(x)$ are an orthogonal set normalized so that

$$\boxed{\int_{-1}^{+1} dx \ P_l(x) P_m(x) = \delta_{lm} \frac{2}{2l + 1}.} \tag{3.35}$$

- As the eigenfunctions of a Sturm-Liouville problem the $P_l(x)$ form a complete set, and any "reasonable" function $f(x)$ can be expanded in the interval $-1 \leq x \leq +1$ in the series[2]

$$f(x) = \sum_{l=0}^{\infty} a_l P_l(x). \tag{3.36}$$

Using the orthogonality relation (3.35) we can then find the a_l:

$$a_l = \frac{2l + 1}{2} \int_{-1}^{+1} dx' \ P_l(x') f(x'). \tag{3.37}$$

Substituting this back in (3.36) it is easy to establish the completeness relation

$$\boxed{\frac{1}{2} \sum_l (2l + 1) P_l(x) P_l(x') = \delta(x - x').} \tag{3.38}$$

---

[2]Various precise mathematical statements can be found in the literature: Courant and Hilbert op. cit., p. 65, 82, 513; Whittaker and Watson, op. cit p. 322; E. W. Hobson, The Theory of Spherical and Ellipsoidal Harmonics (Cambridge University Press, London, 1931), p. 318.

- The $P_l(x)$ are a set of polynomials of degree l, orthogonal in the interval $-1 \leq x \leq +1$. In fact this statement characterizes them uniquely, up to normalization. If one starts with the sequence of powers, $1, x, x^2, x^3, \ldots$ and constructs from them successively polynomials of degree l orthogonal to the polynomials of lower order $1, 2, \ldots l-1$ (the Gram Schmidt orthogonalization procedure), one obtains a unique set of polynomials which must be the Legendre polynomials up to normalization. We can use this fact to prove the most famous expansion in Legendre polynomials.

Proof. Expanding in a power series in u we find

$$\frac{1}{\sqrt{1 - 2ux + u^2}} = \sum_{l=0}^{\infty} R_l(x) u^l, \quad |u| < 1, \tag{3.39}$$

where $R_l(x)$ is some polynomial of degree l in x. Then we have

$$\frac{1}{\sqrt{1 - 2ux + u^2}} \frac{1}{\sqrt{1 - 2vx + v^2}} = \sum_{l,m=0}^{\infty} R_l(x) R_m(x) u^l v^m. \tag{3.40}$$

We can integrate the left side of (3.40) with respect to x using elementary techniques and expand the result in a Taylor series to obtain

$$\int_{-l}^{+l} dx \, \frac{1}{\sqrt{1 - 2ux + u^2}} \frac{1}{\sqrt{1 - 2vx + v^2}} = \frac{1}{\sqrt{uv}} \ln\left(\frac{1 + \sqrt{uv}}{1 - \sqrt{uv}}\right)$$

$$= \sum_{l=0}^{\infty} \frac{2}{2l+1} (uv)^l. \tag{3.41}$$

Comparing this with the integral of the right side of (3.40), we find

$$\int_{-1}^{+1} dx \, R_l(x) R_m(x) = \delta_{lm} \frac{2}{2l+1}. \tag{3.42}$$

Since the $R_l(x)$ are polynomials of degree l with exactly the same orthogonality and normalization properties as the $P_l(x)$, we see that in fact $R_l(x) = P_l(x)$. Thus we can rewrite (3.39) as

$$\boxed{F(u, x) = \frac{1}{\sqrt{1 - 2ux + u^2}} = \sum_{l} P_l(x) u^l, \quad |u| < 1.} \tag{3.43}$$

□

Because of this property the function on the left of (3.43) is sometimes called the generating function of the Legendre polynomials.

- The relation (3.43) can be used to derive some useful identities satisfied by the $P_l(x)$. Differentiating the generating function with respect to u we find

$$\frac{\partial F}{\partial u} = \frac{x - u}{1 - 2ux + u^2} \, F \, ,$$

$$(1 - 2ux + u^2)\frac{\partial F}{\partial u} = (x - u)F. \tag{3.44}$$

Substituting the series expansion on the right of (3.43) for F and equating coefficients of u^l yields the recursion relation

$$(l + 1)P_{l+1}(x) - (2l + 1)xP_l(x) + lP_{l-1}(x) = 0. \tag{3.45}$$

Differentiating with respect to x we obtain

$$(1 - 2ux + u^2)\frac{\partial F}{\partial x} = uF \, , \tag{3.46}$$

and substituting the series yields

$$P'_{l+1}(x) - 2xP'_l(x) + P'_{l-1}(x) - P_l(x) = 0. \tag{3.47}$$

Differentiating (3.45) and combining with (3.47) to eliminate P'_{l+1} or P'_{l-1} yields some simpler relations:

$$xP'_l(x) - P'_{l-1}(x) - lP_l(x) = 0 \tag{3.48}$$

$$P'_{l+1}(x) - xP'_l(x) - (l + 1)P_l(x) = 0. \tag{3.49}$$

We have gone to some effort to derive a number of the properties of the Legendre polynomials. The results derived above as well as many other relations for Legendre polynomials and other functions are recorded in various mathematical handbooks.[3] Such books are invaluable to the reader with enough basic knowledge to use them intelligently. A primary objective of the treatment of this book is to provide this basic knowledge.

---

[3]W. Magnus, F. Oberhettinger and R. P. Soni, Formulas and Theorems for the Special Functions of Mathematical Physics (Springer-Verlag, New York, 1966);
E. Jahnke and F. Emde, Tables of Functions (Dover Publications, New York, 1945);
M. Abramowitz and I. Stegun, Handbook of Mathematical Functions (National Bureau of Standards Applied Mathematics Series. 55, Washington, D.C., 1964);
A. Erdélyi, W. Magnus, F. Oberhettinger and F. C. Tricomi, Bateman Manuscript Project: Higher Transcendental Functions, Vols. I-III, Tables of Integral Transforms, Vols. I, II (McGraw-Hill, New York, 1953).

3.4 THE SECOND SOLUTION $Q_l(x)$ OF LEGENDRE'S EQUATION

For l an integer the Legendre polynomials provide one solution of the differential equation

$$\frac{d}{dx}\left[(1-x^2)\frac{dP_l(x)}{dx}\right] + l(l+1)P_l(x) = 0. \tag{3.50}$$

A second solution of this equation exists, although naturally it violates the finiteness boundary condition imposed to obtain the $P_l(x)$. A useful way to write the second solution is

$$\boxed{Q_l(z) = \frac{1}{2}\int_{-1}^{1} dt\,\frac{P_l(t)}{z-t}.} \tag{3.51}$$

The integral in this expression is not well defined for z on the real axis between -1 and $+1$. However, for any other value of z in the whole complex plane $Q_l(z)$ is well defined by (3.51). We can approach the line $-1 \le z = x \le +1$ by a limiting procedure, either from above or below. Another possibility would be to define $Q_l(x)$, $-1 \le x \le +1$, as a Cauchy principal value integral:

$$Q_l(x) = \lim_{\epsilon\to 0}\left\{\frac{1}{2}\int_{-1}^{x-\epsilon} dt\,\frac{P_l(t)}{x-t} + \frac{1}{2}\int_{x+\epsilon}^{+1} dt\,\frac{P_l(t)}{x-t}\right\}. \tag{3.52}$$

To verify that $Q_l(z)$ satisfies Legendre's equation, differentiate under the integral sign in (3.51), integrate by parts, and use (3.25), (3.26):

$$\frac{dQ_l(z)}{dz} = \frac{1}{2}\int_{-1}^{+1} dt\,P_l(t)\frac{d}{dz}\frac{1}{z-t}$$

$$= -\frac{1}{2}\int_{-1}^{+1} dt\,P_l(t)\frac{d}{dt}\frac{1}{z-t}$$

$$= \frac{1}{2}\int_{-1}^{+1} dt\,\left(\frac{\frac{dP_l(t)}{dt}}{z-t}\right) - \frac{1}{2}\cdot\frac{1}{z-1} + \frac{1}{2}\cdot\frac{(-1)^l}{z+1}, \tag{3.53}$$

and

$$\frac{d}{dz}\left[(1-z^2)\frac{dQ_l(z)}{dz}\right] = \frac{1}{2}\int_{-1}^{+1} dt\,\frac{\frac{d}{dt}\left[(1-t^2)\frac{dP_l(t)}{dt}\right]}{z-t}$$

$$+ \frac{1}{2}\cdot\frac{d}{dz}\int_{-1}^{+1} dt\,\frac{t^2-z^2}{z-t}\frac{dP_l(t)}{dt} + \frac{1}{2} - \frac{1}{2}(-1)^l$$

$$= \frac{1}{2}\int_{-1}^{+1} dt\,\frac{\frac{d}{dt}\left[(1-t^2)\frac{dP_l(t)}{dt}\right]}{z-t}. \tag{3.54}$$

Using (3.50) for $P_l(t)$ we then find

$$\frac{d}{dz}\left[(1-z^2)\frac{dQ_l(z)}{dz}\right] + l(l+1)Q_l(z) = 0. \tag{3.55}$$

For $l = 0$ and 1 we find

$$Q_0(z) = \frac{1}{2}\int_{-1}^{+1} dt\,\frac{1}{z-t} = \frac{1}{2}\ln\frac{z+1}{z-1}, \tag{3.56}$$

$$Q_1(z) = \frac{1}{2}\int_{-1}^{+1} dt\,\frac{t}{z-t} = \frac{1}{2}z\ln\frac{z+1}{z-1} - 1. \tag{3.57}$$

More generally we have

$$Q_l(z) = \frac{1}{2}\int_{-1}^{+1} dt\,\frac{P_l(t)}{z-t}$$

$$= \frac{1}{2}P_l(z)\ln\frac{z+1}{z-1} + \frac{1}{2}\int_{-1}^{+1} dt\,\frac{P_l(t) - P_l(z)}{z-t}. \tag{3.58}$$

Now $P_l(t) - P_l(z)$ is of the form

$$P_l(t) - P_l(z) = (t - z) \times [\text{Polynomial of degree } l - 1 \text{ in } z \text{ and } t]. \tag{3.59}$$

so that (3.58) has the form

$$Q_l(z) = \frac{1}{2}P_l(z)\ln\frac{z+1}{z-1} + [\text{Polynomial of degree } l - 1 \text{ in } z]. \tag{3.60}$$

This formula explicitly displays the singular behavior at $z = \pm 1$.

Finally, note that if we had used the definition (3.52) for $-1 \le z = x \le +1$,

$$\lim_{\epsilon \to 0}\left\{\int_{-1}^{x-\epsilon}\frac{dt}{x-t} + \int_{x=\epsilon}^{+1}\frac{dt}{x-t}\right\} = \ln\frac{1+x}{1-x} \tag{3.61}$$

and the factors $\ln[(z+1)/(z-1)]$ in (3.56), (3.57) and (3.60) would be replaced by $\ln[(1+x)/(1-x)]$. On the other hand, if the $Q_l(z)$ are defined in the complex z plane according to (3.51) and we then approach the real line $-1 \le x \le +1$ from above and below, $z = x \pm i\epsilon$, we find that the logarithmic factor is discontinuous:

$$\ln\frac{z+1}{z-1}\bigg|_{z=x\pm i\epsilon} = \ln\frac{x+1}{(1-x)e^{\pm i\pi}} = \ln\frac{1+x}{1-x} \mp i\pi. \tag{3.62}$$

These complications will become much clearer when we come to Part III of the book, where complex variable theory is discussed. We bring them up here only to avoid misunderstanding through oversimplification.

The asymptotic properties of the $Q_l(z)$ are simple:

$$Q_l(z) = \frac{1}{2z} \int_{-1}^{+1} dt \; \frac{P_l(t)}{1 - \frac{t}{z}}$$

$$\xrightarrow[|z| \to \infty]{} \frac{1}{2z} \int_{-1}^{+1} dt \; P_l(t) \left[1 + \frac{t}{z} + \left(\frac{t}{z} \right)^2 + \ldots \right]. \qquad (3.63)$$

The first term in the series which contributes is $(t/z)^l$ since powers of t of lower order than l can be expressed in terms of Legendre polynomials of lower order than l and these are orthogonal to $P_l(t)$. Thus we find

$$Q_l(z) \xrightarrow[|z| \to \infty]{} \frac{1}{2z^{l+1}} \int_{-1}^{1} dt \; P_l(t) t^l$$

$$= \frac{2^l (l!)^2}{(2l+1)!} \frac{1}{z^{l+1}}, \qquad (3.64)$$

where the integral was calculated by using Rodrigues's formula (3.23) and integrating by parts. The asymptotic form (3.64) for $Q_l(z)$ is to be contrasted with that for $P_l(z)$, (3.30).

From (3.51) we can easily find an expansion theorem for $(z-x)^{-1}$ with $-1 \le x \le 1$ and z a real or complex number outside this interval. With x in the specified interval we can expand according to (3.36), (3.37):

$$\frac{1}{z-x} = \sum_{l=0}^{\infty} a_l P_l(x), \qquad (3.65)$$

$$a_l = \frac{2l+1}{2} \int_{-1}^{1} dt \; P_l(t) \frac{1}{z-t}. \qquad (3.66)$$

Using (3.51) we then obtain

$$\frac{1}{z-x} = \sum_{l=0}^{\infty} (2l+1) Q_l(z) P_l(x). \qquad (3.67)$$

3.5 ASSOCIATED LEGENDRE POLYNOMIALS

For $m \ne 0$ Legendre's equation is

$$\frac{d}{dx} \left[(1-x^2) \frac{dP}{dx} \right] + \left[\lambda - \frac{m^2}{1-x^2} \right] P = 0. \qquad (3.68)$$

In (3.7) we found the roots $\alpha = \pm m/2$ of the indicial equation for expansion about $x = \pm 1$. For solutions finite at $x = \pm 1$ we can then separate off the behavior near $x = \pm 1$ by writing

$$P = (1-x)^{m/2} (1+x)^{m/2} R(x) = (1-x^2)^{m/2} R(x), \qquad (3.69)$$

where in these expressions and below we take m to be a positive integer. Substituting (3.69) in (3.68) we find as equation for $R(x)$

$$(1 - x^2)\frac{d^2 R}{dx^2} - 2(m+1)x\frac{dR}{dx} + [\lambda - m(m+1)]R = 0. \tag{3.70}$$

On the other hand if we differentiate m times the equation (3.31) for the Legendre polynomials we find

$$(1 - x^2)\frac{d^2}{dx^2}\left[\frac{d^m P_l(x)}{dx^m}\right] - 2(m+1)x\frac{d}{dx}\left[\frac{d^m P_l(x)}{dx^m}\right]$$

$$+ \{l(l+1) - m(m+1)\}\left[\frac{d^m P_l(x)}{dx^m}\right] = 0. \tag{3.71}$$

Comparing (3.70) and (3.71) we see that for $\lambda = l(l+1)$ we have found a solution of (3.68), namely

$$P = P_l^m(x) = (1 - x^2)^{\frac{m}{2}}\frac{d^m P_l(x)}{dx^m}$$

$$= \frac{(1 - x^2)^{\frac{m}{2}}}{2^l\, l!}\frac{d^{m+l}}{dx^{m+l}}(x^2 - 1)^l, \quad m = 0, 1, 2, \ldots, l. \tag{3.72}$$

Just as for the $P_l(x)$, it can be verified from study of the series solution that only for $\lambda = l(l+1)$ can we find a solution finite at $x = \pm 1$. Thus we have solved the Sturm-Liouville eigenvalue problem.

$$\frac{d}{dx}\left[(1 - x^2)\frac{dP_l^m(x)}{dx}\right] - \frac{m^2}{1 - x^2}P_l^m(x) = -l(l+1)P_l^m(x), \tag{3.73}$$

$$P_l^m(x) \text{ finite in interval } -1 \leq x \leq 1.$$

The eigenfunctions $P_l^m(x)$, $l = m, m+1, \ldots$, of this Sturm-Liouville problem form a complete orthogonal set in the interval $-1 \leq x \leq 1$:

$$\int_{-1}^{+1} dx\, P_l^m(x)P_{l'}^m(x) = 0, \quad l \neq l'. \tag{3.74}$$

In order to calculate the normalization integral, it is convenient to first find a new formula for $P_l^m(x)$. From the old formula (3.72) we find

$$(1 - x^2)^{\frac{1}{2}}\frac{d}{dx}P_l^m(x) = P_l^{m+1}(x) - \frac{mx}{(1 - x^2)^{\frac{1}{2}}}P_l^m(x), \tag{3.75}$$

$$P_l^{m+1}(x) = (1 - x^2)^{\frac{1}{2}}\frac{dP_l^m(x)}{dx} + \frac{mx}{(1 - x^2)^{\frac{1}{2}}}P_l^m(x). \tag{3.76}$$

On the other hand, multiplying (3.71) by $(1 - x^2)^{m/2}$ and using the definition (3.72) leads to the result

$$P_l^{m+2}(x) - 2(m + 1)\frac{x}{(1 - x^2)^{\frac{1}{2}}}P_l^{m+1}(x) + \{l(l + 1) - m(m + 1)\}P_l^m(x) = 0. \quad (3.77)$$

Using the formula (3.76) with $m \to m + 1$ for $P_l^{m+2}(x)$ in (3.77) and solving (3.77) for $P_l^m(x)$ gives

$$P_l^m(x) = \frac{1}{l(l + 1) - m(m + 1)}\left[\frac{(m + 1)x}{(1 - x^2)^{\frac{1}{2}}} - (1 - x^2)^{\frac{1}{2}}\frac{d}{dx}\right]P_l^{m+1}(x)$$

$$= \frac{-1}{(l + m + 1)(l - m)} \cdot \frac{1}{(1 - x^2)^{\frac{m}{2}}} \cdot \frac{d}{dx}\left[(1 - x^2)^{\frac{m+1}{2}}P_l^{m+1}(x)\right]. \quad (3.78)$$

The identity (3.76) enables one to increase the value of m by one; the identity (3.78) enables one to lower m by one. By employing (3.78) repeatedly one can express P_l^m in terms of P_l^l:

$$P_l^m(x) = \frac{(-1)}{(l + m + 1)(l - m)} \cdot \frac{(-1)}{(l + m + 2)(l - m - 1)} \cdot \frac{1}{(1 - x^2)^{m/2}} \times$$

$$\frac{d}{dx}\left[\frac{(1 - x^2)^{\frac{m+1}{2}}}{1} \cdot \frac{1}{(1 - x^2)^{\frac{m+1}{2}}} \cdot \frac{d}{dx}\left[(1 - x^2)^{\frac{m+2}{2}}P_l^{m+2}(x)\right]\right]$$

$$\vdots$$

$$= \frac{(-1)^{l-m}(l + m)!}{(2l)!(l - m)!} \cdot \frac{1}{(1 - x^2)^{m/2}} \cdot \frac{d^{l-m}}{dx^{l-m}}\left[(1 - x^2)^{\frac{1}{2}}P_l^l(x)\right]. \quad (3.79)$$

From the old formula (3.72) we can obtain $P_l^l(x)$:

$$P_l^l(x) = \frac{(1 - x^2)^{\frac{l}{2}}}{2^l\,l!} \cdot \frac{d^{2l}}{dx^{2l}}(x^2 - 1)^l = \frac{(2l)!}{2^l\,l!}(1 - x^2)^{\frac{l}{2}}. \quad (3.80)$$

Substituting this in (3.79) leads to the desired new formula for $P_l^m(x)$:

$$P_l^m(x) = (-1)^m\frac{(l + m)!}{2^l\,l!(l - m)!} \cdot \frac{1}{(1 - x^2)^{\frac{m}{2}}} \cdot \frac{d^{l-m}}{dx^{l-m}}(x^2 - 1)^l. \quad (3.81)$$

This result is to be contrasted with (3.72).

With these two formulas it is easy to calculate the normalization integral:

$$\int_{-1}^{+1} dx \, [P_l^m(x)]^2 = \frac{(-1)^m}{[2^l \, l!]^2} \cdot \frac{(l+m)!}{(l-m)!} \times$$

$$\int_{-1}^{+1} dx \, \left[\frac{d^{l+m}}{dx^{l+m}}(x^2-1)^l \right] \times \left[\frac{d^{l-m}}{dx^{l-m}}(x^2-1)^l \right]$$

$$= \frac{(l+m)!}{(l-m)!} \int_{-1}^{+1} dx \, \left[\frac{1}{2^l \, l!} \frac{d^l}{dx^l}(x^2-1)^l \right]^2$$

$$= \frac{(l+m)!}{(l-m)!} \frac{2}{2l+1}. \tag{3.82}$$

Here we have used our previous result (3.34) for the normalization integral of the $P_l(x)$.

3.6 SPHERICAL HARMONICS

We have now completed the solution of the problem posed in Section 3.1. We have solved the eigenvalue problem

$$\frac{1}{\sin\theta} \frac{\partial}{\partial\theta} \left(\sin\theta \frac{\partial Y}{\partial\theta} \right) + \frac{1}{\sin^2\theta} \frac{\partial^2 Y}{\partial\varphi^2} = -\lambda Y \tag{3.83}$$

with boundary conditions that Y be

    1. single valued

    2. finite $\tag{3.84}$

over the sphere. We have found

$$\lambda = l(l+1), \qquad\qquad l = 0, 1, 2, \ldots \tag{3.85}$$

$$\boxed{Y = P_l^m(\cos\theta)e^{\pm im\varphi}, \qquad m = 0, 1, 2, \ldots, l.} \tag{3.86}$$

At this point it is convenient to change the notation slightly by letting m take on positive and negative values and to use the result (3.82) to introduce normalized eigenfunctions called spherical harmonics:

$$Y_l^m(\theta,\varphi) = \begin{cases} (-1)^m \sqrt{\dfrac{2l+1}{2} \dfrac{(1-m)!}{(l+m)!}} \, P_l^m(\cos\theta) \dfrac{e^{im\varphi}}{\sqrt{2\pi}}, & m \geq 0, \\[4mm] \sqrt{\dfrac{2l+1}{2} \dfrac{(l-|m|)!}{(l+|m|)!}} \, P_l^{|m|}(\cos\theta) \dfrac{e^{im\varphi}}{\sqrt{2\pi}}, & m < 0, \\[4mm] l = 0, 1, 2\ldots, \quad m = -l, -l+1, \ldots, l-1, l, \end{cases} \tag{3.87}$$

where

$$P_l^m(x) = \frac{(1-x^2)^{\frac{m}{2}}}{2^l \, l!} \frac{d^{l+m}}{dx^{l+m}}(x^2-1)^l$$

$$= \frac{(-1)^m(l+m)!}{2^l \, l!(l-m)!} \cdot \frac{1}{(1-x^2)^{\frac{m}{2}}} \frac{d^{l-m}}{dx^{l-m}}(x^2-1)^l. \qquad (3.88)$$

The choice of phase in (3.87) is convenient in the quantum theory of angular momentum and is called the Condon and Shortley choice of phase. Note that with this choice

$$Y_l^{m*}(\theta, \varphi) = (-1)^m Y_l^{-m}(\theta, \varphi). \qquad (3.89)$$

It is easy to calculate the low order spherical harmonics from the explicit formulas (3.87), (3.88):

$$l = 0:$$

$$Y_0^0(\theta, \varphi) = \frac{1}{\sqrt{4\pi}} \qquad (3.90)$$

$$l = 1:$$

$$Y_1^1(\theta, \varphi) = -\sqrt{\frac{3}{8\pi}} \sin\theta e^{i\varphi} \qquad \propto \frac{x+iy}{r}$$

$$Y_1^0(\theta, \varphi) = \sqrt{\frac{3}{4\pi}} \cos\theta \qquad \propto \frac{z}{r}$$

$$Y_1^{-1}(\theta, \varphi) = \sqrt{\frac{3}{8\pi}} \sin\theta e^{-i\varphi} \qquad \propto \frac{x-iy}{r} \qquad (3.91)$$

$$l = 2:$$

$$Y_2^2(\theta, \varphi) = \frac{1}{4}\sqrt{\frac{15}{2\pi}} \sin^2\theta e^{2i\varphi} \qquad \propto \frac{(x+iy)^2}{r^2}$$

$$Y_2^1(\theta, \varphi) = -\sqrt{\frac{15}{8\pi}} \sin\theta \cos\theta e^{i\varphi} \qquad \propto \frac{z(x+iy)}{r^2}$$

$$Y_2^0(\theta, \varphi) = \frac{1}{2}\sqrt{\frac{5}{4\pi}}(3\cos^2\theta - 1) \qquad \propto \frac{(2z^2 - x^2 - y^2)}{r^2}$$

$$Y_2^{-1}(\theta, \varphi) = \sqrt{\frac{15}{8\pi}} \sin\theta \cos\theta e^{-i\varphi} \qquad \propto \frac{z(x-iy)}{r^2}$$

$$Y_2^{-2}(\theta, \varphi) = \frac{1}{4}\sqrt{\frac{15}{2\pi}} \sin^2\theta e^{-2i\varphi} \qquad \propto \frac{(x-iy)^2}{r^2} \qquad (3.92)$$

$l = 3$:

$$Y_3^3(\theta, \varphi) = -\frac{1}{4}\sqrt{\frac{35}{4\pi}} \sin^3\theta e^{3i\varphi} \qquad \propto \frac{(x+iy)^3}{r^3}$$

$$Y_3^2(\theta, \varphi) = \frac{1}{4}\sqrt{\frac{105}{2\pi}} \sin^2\theta \cos\theta e^{2i\varphi} \qquad \propto \frac{z(x+iy)^2}{r^3}$$

$$Y_3^1(\theta, \varphi) = -\frac{1}{4}\sqrt{\frac{21}{4\pi}} \sin\theta(5\cos^2\theta - 1)e^{i\varphi} \qquad \propto \frac{(4z^2 - x^2 - y^2)(x+iy)}{r^3}$$

$$Y_3^0(\theta, \varphi) = \frac{1}{2}\sqrt{\frac{7}{4\pi}} \cos\theta(5\cos^2\theta - 3) \qquad \propto \frac{2z^3 - 3zx^2 - 3zy^2}{r^3}$$

$$Y_3^{-1}(\theta, \varphi) = \frac{1}{4}\sqrt{\frac{21}{4\pi}} \sin\theta(5\cos^2\theta - 1)e^{-i\varphi} \qquad \propto \frac{(4z^2 - x^2 - y^2)(x-iy)}{r^3}$$

$$Y_3^{-2}(\theta, \varphi) = \frac{1}{4}\sqrt{\frac{105}{2\pi}} \sin^2\theta \cos\theta e^{-2i\varphi} \qquad \propto \frac{z(x-iy)^2}{r^3}$$

$$Y_3^{-3}(\theta, \varphi) = \frac{1}{4}\sqrt{\frac{35}{4\pi}} \sin^3\theta e^{-3i\varphi} \qquad \propto \frac{(x-iy)^3}{r^3} \qquad (3.93)$$

For $m = 0$ we see from (3.87) that

$$Y_l^0(\theta, \varphi) = \sqrt{\frac{2l+1}{4\pi}} P_l^0(\cos\theta). \qquad (3.94)$$

The spherical harmonics are orthonormal in the sense

$$\boxed{\int_0^\pi \sin\theta \, d\theta \int_0^{2\pi} d\varphi \, Y_l^{m*}(\theta, \varphi) Y_{l'}^{m'}(\theta, \varphi) = \delta_{ll'}\delta_{mm'}.} \qquad (3.95)$$

Here the φ integration forces $m = m'$ for a nonvanishing result, and the orthogonality property (3.74) then forces $l = l'$ for a nonzero result. The factors in (3.87) have been chosen so that the normalization integral with $l = l'$, $m = m'$ is unity.

Since the $e^{im\varphi}$ form a complete set for functions of φ and the $P_l^m(\cos\theta)$ a complete set for functions of $\cos\theta$, the $Y_l^m(\theta, \varphi)$ form a complete set for functions of angle on the sphere.[4] Thus, we can expand an arbitrary function $f(\theta, \varphi)$ in the series

$$f(\theta, \varphi) = \sum_{l=0}^\infty \sum_{m=-l}^{+l} C_{lm} Y_l^m(\theta, \varphi). \qquad (3.96)$$

Using the orthonormality property (3.95) we can evaluate the C_{lm}:

$$C_{lm} = \int_0^\pi \sin\theta' \, d\theta' \int_0^{2\pi} d\varphi' \, Y_l^{m*}(\theta', \varphi') f(\theta', \varphi'). \qquad (3.97)$$

Substituting (3.97) in (3.96) we derive the completeness relation:

$$\boxed{\sum_{l=0}^\infty \sum_{m=-l}^{+l} Y_l^m(\theta, \varphi) Y_l^{m*}(\theta', \varphi') = \frac{\delta(\theta - \theta')\delta(\varphi - \varphi')}{\sin\theta'}.} \qquad (3.98)$$

---

[4]Courant and Hilbert, op. cit. p. 513; Hobson, op. cit. p. 342.

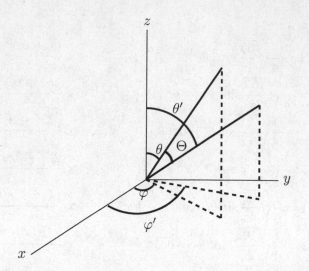

Figure 3.1: Angle between two directions in spherical coordinates .

3.7 THE SPHERICAL HARMONICS ADDITION THEOREM

We complete the formal part of our study of spherical harmonics with a derivation of the addition theorem which expresses $P_l(\cos\Theta)$, Θ being the angle between two directions in space, in terms of $Y_l^m(\theta,\varphi)$ and $Y_l^m(\theta',\varphi')$, θ,φ and θ',φ' being the spherical coordinates locating the two directions in question. The angles are indicated in Figure 3.1.

The proof is based on the invariance of the differential operator

$$\mathcal{L} = \frac{1}{\sin\theta}\frac{\partial}{\partial\theta}\left[\sin\theta\frac{\partial}{\partial\theta}\right] + \frac{1}{\sin^2\theta}\frac{\partial^2}{\partial\varphi^2}$$

$$= \mathfrak{L} = \frac{1}{\sin\Theta}\frac{\partial}{\partial\Theta}\left[\sin\Theta\frac{\partial}{\partial\Theta}\right] + \frac{1}{\sin^2\Theta}\frac{\partial^2}{\partial\Phi^2} \tag{3.99}$$

with respect to the rotations of coordinates $(x,y,z) \to (X,Y,Z)$ as shown in Figure 3.2. This follows from the invariance of the Laplacian

$$\boldsymbol{\nabla}^2 = \frac{\partial^2}{\partial x^2} + \frac{\partial^2}{\partial y^2} + \frac{\partial^2}{\partial z^2} = \boldsymbol{\nabla}'^2$$

$$= \frac{\partial^2}{\partial X^2} + \frac{\partial^2}{\partial Y^2} + \frac{\partial^2}{\partial Z^2}, \tag{3.100}$$

since $\mathcal{L}$ is just the angular part of the Laplacian.

Proof. Now the $Y_l^m(\theta,\varphi)$ are the eigenfunctions of $\mathcal{L}$,

$$\mathcal{L}Y_l^m(\theta,\varphi) = -l(l+1)Y_l^m(\theta,\varphi), \tag{3.101}$$

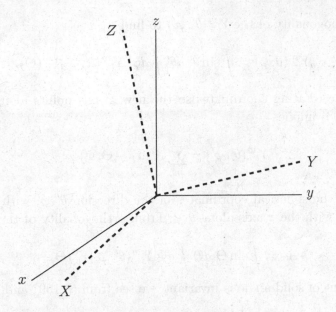

Figure 3.2: Rotation of coordinate axis

and since $\mathcal{L} = \mathfrak{L}$, we have

$$\mathfrak{L} Y_l^m(\theta, \varphi) = -l(l+1) Y_l^m(\theta, \varphi). \tag{3.102}$$

On the other hand, the eigenfunctions of $\mathfrak{L}$ belonging to the eigenvalue l are the $Y_l^m(\Theta, \Phi)$, so we must have

$$Y_l^m(\theta, \varphi) = \sum_{m'} a_{mm'} Y_l^{m'}(\Theta, \Phi). \tag{3.103}$$

Applying this result to the problem in hand, by taking the new z axis along one or the other of the two directions θ, φ or θ', φ', we see that

$$Y_l^0(\Theta) = \sum_m a_m(\theta', \varphi') Y_l^m(\theta, \varphi)$$

$$= \sum_{m'} b_{m'}(\theta, \varphi) Y_l^{m'}(\theta', \varphi'). \tag{3.104}$$

Together these imply

$$Y_l^0(\Theta) = \sum_{m,m'} c_{mm'} Y_l^{m'}(\theta', \varphi') Y_l^m(\theta, \varphi). \tag{3.105}$$

Consider now a rotation around the z axis. Neither Θ nor $\varphi - \varphi'$ changes, but in the terms in (3.105), containing the factors $e^{im'\varphi' + im\varphi}$, we must have $m' = m$. Thus (3.105) reduces to the form

$$Y_l^0(\Theta) = \sum_m c_m Y_l^{m*}(\theta', \varphi') Y_l^m(\theta, \varphi). \tag{3.106}$$

We can use orthogonality of the $Y_l^m(\theta', \varphi')$ to find c_m

$$c_m Y_l^m(\theta, \varphi) = \int \sin\theta' \, d\theta' \int d\varphi' \, Y_l^m(\theta', \varphi') Y_l^0(\Theta). \tag{3.107}$$

Now imagine rotating coordinates so the new z axis points along the direction θ, φ. Then (3.103) implies

$$Y_l^m(\theta', \varphi') = \sum_{m'} A'_m Y_l^{m'}(\Theta, \Phi), \tag{3.108}$$

where Θ, Φ are the spherical coordinates of the direction θ', φ' with respect to the new set of axes with the z axis along θ, φ. From orthogonality of the $Y_l^{m'}(\Theta, \Phi)$ we obtain

$$A_0 = \int \sin\Theta \, d\Theta \int d\Phi \, Y_l^m(\theta', \varphi') Y_l^0(\Theta). \tag{3.109}$$

Since the element of solid angle is invariant, we see from (3.109) and (3.107) that

$$c_m Y_l^m(\theta, \varphi) = A_0. \tag{3.110}$$

But if we set $\Theta = 0$ in (3.108), $\theta, \varphi \to \theta', \varphi'$ and all $Y_l^{m'}(\Theta, \Phi)$ vanish except when $m' = 0$ [see (3.88)], so we find

$$Y_l^m(\theta, \varphi) = A_0 Y_l^0(0). \tag{3.111}$$

Combining (3.110) and (3.111) yields

$$c_m = \frac{1}{Y_l^0(0)}, \tag{3.112}$$

so that (3.106) becomes

$$Y_l^0(\Theta) Y_l^0(0) = \sum_m Y_l^{m*}(\theta', \varphi') Y_l^m(\theta, \varphi). \tag{3.113}$$

Using (3.94) and (3.25) this can be put finally in the form

$$\boxed{P_l(\cos\Theta) = \frac{4\pi}{2l+1} \sum_{m=-l}^{l} Y_l^{m*}(\theta', \varphi') Y_l^m(\theta, \varphi),} \tag{3.114}$$

where

$$\boxed{\cos\Theta = \cos\theta \cos\theta' + \sin\theta \sin\theta' \cos(\varphi - \varphi').} \tag{3.115}$$

□

The remarkable addition theorem (3.114) enables one to separate the dependence on the two direction θ, φ and θ', φ' between which Θ is the angle.

3.8 MULTIPOLE EXPANSIONS

In the remainder of this chapter we study some basic applications of spherical harmonics. First we consider the multipole expansion of the electrostatic potential due to a charge distribution $\rho(\mathbf{r}')$. For a point charge e at $\mathbf{r}'$ the electrostatic potential at $\mathbf{r}$ is given by

$$\frac{e}{|\mathbf{r} - \mathbf{r}'|}. \tag{3.116}$$

For a distribution with charge density $\rho(\mathbf{r}')$ the electrostatic potential at $\mathbf{r}$ is given by

$$\phi(r) = \int d^3x' \, \frac{\rho(\mathbf{r}')}{|\mathbf{r} - \mathbf{r}'|}. \tag{3.117}$$

This is the solution of the Poisson equation

$$\nabla^2 \phi(\mathbf{r}) = -4\pi \rho(\mathbf{r}). \tag{3.118}$$

In terms of $\phi(\mathbf{r})$, the electric field $\mathbf{E}$ is given by

$$\mathbf{E}(\mathbf{r}) = -\nabla \phi(\mathbf{r}). \tag{3.119}$$

Suppose all of the charge is located in a finite region $r' < R$, and we ask for the potential outside this region, $r > R$, as shown in Figure 3.3. We can use our results for Legendre polynomials and spherical harmonics. The expansion (3.43) gives

$$\frac{1}{|\mathbf{r} - \mathbf{r}'|} = \frac{1}{\sqrt{(\mathbf{r} - \mathbf{r}')^2}}$$

$$= \frac{1}{\sqrt{r^2 - 2rr'\cos\Theta + r'^2}}$$

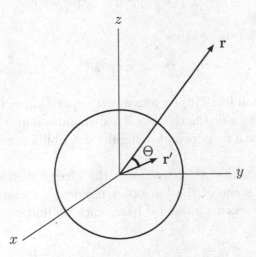

Figure 3.3: Vectors to observation point, $\mathbf{r}$, and to sources, $\mathbf{r}'$, for multipole expansion

or

$$\frac{1}{|\mathbf{r} - \mathbf{r}'|} = \frac{1}{r\sqrt{1 - 2\frac{r'}{r}\cos\Theta + \left(\frac{r'}{r}\right)^2}}$$

$$\boxed{\frac{1}{|\mathbf{r} - \mathbf{r}'|} = \frac{1}{r}\sum_l \left(\frac{r'}{r}\right)^l P_l(\cos\Theta), \quad r' < r.}$$ (3.120)

We emphasize that this series converges only for $r' < r$. The spherical harmonics addition theorem (3.114) can now be employed for $P_l(\cos\Theta)$:

$$P_l(\cos\Theta) = \frac{4\pi}{2l+1}\sum_m Y_l^{m*}(\theta', \varphi')Y_l^m(\theta, \varphi),$$ (3.121)

$$\frac{1}{|\mathbf{r} - \mathbf{r}'|} = \sum_{l,m} \frac{4\pi}{2l+1} r'^l Y_l^{m*}(\theta', \varphi')\frac{1}{r^{l+1}} Y_l^m(\theta, \varphi).$$ (3.122)

In this expression θ, φ are the spherical angles of $\mathbf{r}$, and θ', φ' are the spherical angles of $\mathbf{r}'$. The formula (3.122) separates the dependence on $\mathbf{r}$ from the dependence on $\mathbf{r}'$. The price paid is that there is an infinite sum of such separated terms. Substituting (3.122) in (3.117) we find

$$\boxed{\phi(\mathbf{r}) = \sum_{l=0}^{\infty} \sum_{m=-l}^{l} \frac{4\pi}{2l+1} q_l^m \frac{Y_l^m(\theta, \varphi)}{r^{l+1}},}$$ (3.123)

where

$$q_l^m = \int d^3x' \, Y_l^{m*}(\theta', \varphi')r'^l\rho(\mathbf{r}').$$ (3.124)

From the identity (3.89) we see that

$$q_l^{-m} = (-1)^m q_l^{m*}.$$ (3.125)

The multipole expansion (3.123) provides a nice separation of the potential $\phi(\mathbf{r})$ into pieces each of which has a distinctive angular distribution and dependence on r. The terms with larger l have a more complex angular distribution and fall off more rapidly with increasing r.

The multipole moments q_l^m characterize the charge distribution $\rho(\mathbf{r}')$. It is instructive to work out some of the low order moments. Using the explicit formulas (3.90)-(3.92) for the low order spherical harmonics we find

$$q_0^0 = \frac{1}{\sqrt{4\pi}} \int d^3x' \, \rho(\mathbf{r}') = \frac{1}{\sqrt{4\pi}}q,$$ (3.126)

where q is the total charge.

We also find

$$q_1^1 = -\sqrt{\frac{3}{8\pi}} \int d^3x' \; r' \sin\theta' e^{-i\varphi'} \rho(\mathbf{r'})$$

$$= -\sqrt{\frac{3}{8\pi}} \int d^3x' (x' - iy') \rho(\mathbf{r'})$$

$$= -\sqrt{\frac{3}{8\pi}} \, (p_x - i p_y), \tag{3.127}$$

$$q_1^0 = \sqrt{\frac{3}{4\pi}} \int d^3x' \; r' \cos\theta' \rho(\mathbf{r'})$$

$$= \sqrt{\frac{3}{4\pi}} \; p_z,$$

where the dipole moment is

$$\mathbf{p} = \int d^3x' \; \mathbf{r'} \rho(\mathbf{r'}), \tag{3.128}$$

and

$$q_2^2 = \frac{1}{4}\sqrt{\frac{15}{2\pi}} \int d^3x' \; r'^2 \sin^2\theta' e^{-2i\varphi'} \rho(\mathbf{r'})$$

$$= \frac{1}{4}\sqrt{\frac{15}{2\pi}} \int d^3x' \; (x' - iy')^2 \rho(\mathbf{r'})$$

$$= \frac{1}{12}\sqrt{\frac{15}{2\pi}} \, (Q_{11} - 2iQ_{12} - Q_{22}),$$

$$q_2^1 = -\sqrt{\frac{15}{8\pi}} \int d^3x' \; r'^2 \sin\theta' \cos\theta' e^{-i\varphi'} \rho(\mathbf{r'})$$

$$= -\sqrt{\frac{15}{8\pi}} \int d^3x' \; (x' - iy') z' \rho(\mathbf{r'}) \tag{3.129}$$

$$= -\frac{1}{3}\sqrt{\frac{15}{8\pi}} \, (Q_{13} - iQ_{23}),$$

$$q_2^0 = \frac{1}{2}\sqrt{\frac{5}{4\pi}} \int d^3x' \; r'^2 (3\cos^2\theta' - 1) \rho(\mathbf{r'})$$

$$= \frac{1}{2}\sqrt{\frac{5}{4\pi}} \int d^3x' \; (3z'^2 - r'^2) \rho(\mathbf{r'})$$

$$= \frac{1}{2}\sqrt{\frac{5}{4\pi}} \, Q_{33},$$

where

$$Q_{ij} = \int d^3x' (3x'_i x'_j - r'^2 \delta_{ij}) \rho(\mathbf{r}') \tag{3.130}$$

is the quadrupole moment tensor satisfying

$$\sum_{i=1}^{3} Q_{ii} = 0. \tag{3.131}$$

We note that q_1^m and $\mathbf{p}$ are spherical and cartesian components for the same quantity as are q_2^m and Q_{ij}. Note in particular that there are three independent real numbers in q_1^m or $\mathbf{p}$ and five independent real numbers in q_2^m or Q_{ij}.

Substituting (3.126), (3.127), and (3.129) in (3.123) and using (3.131) we find

$$\boxed{\phi(\mathbf{r}) = \frac{q}{r} + \frac{\mathbf{p} \cdot \mathbf{r}}{r^3} + \frac{1}{2} \sum_{i,j} Q_{ij} \frac{x_i x_j}{r^5} + \cdots.} \tag{3.132}$$

This can also be obtained by expanding $|\mathbf{r} - \mathbf{r}'|^{-1}$ in a Taylor series, but the expansion in cartesian coordinates becomes more and more awkward for the higher terms. The general case is handled much more easily by the expansion in spherical harmonics (3.123).

We can also expand in multipole moments the energy of a charge distribution $\rho(\mathbf{r})$:

$$W = \int d^3x \, \rho(\mathbf{r}) \phi(\mathbf{r}). \tag{3.133}$$

Expand $\phi(\mathbf{r})$ in a Taylor series around the point $\mathbf{r} = 0$, assumed located inside the charge distribution:

$$\phi(\mathbf{r}) = \phi(0) + \sum_i x_i \frac{\partial \phi}{\partial x_i} + \frac{1}{2} \sum_{i,j} x_i x_j \frac{\partial^2 \phi}{\partial x_i \partial x_j} + \cdots$$

$$= \phi(0) - \mathbf{r} \cdot \mathbf{E}(0) - \frac{1}{2} \sum_{i,j} x_i x_j \frac{\partial E_j}{\partial x_i} + \cdots$$

$$= \phi(0) - \mathbf{r} \cdot \mathbf{E}(0) - \frac{1}{6} \sum_{i,j} (3x_i x_j - \delta_{ij} r^2) \frac{\partial E_j}{\partial x_i} + \cdots. \tag{3.134}$$

Here the last step depends on the assumption

$$\nabla \cdot E = 0 \tag{3.135}$$

valid for an electric field $\mathbf{E}$ produced by sources outside the region in question. Substituting (3.134) in (3.133) and using the definitions (3.128), (3.130) we find

$$W = q\phi(0) - \mathbf{p} \cdot \mathbf{E}(0) - \frac{1}{6} \sum_{i,j} Q_{ij} \frac{\partial E_j}{\partial x_i} + \cdots. \tag{3.136}$$

We can use this to calculate the interaction between two dipoles $\mathbf{p}_1$, $\mathbf{p}_2$. The potential due to $\mathbf{p}_1$ at distance $\mathbf{r}$ relative to $\mathbf{p}_1$ is then

$$\phi(\mathbf{r}) = \frac{\mathbf{p}_1 \cdot \mathbf{r}}{r^3}, \tag{3.137}$$

and the electric field is

$$\mathbf{E} = -\boldsymbol{\nabla}\phi = -\frac{\mathbf{p}_1}{r^3} + \frac{3\mathbf{p}_1 \cdot \mathbf{r}}{r^4} \cdot \frac{\mathbf{r}}{r}. \tag{3.138}$$

The interaction with the second dipole is

$$W = -\mathbf{p}_2 \cdot \mathbf{E} = \frac{\mathbf{p}_1 \cdot \mathbf{p}_2}{r^3} - 3\frac{(\mathbf{p}_1 \cdot \mathbf{r})(\mathbf{p}_2 \cdot \mathbf{r})}{r^5}. \tag{3.139}$$

Since atoms and nuclei have multipole moments, the expressions derived in this section are important in various atomic and nuclear problems.

3.9 LAPLACE'S EQUATION IN SPHERICAL COORDINATES

We have already done all the work for this problem. It is only necessary to assemble the pieces. The separation of variables was carried out in Section 2.5. The appropriate single valued, everywhere finite solutions of the angular equation (2.35) are the spherical harmonics $Y_l^m(\theta, \varphi)$, and the separation constant λ is the eigenvalue $\lambda = l(l+1)$ according to Section 3.6. With this value of λ we find from (2.44), (2.45) as solution of the radial equation

$$R = Ar^l + B\frac{1}{r^{l+1}}. \tag{3.140}$$

Multiplying together the radial and angular parts we obtain a solution of Laplace's equation

$$\left[Ar^l + B\frac{1}{r^{l+1}}\right] Y_l^m(\theta, \varphi). \tag{3.141}$$

A general solution of Laplace's equation is a superposition of such solutions:

$$\boxed{\psi(\mathbf{r}) = \sum_{l=0}^{\infty} \sum_{m=-l}^{l} \left[A_{lm}r^l + B_{lm}\frac{1}{r^{l+1}}\right] Y_l^m(\theta, \varphi).} \tag{3.142}$$

That a superposition of solutions is also a solution follows from the linear homogeneous form of Laplace's equation. To see that (3.142) is a general solution requires more effort. First we note that the form (3.142) is convenient for problems involving boundary conditions given on the surfaces of spheres, and convenient only for such problems, so we restrict ourselves to that case. As we show in the examples below, the arbitrary constants A_{lm} and B_{lm} can then be chosen to satisfy arbitrary boundary conditions for ψ on the surfaces of spheres or arbitrary boundary conditions for $\partial\psi/\partial r$ on the surfaces of spheres. On the other hand with either ψ or $\partial\psi/\partial r$ given on the surfaces of spheres, ψ is uniquely determined. To see this, suppose we have two solutions ψ_1, ψ_2, both satisfying Laplace's equation and both satisfying the same boundary conditions for ψ or the normal derivative $\partial\psi/\partial n$ on some surface S (not necessarily spheres) of a volume V. Then $u = \psi_1 - \psi_2$ satisfies Laplace's equation and the boundary condition $u = 0$ or $\partial u/\partial n = 0$ on S. Applying Gauss's theorem to $\nabla \cdot (u\nabla u) = u\nabla^2 u + (\nabla u)^2$ we find

$$\int_S d\mathbf{A} \cdot u\nabla u = \int_V d^3x \left[u\nabla^2 u + (\nabla u)^2 \right], \tag{3.143}$$

from which for either case we obtain

$$\int_V d^3x \, (\nabla^2 u)^2 = 0. \tag{3.144}$$

Since the integrand is always positive, we find

$$\nabla u = 0. \tag{3.145}$$

For the case $u = 0$ on the boundary (3.145) implies $u = \psi_1 - \psi_2 = 0$ everywhere. For the case $\partial u/\partial n = 0$ on the boundary ψ_1 and ψ_2 can differ only by a constant.

We can now apply our general solution (3.142) to various examples of boundary value problems:

3.9.1 Interior Problem I, $r \leq a$ with $\psi(a,\theta,\varphi) = u(\theta,\varphi)$ given

If we specify in addition that ψ be finite everywhere, then we have

$$B_{lm} = 0, \tag{3.146}$$

$$\psi(r,\theta,\varphi) = \sum_{l,m} A_{lm} r^l Y_l^m(\theta,\varphi). \tag{3.147}$$

On the surface of the sphere this becomes

$$\psi(a,\theta,\varphi) = u(\theta,\varphi) = \sum_{l,m} A_{lm} a^l Y_l^m(\theta,\varphi). \tag{3.148}$$

Since the $Y_l^m(\theta,\varphi)$ form a complete set, this expansion of the arbitrary function $u(\theta,\varphi)$ is possible. Using the orthonormality of the $Y_l^m(\theta,\varphi)$ we can solve for the coefficients A_{lm}:

$$A_{lm} = \frac{1}{a^l} \int_0^\pi \sin\theta d\theta \int_0^{2\pi} d\varphi \, Y_l^{m*}(\theta,\varphi) u(\theta,\varphi). \tag{3.149}$$

3.9.2 Interior Problem II, $r \leq a$ with $\left.\frac{\partial\psi}{\partial r}\right|_{r=a} = v(\theta,\varphi)$ given

Specifying ψ finite at $r = 0$ eliminates the B_{lm} as above and we find

$$\psi(r,\theta,\varphi) = \sum_{l,m} A_{lm} r^l Y_l^m(\theta,\varphi), \tag{3.150}$$

$$\frac{\partial \psi}{\partial r} = \sum_{l,m} l A_{lm} r^{l-1} Y_l^m(\theta, \varphi), \tag{3.151}$$

$$v(\theta, \varphi) = \sum_{l,m} l A_{lm} a^{l-1} Y_l^m(\theta, \varphi), \tag{3.152}$$

$$A_{lm} = \frac{1}{l a^{l-1}} \int d\Omega \, Y_l^{m*}(\theta, \varphi) v(\theta, \varphi), \quad l \neq 0. \tag{3.153}$$

A_{00} arbitrary.

3.9.3 Exterior Problem, $r \geq a$ with $\psi(a, \theta, \varphi) = u(\theta, \varphi)$ given

If we specify in addition that $\psi \to 0$ as $r \to \infty$, then we have

$$A_{lm} = 0, \tag{3.154}$$

$$\psi(r, \theta, \varphi) = \sum_{l,m} B_{lm} \frac{1}{r^{l+1}} Y_l^m(\theta, \varphi), \tag{3.155}$$

$$B_{lm} = a^{l+1} \int d\Omega \, Y_l^{m*}(\theta, \varphi) u(\theta, \varphi). \tag{3.156}$$

3.9.4 Exterior Problem, $r \geq a$ with $\left.\frac{\partial \psi}{\partial r}\right|_{r=a} = v(\theta, \varphi)$ given

Specifying $\psi \to 0$ at $r \to \infty$ eliminates A_{lm} as above and we find

$$\psi(r, \theta, \varphi) = \sum_{l,m} B_{lm} \frac{1}{r^{l+1}} Y_l^m(\theta, \varphi), \tag{3.157}$$

$$B_{lm} = -\frac{a^{l+2}}{l+1} \int d\Omega \, Y_l^{m*}(\theta, \varphi) v(\theta, \varphi) \tag{3.158}$$

3.9.5 Region Between Two Spheres, $a \leq r \leq b$ with $\psi(a, \theta, \varphi) = u(\theta, \varphi)$ and $\psi(b, \theta, \varphi) = v(\theta, \varphi)$ given

We need all the terms in the expansion:

$$\psi(r, \theta, \varphi) = \sum_{l,m} \left[A_{lm} r^l + B_{lm} \frac{1}{r^{l+1}} \right] Y_l^m(\theta, \varphi) \tag{3.159}$$

The boundary conditions now read

$$\psi(a, \theta, \varphi) = u(\theta, \varphi) = \sum_{l,m} \left[A_{lm} a^l + B_{lm} \frac{1}{a^{l+1}} \right] Y_l^m(\theta, \varphi), \tag{3.160}$$

$$\psi(b,\theta,\varphi) = v(\theta,\varphi) = \sum_{l,m} \left[A_{lm}b^l + B_{lm}\frac{1}{b^{l+1}} \right] Y_l^m(\theta,\varphi). \qquad (3.161)$$

Using the orthonormality of the $Y_l^m(\theta,\varphi)$ we obtain

$$A_{lm}a^l + B_{lm}\frac{1}{a^{l+1}} = \int d\Omega \, Y_l^{m*}(\theta,\varphi)u(\theta,\varphi), \qquad (3.162)$$

$$A_{lm}b^l + B_{lm}\frac{1}{b^{l+1}} = \int d\Omega \, Y_l^{m*}(\theta,\varphi)v(\theta,\varphi), \qquad (3.163)$$

These two equations can be solved for A_{lm} and B_{lm}.

3.9.6 Notes on Solving Other Boundary Conditions on Regions Between Two Spheres

Using obvious extensions of the above techniques, we can also solve the cases in which $\partial\psi/\partial r$ is given on both surfaces or ψ is given on one surface, $\partial\psi/\partial r$ on the other.

3.10 CONDUCTING SPHERE IN A UNIFORM EXTERNAL ELECTRIC FIELD

This is a favorite exterior problem. $\psi(\mathbf{r})$ is now the electrostatic potential, and the boundary conditions are

$$\psi(a,\theta,\varphi) = \text{const.}, \qquad (3.164)$$

$$\psi(r,\theta,\varphi) \xrightarrow[r\to+\infty]{} -E_0 r \cos\theta, \qquad (3.165)$$

where E_0 is the external electric field, assumed in the z direction:

$$-\boldsymbol{\nabla}(-E_0 r \cos\theta) = \boldsymbol{\nabla}E_o z = E_0\hat{\mathbf{k}}. \qquad (3.166)$$

The general solution of the Laplace's equation for the electrostatic potential outside the sphere can be written in the form (3.142):

$$\psi(\mathbf{r}) = \sum_{l,m} \left[A_{lm}r^l + B_{lm}\frac{1}{r^{l+1}} \right] Y_l^m(\theta,\varphi). \qquad (3.167)$$

Using $Y_1^0(\theta) = \sqrt{\frac{3}{4\pi}}\cos\theta$ we see that to satisfy the boundary condition (3.165) we must take

$$A_{10} = -\sqrt{\frac{4\pi}{3}}E_0,$$

$$A_{11} = A_{1-1} = 0,$$

$$A_{lm} = 0, \quad l \geq 2. \qquad (3.168)$$

Thus $\psi(\mathbf{r})$ has the form

$$\psi(\mathbf{r}) = \left[A_{00} + \frac{B_{00}}{r}\right]\frac{1}{\sqrt{4\pi}} + \left[-\sqrt{\frac{4\pi}{3}}E_0 r + \frac{B_{10}}{r^2}\right]Y_1^0(\theta) + \sum_{l,m}{}' \frac{B_{lm}}{r^{l+1}}\,Y_l^m(\theta,\varphi), \quad (3.169)$$

where $\sum'$ is a sum over all l, m pairs except 0,0 and 1,0.

To satisfy the boundary condition (3.164) we must have

$$-\sqrt{\frac{4\pi}{3}}E_0 a + \frac{B_{10}}{a^2} = 0,$$

$$B_{11} = B_{1-1} = 0$$
$$B_{lm} = 0, \quad l \geq 2. \quad (3.170)$$

Since only $\boldsymbol{\nabla}\psi = -\mathbf{E}$ has physical significance, the constant A_{00} can be dropped. On the other hand $B_{00}/\sqrt{4\pi}$ is the charge Q on the sphere, so finally we obtain

$$\psi(\mathbf{r}) = \frac{Q}{r} - E_0\left[r - \frac{a^3}{r^2}\right]\cos\theta. \quad (3.171)$$

Comparing this formula with (3.132) we see that the external field $\mathbf{E}$ has induced a dipole moment $p = a^3 E_0$ in the conducting sphere.

3.11   FLOW OF AN INCOMPRESSIBLE FLUID AROUND A SPHERICAL OBSTACLE

This is a very similar to the problem just worked, but with a different boundary condition at the surface of the sphere $r = a$. The fluid velocity is

$$\mathbf{v} = -\boldsymbol{\nabla}\psi, \quad (3.172)$$

where

$$\boldsymbol{\nabla}^2\psi = 0. \quad (3.173)$$

The boundary conditions are

$$v_r = -\frac{\partial\psi}{\partial r} = 0 \quad \text{at } r = a, \quad (3.174)$$

$$\psi \xrightarrow[r\to+\infty]{} -v_0 r\cos\theta. \quad (3.175)$$

The boundary condition (3.174) says that the flow at the surface of the obstacle must be tangent to the surface. The asymptotic boundary condition (3.175) is essentially identical to the boundary condition for the electrostatics problem above. By the same reasoning as used there, we conclude that the only nonvanishing terms are

$$\psi(\mathbf{r}) = A + \frac{B}{r} + \left(-v_0 r + \frac{C}{r^2}\right)\cos\theta. \quad (3.176)$$

The boundary condition (3.174) then leads to

$$B = 0$$

$$-v_0 - \frac{2C}{a^3} = 0. \tag{3.177}$$

The constant A has no significance since $\mathbf{v} = -\boldsymbol{\nabla}\psi$, so we finally obtain

$$\psi(\mathbf{r}) = -v_0 \left[r + \frac{a^3}{2r^2} \right] \cos\theta, \tag{3.178}$$

$$\mathbf{v} = -\boldsymbol{\nabla}\psi = v_0 \left[1 - \frac{a^3}{r^3} \right] \cos\theta\, \hat{\mathbf{r}} - v_0 \left[1 + \frac{a^3}{2r^3} \right] \sin\theta\, \hat{\boldsymbol{\theta}}. \tag{3.179}$$

There are a large number of possible variations of the examples in Sections 3.10 and 3.11. We consider some of these variations later when we study boundary value problems in dielectric and magnetic media. In all cases only the one term

$$\boxed{\left[Ar + \frac{B}{r^2} \right] \cos\theta} \tag{3.180}$$

from the general solution (3.142) is needed to satisfy the boundary conditions.

PROBLEMS

3.1 Expand the step function

$$f(x) = \begin{cases} -1, & -1 < x < 0 \\ +1, & 0 < x < 1 \end{cases}$$

in a series of Legendre polynomials $P_l(x)$. Obtain an explicit formula for the expansion coefficients.

3.2 Consider the integral

$$I = \int d\Omega \ f(\cos\alpha)g(\cos\beta) = \int \sin\theta \ d\theta d\varphi \ f(\cos\alpha)g(\cos\beta),$$

where f and g are two arbitrary functions and $\alpha = \theta$ and β are the angles between the integration direction (θ, φ) and the two fixed axes (the z axis and the z' axis at an angle γ with respect to the z axis in the x, z plane) as shown in Figure 3.4.

a) By transforming integration variables show that I can be written in the form

$$I = 2\int_{-1}^{1} d(\cos\alpha) \left[\int_{-1}^{1} d(\cos\beta) f(\cos\alpha)g(\cos\beta) \times \right.$$

$$\left. \frac{\theta(1 - \cos^2\alpha - \cos^2\beta - \cos^2\gamma + 2\cos\alpha\cos\beta\cos\gamma)}{\sqrt{1 - \cos^2\alpha - \cos^2\beta - \cos^2\gamma + 2\cos\alpha\cos\beta\cos\gamma}} \right],$$

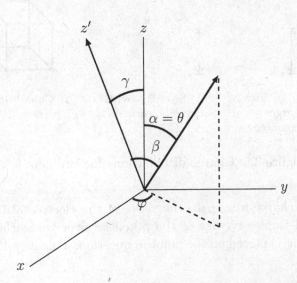

Figure 3.4: Angles of consideration for Problem 3.2

where $\theta(x)$ is the step function

$$\theta(x) = \begin{cases} 1, & x > 0 \\ 0, & x < 0. \end{cases}$$

b) By expanding $f(\cos\alpha)$ and $g(\cos\beta)$ in a series of Legendre polynomials and using the spherical harmonics addition theorem, show that I can be written in the form

$$I = \int_{-1}^{1} d(\cos\alpha) \left[\int_{-1}^{1} d(\cos\beta) \times \right.$$

$$\left. \pi \sum_{l=0}^{\infty} (2l+1) P_l(\cos\alpha) P_l(\cos\beta) P_l(\cos\gamma) f(\cos\alpha) g(\cos\beta) \right].$$

c) Derive the identity

$$\pi \sum_{l=0}^{\infty} (2l+1) P_l(x) P_l(y) P_l(z) = \frac{2\theta(1 - x^2 - y^2 - z^2 + 2xyz)}{\sqrt{1 - x^2 - y^2 - z^2 + 2xyz}}.$$

3.3 Find the electrostatic potential at large distances due to the following arrangements of point charges. The charges have equal magnitude q and signs $(\pm)$ as indicated in Figure 3.5. What is the lowest order multipole moment in each case?

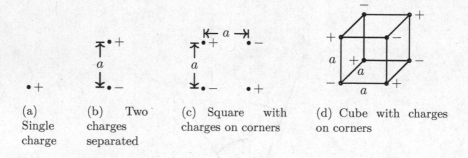

(a) Single charge

(b) Two charges separated

(c) Square with charges on corners

(d) Cube with charges on corners

Figure 3.5: Charge distributions for Problem 3.3

3.4 Show that in charge-free space the value of the electrostatic potential at any point is equal to the average of the potential over the surface of a sphere of arbitrary radius centered on the point in question. This is called the mean value theorem.

3.5 Two concentric conducting spheres of radii a and b are separated into four hemispheres by a nonconducting sheet.

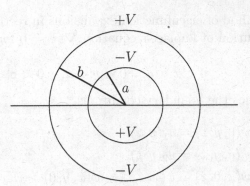

Figure 3.6: Conducting spheres separated into hemispheres for Problem 3.5

The upper outer and lower inner hemispheres are kept at potential $+V$ while the upper inner and lower outer hemispheres are kept at potential $-V$, as identified in Figure 3.6. Find the electrostatic potential in the region between the two spheres. Evaluate the expansion coefficients explicitly for terms up to $l = 4$.

3.6 The electrostatic potential $\psi(r, \theta, \varphi)$ inside a sphere of radius a satisfies Laplace's equation $\nabla^2 \psi = 0$ and the boundary condition $\psi(a, \theta, \varphi) = V(\theta, \varphi)$. Show that the potential inside the sphere is given by

$$\psi(r, \theta, \varphi) = \frac{a(a^2 - r^2)}{4\pi} \int d\Omega' \frac{V(\theta', \varphi')}{(r^2 + a^2 - 2ar\cos\gamma)^{\frac{3}{2}}},$$

where $\cos\gamma = \cos\theta\cos\theta' + \sin\theta\sin\theta'\cos(\varphi - \varphi')$.

To carry out the derivation of this formula sum the series (3.147), (3.149), using the spherical harmonics addition theorem, the generating function

$$\frac{1}{(r^2 + a^2 - 2ar\cos\gamma)^{\frac{1}{2}}} = \sum_l \frac{r^l}{a^{l+1}} P_l(\cos\gamma),$$

and its derivative with respect to a.

3.7 The z independent solutions of Laplace's equation in cylindrical coordinates are

$$\psi = (A_0 + B_0 \ln r)(C_0 + D_0 \theta)$$
$$+ \sum_{m=1}^{\infty} (A_m r^m + B_m r^{-m})(C_m e^{im\theta} + D_m e^{-im\theta}).$$

Study the (z independent) interior problem for the cylinder. Find the finite solution which satisfies the boundary condition $\psi(a, \theta) = u(\theta)$. Sum the series and show that your solution can be written in the form

$$\psi(r, \theta) = \frac{a^2 - r^2}{2\pi} \int_0^{2\pi} d\theta' \, u(\theta') \frac{1}{a^2 + r^2 - 2ar\cos(\theta - \theta')}.$$

3.8 a) Use the method of eigenfunction expansions in rectangular coordinates to find the solution of Laplace's equation $\nabla^2 \psi = 0$ for the region inside the cube

$$0 < x < a, \quad 0 < y < a, \quad 0 < z < a,$$

which satisfies the boundary conditions

$$\psi(0, y, z) = u_1(y, z),$$
$$\psi(a, y, z) = u_2(y, z),$$
$$\psi(x, 0, z) = \psi(x, a, z) = \psi(x, y, 0) = \psi(x, y, a) = 0.$$

b) Apply this solution to find the electrostatic potential inside a hollow cube with conducting walls when the right and left sides of the cube are kept at constant potential V and the other sides are kept at zero potential. Evaluate the expansion coefficients explicitly.

c) Evaluate numerically the potential at the center of the cube. How many terms of the series are necessary to obtain a result to three significant figures?

d) Find the solution of Laplace's equation for the inside of the cube which satisfies nonzero boundary conditions on all six sides:

$$\psi(0, y, z) = u_1(y, z),$$
$$\psi(a, y, z) = u_2(y, z),$$
$$\psi(x, 0, z) = v_1(x, z),$$
$$\psi(x, a, z) = v_2(x, z),$$
$$\psi(x, y, 0) = w_1(x, y),$$
$$\psi(x, y, a) = w_2(x, y).$$

CHAPTER 4

Bessel Functions and Applications

4.1 INTRODUCTION

When one separates variables in the Helmholtz equation in cylindrical or spherical coordinates the radial equation is Bessel's equation or can be reduced to Bessel's equation (see Sections 2.4 and 2.5). Bessel functions come up in many calculations in physics. Many important integrals, for example, can be expressed in terms of them. In fact, Bessel functions are similar in many respects to trigonometric functions and it is necessary to have a working knowledge of them for the same sorts of reasons that it is necessary to have a working knowledge of trigonometric functions.

4.2 SERIES SOLUTIONS OF BESSEL'S EQUATION; BESSEL FUNCTIONS

As we saw in Section 2.4 Bessel's equation is

$$\frac{d^2u}{dx^2} + \frac{1}{x}\frac{du}{dx} + \left[1 - \frac{m^2}{x^2}\right]u = 0. \tag{4.1}$$

We shall obtain series solutions according to the method of Section 2.7 by expanding around the regular singular point $x = 0$:

$$u(x) = x^\alpha \left[1 + \sum_{j=1}^{\infty} u_j x^j\right]. \tag{4.2}$$

Substituting this series in the differential equation (4.1), after multiplication by x^2, we obtain

$$x^\alpha \left[\alpha(\alpha-1) + \sum_{j=1}^{\infty}(\alpha+j)(\alpha+j-1)u_j x^j\right] + x^\alpha \left[\alpha + \sum_{j=1}^{\infty}(\alpha+j)u_j x^j\right]$$

$$+ x^\alpha \left[x^2 + \sum_{j=1}^{\infty} u_j x^{j+2} - m^2 - m^2 \sum_{j=1}^{\infty} u_j x^j\right] = 0. \tag{4.3}$$

Equating coefficients of successive powers of x to zero we obtain first the indicial equation

$$\alpha(\alpha - 1) + \alpha - m^2 = 0, \quad \alpha = \pm m, \tag{4.4}$$

then

$$[(\alpha + 1)\alpha + \alpha + 1 - m^2]u_1 = 0, \tag{4.5}$$
$$[(\alpha + 2)(\alpha + 1) + \alpha + 2 - m^2]u_2 + 1 = 0 \tag{4.6}$$
$$\vdots$$

$$[(\alpha + j)(\alpha + j - 1) + \alpha + j - m^2]u_j + u_{j-2} = 0. \tag{4.7}$$

Using the indicial equation, (4.5) reduces to

$$(2\alpha + 1)u_1 = 0. \tag{4.8}$$

For $\alpha \neq -1/2$ this implies $u_1 = 0$; in fact we can and do take $u_1 = 0$ even for $\alpha = -1/2$. The recursion relation (4.7) then implies

$$u_1 = u_3 = u_5 = \ldots = u_{2n+1} = \ldots = 0. \tag{4.9}$$

Using the indicial equation $\alpha^2 = m^2$ in the recursion relation (4.7), we find it can be rewritten in the form

$$u_{2j} = -\frac{u_{2(j-1)}}{(2j + \alpha)^2 - m^2} = -\frac{u_{2(j-1)}}{4j(j + \alpha)}. \tag{4.10}$$

Applying (4.10) repeatedly we can express u_{2j} in terms of $u_0 = 1$:

$$u_{2j} = (-1)^2 \frac{u_{2(j-2)}}{4j(j - 1)(j + \alpha)(j + \alpha - 1)} = \frac{(-1)^j \Gamma(1 + \alpha)}{2^{2j} j! \, \Gamma(j + \alpha + 1)} u_0, \tag{4.11}$$

where Γ is the gamma function. Substituting (4.9) and (4.11) in (4.2) we find two solutions corresponding to the two roots of the indicial equation (4.4):

$$u(x) = \begin{cases} x^m \displaystyle\sum_{r=0}^{\infty} \frac{\Gamma(m + 1)(-1)^r}{2^{2r} r! \, \Gamma(r + m + 1)} x^{2r} \\[2em] x^{-m} \displaystyle\sum_{r=0}^{\infty} \frac{\Gamma(-m + 1)(-1)^r}{2^{2r} r! \, \Gamma(r - m + 1)} x^{2r} \end{cases} \tag{4.12}$$

Multiplying by $2^{\mp m}/\Gamma(\pm m + 1)$ we obtain the conventionally defined Bessel functions:

$$J_m(x) = \left(\frac{x}{2}\right)^m \sum_{r=0}^{\infty} \frac{(-1)^r}{r! \, \Gamma(r + m + 1)} \left(\frac{x}{2}\right)^{2r}, \tag{4.13}$$

$$J_{-m}(x) = \left(\frac{x}{2}\right)^{-m} \sum_{r=0}^{\infty} \frac{(-1)^r}{r! \, \Gamma(r - m + 1)} \left(\frac{x}{2}\right)^{2r}. \tag{4.14}$$

These series converge for all x according to the ratio test,

$$\frac{\left(\frac{x}{2}\right)^2}{r(r \pm m)} \xrightarrow[r \to +\infty]{} 0, \qquad (4.15)$$

and so define two solutions to Bessel's differential equation (4.1).

The two solutions obtained above are well defined and independent except for the difficult case in which the roots of the indicial equation differ by an integer, $2m =$ integer. Even for $2m =$ odd integer the two solutions are well defined and independent – the vanishing denominators appear in the terms $u_1, u_3, u_5, \ldots$, which have been set to zero. For $m =$ positive integer the first few terms in the series (4.14) vanish because the gamma functions in the denominator are infinite for negative arguments:

$$\Gamma(r - m + 1) = \infty, \quad \begin{cases} r + 1 - m = 0, -1, -2, \ldots, \text{ or} \\ r = m - 1, m - 2, \ldots, 0. \end{cases} \qquad (4.16)$$

Leaving out these terms and changing the dummy summation index we find from (4.14)

$$J_{-m}(x) = \left(\frac{x}{2}\right)^{-m} \sum_{r=m}^{\infty} \frac{(-1)^r}{r! \, \Gamma(r - m + 1)} \left(\frac{x}{2}\right)^{2r}$$

$$= \left(\frac{x}{2}\right)^{-m} \sum_{s=0}^{\infty} \frac{(-1)^{m+s}}{(m+s)! \, \Gamma(s+1)} \left(\frac{x}{2}\right)^{2(m+s)}$$

$$= (-1)^m \left(\frac{x}{2}\right)^{m} \sum_{s=0}^{\infty} \frac{(-1)^s}{s! \, \Gamma(s+m+1)} \left(\frac{x}{2}\right)^{2s}$$

$$\boxed{J_{-m}(x) = (-1)^m J_m(x), \quad m = 0, 1, 2, \ldots .} \qquad (4.17)$$

Thus for the case $m =$ integer we have found only one solution.

4.3 NEUMANN FUNCTIONS

For the case $m =$ integer the general theory discussed in Section 2.7 tells us there will be a second solution of the form

$$J_m(x) \ln x + x^{-m} \sum_{r=0}^{\infty} a_r x^r. \qquad (4.18)$$

We could substitute this in Bessel's equation and determine the a_r so as to obtain a solution. It is conventional to proceed differently. The second solution of Bessel's equation (the Neumann function) is defined by

$$\boxed{N_m(x) = \frac{J_m(x) \cos m\pi - J_{-m}(x)}{\sin m\pi}.} \qquad (4.19)$$

For $m \neq$ integer this is a perfectly well defined linear combination of $J_m(x)$ and $J_{-m}(x)$. For $m =$ integer the formula (4.19) reduces to 0/0 in virtue of (4.17). The Neumann function $N_m(x)$ is then defined for m integral by de l'Hospital's rule:

$$N_m(x) = \lim_{m \to \text{integer}} \frac{J_m(x) \cos m\pi - J_{-m}(x)}{\sin m\pi}$$

$$= \lim_{m \to \text{integer}} \frac{\frac{dJ_m(x)}{dm} \cos m\pi - \pi \sin m\pi J_m(x) - \frac{dJ_{-m}(x)}{dm}}{\pi \cos m\pi}$$

$$= \frac{1}{\pi} \left[\frac{dJ_m(x)}{dm} - (-1)^m \frac{dJ_{-m}(x)}{dm} \right], \quad m = \text{integer}. \tag{4.20}$$

We can see that this actually satisfies Bessel's equation for $m =$ integer by differentiating Bessel's equation with respect to m:

$$\frac{d^2 J_m(x)}{dx^2} + \frac{1}{x} \frac{dJ_m(x)}{dx} + \left[1 - \frac{m^2}{x^2} \right] J_m(x) = 0, \tag{4.21}$$

$$\frac{d^2}{dx^2} \left[\frac{dJ_m(x)}{dm} \right] + \frac{1}{x} \frac{d}{dx} \left[\frac{dJ_m(x)}{dm} \right] + \left[1 - \frac{m^2}{x^2} \right] \frac{dJ_m(x)}{dm} = \frac{2m}{x^2} J_m(x), \tag{4.22}$$

$$\frac{d^2}{dx^2} \left[\frac{dJ_{-m}(x)}{dm} \right] + \frac{1}{x} \left[\frac{dJ_{-m}(x)}{dm} \right] + \left[1 - \frac{m^2}{x^2} \right] \frac{dJ_{-m}(x)}{dm} = \frac{2m}{x^2} J_{-m}(x). \tag{4.23}$$

Multiplying (4.23) by $(-1)^m$, subtracting the result from (4.22) and using (4.17), we immediately find that (4.20) is a solution of Bessel's equation.

To calculate the derivatives with respect to m which appear in (4.20) we use the series expansions (4.13) and (4.14):

$$\frac{dJ_m(x)}{dm} = J_m(x) \ln \frac{x}{2} + \left(\frac{x}{2} \right)^m \sum_{r=0}^{\infty} \frac{(-1)^r}{r!} \left(\frac{x}{2} \right)^{2r} \frac{d}{dt} \frac{1}{\Gamma(t)} \Big|_{t=m+r+1}, \tag{4.24}$$

$$\frac{dJ_{-m}(x)}{dm} = -J_{-m}(x) \ln \frac{x}{2} - \left(\frac{x}{2} \right)^{-m} \sum_{r=0}^{\infty} \frac{(-1)^r}{r!} \left(\frac{x}{2} \right)^{2r} \frac{d}{dt} \frac{1}{\Gamma(t)} \Big|_{t=-m+r+1}. \tag{4.25}$$

We need the derivatives of the gamma function:

$$\frac{d}{dt} \frac{1}{\Gamma(t)} = -\frac{1}{\Gamma(t)} \frac{\Gamma'(t)}{\Gamma(t)} = -\frac{1}{\Gamma(t)} \frac{d}{dt} \ln \Gamma(t). \tag{4.26}$$

Repeated use of the recursion relation for the gamma function yields

$$\Gamma(t) = (t-1)\Gamma(t-1) = (t-1)(t-2)\ldots(t-n+1)\Gamma(t-n+.+1). \tag{4.27}$$

Taking the logarithm of (4.27) and differentiating, we obtain

$$\frac{\Gamma'(t)}{\Gamma(t)} = \frac{1}{t-1} + \frac{1}{t-2} + \ldots + \frac{1}{t-n+1} + \frac{\Gamma'(t-n+1)}{\Gamma(t-n+1)}. \tag{4.28}$$

If we now let $t = n =$ integer in (4.28), we obtain

$$\frac{\Gamma'(n)}{\Gamma(n)} = \frac{1}{n-1} + \frac{1}{n-2} + \ldots + \frac{1}{1} + \frac{\Gamma'(1)}{\Gamma(1)}. \tag{4.29}$$

The quantity $\Gamma'(1)/\Gamma(1)$ is by definition the negative of Euler's constant γ. A formula for γ can be obtained by taking the limit of (4.29) as $n \to \infty$ and using Stirling's formula for $\Gamma(n)$ for large n:

$$\Gamma(n+1) \xrightarrow[n \to +\infty]{} \sqrt{2\pi} \; n^{n+\frac{1}{2}} \; e^{-n}, \tag{4.30}$$

$$\frac{\Gamma'(n+1)}{\Gamma(n+1)} \xrightarrow[n \to +\infty]{} \ln n, \tag{4.31}$$

$$\frac{\Gamma'(1)}{\Gamma(1)} = -\lim_{n \to \infty} \left[\frac{1}{1} + \frac{1}{2} + \ldots + \frac{1}{n} - \frac{\Gamma'(n+1)}{\Gamma(n+1)} \right]$$

$$= -\lim_{n \to \infty} \left[\frac{1}{1} + \frac{1}{2} + \ldots + \frac{1}{n} - \ln n \right]$$

$$= -\gamma$$

$$= -0.5772\ldots. \tag{4.32}$$

Combining our results we find

$$\frac{d}{dt} \frac{1}{\Gamma(t)} \bigg|_{t=n} = \begin{cases} -\dfrac{1}{(n-1)!} \left[\dfrac{1}{n-1} + \dfrac{1}{n-2} + \ldots + \dfrac{1}{1} - \gamma \right], & n \geq 2, \\[2em] \gamma, & n = 1. \end{cases} \tag{4.33}$$

For some of the terms in (4.25) we need this derivative when t is 0 or a negative integer. To handle this case use (4.28) with $t \to t + k + 1$, $n \to k + 2$:

$$\frac{\Gamma'(t+k+1)}{\Gamma(t+k+1)} = \frac{1}{t+k} + \frac{1}{t+k-1} + \ldots + \frac{1}{t} + \frac{\Gamma'(t)}{\Gamma(t)}, \tag{4.34}$$

$$\frac{d}{dt} \frac{1}{\Gamma(t)} = -\frac{\Gamma'(t)}{\Gamma^2(t)}$$

$$= \frac{1}{\Gamma(t)} \left[\frac{1}{t+k} + \frac{1}{t+k-1} + \ldots + \frac{1}{t} - \frac{\Gamma'(t+k+1)}{\Gamma(t+k+1)} \right]. \tag{4.35}$$

Now set $t = -k$, $k = 0, 1, 2, \ldots$. For these values of k, $\Gamma(-k) = \infty$, so we obtain from (4.35)

$$\frac{d}{dt} \frac{1}{\Gamma(t)}\bigg|_{t=-k} = \lim_{t \to -k} \frac{1}{\Gamma(t)} \frac{1}{t+k}$$

$$= \lim_{t \to -k} \frac{t(t+1)\ldots(t+k-1)}{\Gamma(t)t(t+1)\ldots(t+k)}$$

$$= \lim_{t \to -k} \frac{t(t+1)\ldots(t+k-1)}{\Gamma(t+k+1)}$$

$$= (-1)^k \, k!, \quad k = 0, 1, 2, \ldots . \tag{4.36}$$

We can now combine formulas (4.20), (4.24), (4.25), (4.33), (4.36) to obtain the Neumann function $N_m(x)$ for m an integer:

$$N_m(x) = \frac{2}{\pi} J_m(x) \ln \frac{x}{2}$$

$$+ \frac{1}{\pi} \left(\frac{x}{2}\right)^m \sum_{r=0}^{\infty} \frac{(-1)^r}{r!} \left(\frac{x}{2}\right)^{2r} \frac{(-1)}{(m+r)!} \times$$

$$\left\{\frac{1}{m+r} + \frac{1}{m+r-1} + \ldots + \frac{1}{1} - \gamma\right\}$$

$$+ \frac{1}{\pi}(-1)^m \left(\frac{x}{2}\right)^{-m} \sum_{r=0}^{m-1} \frac{(-1)^r}{r!} \left(\frac{x}{2}\right)^{2r} (-1)^{m-r-1}(m-r-1)!$$

$$+ \frac{1}{\pi}(-1)^m \left(\frac{x}{2}\right)^{-m} \sum_{r=m}^{\infty} \frac{(-1)^r}{r!} \left(\frac{x}{2}\right)^{2r} \frac{(-1)}{(r-m)!} \times$$

$$\left\{\frac{1}{r-m} + \frac{1}{r-m-1} + \ldots + \frac{1}{1} - \gamma\right\}$$

continuing further

$$N_m(x) = \frac{2}{\pi} J_m(x) \left[\ln \frac{x}{2} + \gamma\right]$$

$$- \frac{1}{\pi} \left(\frac{x}{2}\right)^{-m} \sum_{r=0}^{m-1} \frac{(m-r-1)!}{r!} \left(\frac{x}{2}\right)^{2r}$$

$$- \frac{1}{\pi} \left(\frac{x}{2} \right)^m \frac{1}{m!} \left\{ \frac{1}{m} + \frac{1}{m-1} + \ldots + \frac{1}{1} \right\}$$

$$- \frac{1}{\pi} \left(\frac{x}{2} \right)^m \sum_{r=1}^{\infty} \frac{(-1)^r}{r! \, (m+r)!} \left(\frac{x}{2} \right)^{2r} \times$$

$$\left\{ \frac{1}{m+r} + \frac{1}{m+r-1} + \ldots + \frac{1}{1} + \frac{1}{r} + \frac{1}{r-1} + \ldots + \frac{1}{1} \right\}. \qquad (4.37)$$

This function has the expected singular behavior near $x = 0$ – see (4.18).

4.4 SMALL ARGUMENT AND ASYMPTOTIC EXPANSIONS

We have now explicitly constructed the two solutions of Bessel's equation, the Bessel function $J_m(x)$, $m \geq 0$, and the Neumann function $N_m(x)$. $J_m(x)$, $m \geq 0$, is finite at $x = 0$, and $N_m(x)$ is infinite at $x = 0$. Our power series solutions give explicitly the behavior near $x = 0$:

$$J_m(x) \xrightarrow[x \to 0]{} \frac{1}{\Gamma(m+1)} \left(\frac{x}{2} \right)^m, \qquad (4.38)$$

$$N_m(x) \xrightarrow[x \to 0]{} \begin{cases} -\dfrac{1}{\sin m\pi} J_{-m}(x) = -\dfrac{1}{\sin m\pi \, \Gamma(-m+1)} \left(\dfrac{x}{2} \right)^{-m}, & m \neq \text{integer} \\[3mm] \dfrac{2}{\pi} \left(\ln \dfrac{x}{2} + \gamma \right), & m = 0 \\[3mm] -\dfrac{(m-1)!}{\pi} \left(\dfrac{x}{2} \right)^{-m}, & m = \text{integer} \neq 0. \end{cases}$$

$$(4.39)$$

Using the known result [see (14.12)]

$$\Gamma(m)\Gamma(1-m) = \frac{\pi}{\sin m\pi} \qquad (4.40)$$

for the Γ function we see that the $m \neq 0$ cases can be described by the same formula whether m is an integer or not:

$$N_m(x) \xrightarrow[x \to 0]{} \begin{cases} -\dfrac{\Gamma(m)}{\pi} \left(\dfrac{x}{2} \right)^{-m}, & m \neq 0, \\[3mm] \dfrac{2}{\pi} \left(\ln \dfrac{x}{2} + \gamma \right), & m = 0, \end{cases} \qquad (4.41)$$

$$\text{with } \gamma = 0.5772\ldots$$

Formulas for the asymptotic behavior of $J_m(x)$ and $N_m(x)$ for $x \to \infty$ can be derived by using complex variable techniques and will be derived in that part of this book dealing with such methods – see Chapter 13. Here we quote the results:

$$J_m(x) \xrightarrow[x \to +\infty]{} \sqrt{\frac{2}{\pi x}} \cos\left[x - \frac{m\pi}{2} - \frac{\pi}{4}\right], \tag{4.42}$$

$$N_m(x) \xrightarrow[x \to +\infty]{} \sqrt{\frac{2}{\pi x}} \sin\left[x - \frac{m\pi}{2} - \frac{\pi}{4}\right]. \tag{4.43}$$

The formulas (4.38), (4.42) for $J_m(x)$ and (4.41), (4.43) for $N_m(x)$ together give a good picture of the general behavior of these functions. The small argument behavior changes to the large argument behavior in the region $x \sim m$.

The asymptotic formulas (4.42) and (4.43) show the similarity of Bessel functions to trigonometric functions. As in the trigonometric case, it is convenient to introduce linear combinations corresponding to complex exponentials. The Hankel functions are defined by

$$\boxed{H_m^{(1)}(x) = J_m(x) + iN_m(x),} \tag{4.44}$$

$$\boxed{H_m^{(2)}(x) = J_m(x) - iN_m(x).} \tag{4.45}$$

Referring to (4.42), (4.43) we see that these functions have the asymptotic form

$$H_m^{(1)}(x) \xrightarrow[x \to +\infty]{} \sqrt{\frac{2}{\pi x}} \exp\left[+i\left(x - \frac{m\pi}{2} - \frac{\pi}{4}\right)\right], \tag{4.46}$$

$$H_m^{(2)}(x) \xrightarrow[x \to +\infty]{} \sqrt{\frac{2}{\pi x}} \exp\left[-i\left(x - \frac{m\pi}{2} - \frac{\pi}{4}\right)\right]. \tag{4.47}$$

For small argument, $x \to 0$, the Hankel functions are dominated by the singular behavior of the Neumann function $N_m(x)$.

4.5   BESSEL FUNCTIONS OF IMAGINARY ARGUMENT

Bessel functions of imaginary argument often occur in practice and have special names and symbols. The differential equation

$$\frac{d^2u}{dx^2} + \frac{1}{x}\frac{du}{dx} - \left[1 + \frac{\lambda^2}{x^2}\right]u = 0 \tag{4.48}$$

can be written in the form

$$\frac{d^2u}{d(ix)^2} + \frac{1}{ix}\frac{du}{d(ix)} + \left[1 - \frac{\lambda^2}{(ix)^2}\right]u = 0 \tag{4.49}$$

and has solutions

$$u = \begin{cases} J_\lambda(ix) \\ . \qquad . \\ N_\lambda(ix) \end{cases} \tag{4.50}$$

It is convenient to define new functions which are real for real x:

$$I_\lambda(x) = e^{-\frac{1}{2}\lambda\pi i} J_\lambda(ix)$$

$$= \left(\frac{x}{2}\right)^\lambda \sum_{r=0}^{\infty} \frac{1}{r! \, \Gamma(r+\lambda+1)} \left(\frac{x}{2}\right)^{2r}, \tag{4.51}$$

$$K_\lambda(x) = \frac{\pi}{2} \frac{I_{-\lambda}(x) - I_\lambda(x)}{\sin\lambda\pi}$$

$$= \frac{\pi}{2} \frac{e^{\frac{1}{2}\lambda\pi i} J_{-\lambda}(ix) - e^{-\frac{1}{2}\lambda\pi i} J_\lambda(ix)}{\sin\lambda\pi}. \tag{4.52}$$

Solving (4.19) for $J_{-\lambda}(ix)$, substituting in (4.52) and using the definition (4.44) of $H_\lambda^{(1)}(ix)$, this becomes

$$K_\lambda(x) = \frac{\pi}{2} \, i \, e^{\frac{1}{2}\lambda\pi i} H_\lambda^{(1)}(ix). \tag{4.53}$$

Using the small and large argument expansions given above for $J_\lambda(x)$, $N_\lambda(x)$, $H_\lambda^{(1)}(x)$ and $H_\lambda^{(2)}(x)$ it is easy to find the small and large argument expansions for $I_\lambda(x)$, $K_\lambda(x)$:

$$I_\lambda(x) \xrightarrow[x\to 0]{} \frac{1}{\Gamma(\lambda+1)} \left(\frac{x}{2}\right)^\lambda, \tag{4.54}$$

$$I_\lambda(x) \xrightarrow[x\to\infty]{} \frac{1}{\sqrt{2\pi x}} \, e^x, \tag{4.55}$$

$$K_\lambda(x) \xrightarrow[x\to 0]{} \begin{cases} \frac{1}{2}\Gamma(\lambda) \left(\frac{x}{2}\right)^{-\lambda}, & \lambda \neq 0, \\[2mm] -\left(\ln\frac{x}{2} + \gamma\right), & \lambda = 0, \end{cases} \tag{4.56}$$

$$K_\lambda(x) \xrightarrow[x\to\infty]{} \sqrt{\frac{\pi}{2x}} \, e^{-x}, \tag{4.57}$$

4.6 LAPLACE'S EQUATION IN CYLINDRICAL COORDINATES

We have already carried through the separation of variables for the Helmholtz equation in cylindrical coordinates in Section 2.4. Setting $k^2 = 0$ so as to reduce Helmholtz's equation to Laplace's equation and forming a superposition of the solutions of Section 2.4 so as to obtain a general solution of Laplace's equation, we find

$$\nabla^2 \psi(r, \theta, z) = 0, \tag{4.58}$$

$$\psi(r, \theta, z) = \sum_{\alpha, m} \left[A J_m(\alpha r) + B N_m(\alpha r) \right] \left[C e^{im\theta} + D e^{-im\theta} \right] \left[E e^{\alpha z} + F e^{-\alpha z} \right]. \tag{4.59}$$

In (4.59) the constants A, B, C, D, E, F are functions of the separation constants α, m. We choose these constants and sum or integrate over α and m in whatever way is dictated by the boundary conditions imposed on $\psi(r, \theta, z)$. For most physics problems we will want to impose a boundary condition of a single valuedness,

$$\psi(r, \theta, z) = \psi(r, \theta + 2\pi, z), \tag{4.60}$$

which requires that m be integral,

$$m = 0, 1, 2 \ldots . \tag{4.61}$$

4.7 INTERIOR OF A CYLINDER OF FINITE LENGTH

As a first example suppose we wish to solve Laplace's equation for the interior of a cylinder of finite length, as seen in Figure 4.1, with $\psi(r, \theta, z)$ given on the surfaces:

$$\left. \begin{array}{c} \psi(r, \theta, b) \\ \psi(r, \theta, 0) \\ \psi(a, \theta, z) \end{array} \right\} \quad \text{given.} \tag{4.62}$$

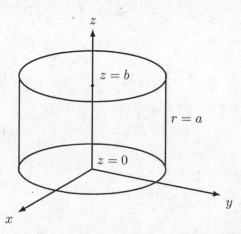

Figure 4.1: Finite length cylinder

We can divide this problem into three parts, in each of which ψ reduces to the given function on one of the surfaces and 0 on the other two surfaces:

$$\psi = \psi_1 + \psi_2 + \psi_3, \tag{4.63}$$

$$\nabla^2\psi_1 = \nabla^2\psi_2 = \nabla^2\psi_3 = 0, \tag{4.64}$$

$$\psi_1(r,\theta,b) = \psi(r,\theta,b), \quad \psi_1(r,\theta,0) = 0, \quad \psi_1(a,\theta,z) = 0, \tag{4.65}$$

$$\psi_2(r,\theta,b) = 0, \quad \psi_2(r,\theta,0) = \psi(r,\theta,0), \quad \psi_2(a,\theta,z) = 0, \tag{4.66}$$

$$\psi_3(r,\theta,b) = 0, \quad \psi_3(r,\theta,0) = 0, \quad \psi_3(a,\theta,z) = \psi(a,\theta,z). \tag{4.67}$$

Consider first the boundary value problem for ψ_1. We use the general expansion (4.59) for ψ_1. To ensure that ψ_1 vanishes on the surface $z = 0$, choose $F = -E$ so that

$$Ee^{\alpha z} + Fe^{-\alpha z} = E(e^{\alpha z} - e^{-\alpha z}) = 2E \sinh \alpha z. \tag{4.68}$$

As for the r dependence, we first impose the condition that ψ_1 be finite at $r = 0$. Because of the singular behavior of $N_m(\alpha r)$ for $r \to 0$ [see (4.41)] this implies $B = 0$. We have also imposed on ψ_1 the boundary condition $\psi_1 = 0$ at $r = a$. With $B = 0$ this implies

$$J_m(\alpha a) = 0. \tag{4.69}$$

Now, $J_m(x)$ is an oscillatory function with infinitely many zeros [see for example the asymptotic form (4.42)], so there is an infinite set of solutions of (4.69). Let x_{mn} be the nth zero (counting out from small x) of $J_m(x)$,

$$J_m(x_{mn}) = 0. \tag{4.70}$$

Then the solutions of (4.69) are given by

$$\alpha = \alpha_{mn} = \frac{x_{mn}}{a}. \tag{4.71}$$

We have now satisfied the $\psi_1 = 0$ boundary conditions on the bottom and sides of the cylinder and in so doing reduced the general form (4.59) to

$$\psi_1(r,\theta,z) = \sum_{m=0}^{\infty} \sum_{n=1}^{\infty} J_m(\alpha_{mn}r) \sinh(\alpha_{mn}z)\left(C_{mn}e^{im\theta} + D_{mn}e^{-im\theta}\right)$$

$$= \sum_{m=-\infty}^{\infty} \sum_{n=1}^{\infty} C_{mn} J_{|m|}(\alpha_{|m|n}r) \sinh(\alpha_{|m|n}z)e^{im\theta}. \tag{4.72}$$

Notice in particular how the form of summation over α in (4.59) has been determined by the boundary condition at $r = a$. We are left with the arbitrary constants C_{mn} in (4.72) which we can use to satisfy the boundary condition on the top of the cylinder, $\psi_1(r,\theta,b) = \psi(r,\theta,b)$.

4.8 THE STURM-LIOUVILLE EIGENVALUE PROBLEM AND APPLICATION OF THE EXPANSION THEOREM

With regard to the expansion problem just discussed in Section 4.7 we note that the $e^{im\theta}$ form a complete set for functions of θ, so that we have no difficulty expanding an arbitrary function of θ in these functions. As for the radial functions we note that the set of functions

$$u_n(r) = J_m(\alpha_{mn}r), \quad n = 1, 2, \ldots, \tag{4.73}$$

form the solution of the Sturm-Liouville eigenvalue problem

$$\frac{d^2 u_n(r)}{dr^2} + \frac{1}{r}\frac{du_n(r)}{dr} - \frac{m^2}{r^2}u_n(r) = -\alpha_{mn}^2 u_n(r) \tag{4.74}$$

with boundary conditions

$$u_n(r) \text{ finite at } r = 0, \quad u_n(a) = 0. \tag{4.75}$$

This can be put in the self-adjoint form

$$\mathcal{L}u_n(r) \equiv \frac{d}{dr}\left[r\frac{du_n}{dr}\right] - \frac{m^2}{r}u_n = -\alpha_{mn}^2 r u_n. \tag{4.76}$$

It is easy to check that for functions u and v satisfying the boundary conditions (4.75)

$$\int_0^a dr\, u\mathcal{L}v = \int_0^a dr\, u\left[\frac{d}{dr}\left(r\frac{dv}{dr}\right) - \frac{m^2}{r}v\right]$$

$$= \int_0^a dr\, (\mathcal{L}u)v. \tag{4.77}$$

In this connection we note that for $m \neq 0$ the eigenfunctions (4.73) vanish at $r = 0$ [see (4.13)] so that the term involving m^2 converges when integrated in (4.77). The relation (4.77) implies by the standard argument (2.110) the orthogonality of the eigenfunctions:

$$\int_0^a r\, dr\, J_m(\alpha_{mn}r)J_m(\alpha_{mp}r) = 0, \quad p \neq n. \tag{4.78}$$

We note in (4.76) and (4.78) the appearance of the weight function $\rho = r$, which has its origin in the volume element $d^3x = r\, dr\, d\theta\, dz$ in cylindrical coordinates.

As usual our eigenfunctions $J_m(\alpha_{mn}r)$ form a complete set, so that an arbitrary function $f(r)$ in the range $0 \le r \le a$ can be expanded

$$f(r) = \sum_{n=1}^{\infty} c_n J_m(\alpha_{mn}r). \tag{4.79}$$

Using the orthogonality relation (4.78) we can solve for the expansion coefficients

$$c_n = \frac{\int_0^a r\, dr\, J_m(\alpha_{mn}r)f(r)}{\int_0^a r\, dr\, [J_m(\alpha_{mn}r)]^2}. \tag{4.80}$$

To evaluate the integral in the denominator of (4.80) we note that

$$\frac{d}{dr}\left[\left(\alpha_{mn}^2 r^2 - m^2\right) J_m^2(\alpha_{mn}r) + \left\{r\frac{d}{dr}J_m(\alpha_{mn}r)\right\}^2\right]$$

$$= 2\alpha_{mn}^2 r J_m^2 + 2(\alpha_{mn}^2 r^2 - m^2)J_m\frac{dJ_m}{dr} + 2r\frac{dJ_m}{dr}\frac{d}{dr}\left(r\frac{dJ_m}{dr}\right)$$

$$= 2\alpha_{mn}^2 r J_m^2, \tag{4.81}$$

where we have used the differential equation (4.76) satisfied by the $J_m(\alpha_{mn}r)$. Integrating (4.81) we find

$$\int_0^a rdr\ [J_m(\alpha_{mn}r)]^2 = \frac{1}{2\alpha_{mn}^2}\left[\left(\alpha_{mn}^2 r^2 - m^2\right)J_m^2(\alpha_{mn}r)\right.$$

$$\left. + \left\{r\frac{d}{dr}J_m(\alpha_{mn}r)\right\}^2\right]_0^a$$

$$\boxed{\int_0^a rdr\ [J_m(\alpha_{mn}r)]^2 = \frac{a^2}{2}\left[J_m'(\alpha_{mn}a)\right]^2.} \tag{4.82}$$

Here we have used $J_m(0) = 0$, $m \neq 0$.

4.9 INTERIOR OF A CYLINDER OF FINITE LENGTH - CONTINUED

Return now to the expansion problem (4.72). Setting $z = b$ and using the boundary condition $\psi_1(r,\theta,b) = \psi(r,\theta,b)$ we have

$$\psi(r,\theta,b) = \sum_{m=-\infty}^{\infty}\sum_{n=1}^{\infty} C_{mn}J_{|m|}(\alpha_{|m|n}r)\sinh(\alpha_{|m|n}b)e^{im\theta}. \tag{4.83}$$

Together the Bessel functions $J_{|m|}(\alpha_{|m|n}r)$ and the trigonometric functions $e^{im\theta}$ form a complete set for functions of r and θ, so the expansion (4.83) is possible. Using the orthogonality and normalization relations (4.78), (4.80), (4.82) for the Bessel functions and the corresponding relations for the trigonometric functions $e^{im\theta}$, we can solve for the expansion coefficients C_{mn}:

$$C_{mn} = \frac{\int_0^{2\pi} d\theta \int_0^a rdr\ e^{-im\theta}J_{|m|}(\alpha_{|m|n}r)\psi(r,\theta,b)}{2\pi\frac{a^2}{2}\left[J_{|m|}'(\alpha_{|m|n}a)\right]^2\sinh(\alpha_{|m|n}b)}. \tag{4.84}$$

We can solve the boundary value problem (4.64), (4.66) for $\psi_2(r,\theta,z)$ by similar means:

$$\psi_2(r,\theta,z) = \sum_{m=-\infty}^{\infty}\sum_{n=1}^{\infty} D_{mn}J_{|m|}(\alpha_{|m|n}r)\sinh[\alpha_{|m|n}(b-z)]e^{im\theta}, \tag{4.85}$$

with

$$D_{mn} = \frac{\int_0^{2\pi} d\theta \int_0^a r\,dr\; e^{-im\theta} J_{|m|}(\alpha_{|m|n}r)\psi(r,\theta,0)}{2\pi \frac{a^2}{2} \left[J'_{|m|}(\alpha_{|m|n}a)\right]^2 \sinh(\alpha_{|m|n}b)}. \tag{4.86}$$

To complete our job we must solve the boundary value problem (4.64), (4.67) for ψ_3. In applying the general solution (4.59) to this case we see that in order to have ψ_3 vanish on both ends, $z = 0, b$, we must take

$$F = -E, \quad \alpha = i\beta, \quad \beta = \frac{n\pi}{b}, \quad n = 1, 2, \dots, \tag{4.87}$$

so that

$$Ee^{\alpha z} + Fe^{-\alpha z} = E\left[e^{i\beta z} - e^{-i\beta z}\right] = 2iE\sin\beta z$$

$$= 2iE\sin\left(\frac{n\pi z}{b}\right). \tag{4.88}$$

Again we impose the boundary condition that ψ_3 be finite for $r \to 0$ so that the coefficient of the Neumann function in (4.59) must vanish. Thus we obtain a series of the form

$$\psi_3(r,\theta,z) = \sum_{m=-\infty}^{\infty} \sum_{n=1}^{\infty} E_{mn}\, I_{|m|}\left(\frac{n\pi r}{b}\right)\sin\left(\frac{n\pi z}{b}\right)e^{im\theta}, \tag{4.89}$$

where we have used (4.51) to replace $J_m(\alpha r) = J_m(i\beta r)$ by an I function with a real argument. The double Fourier series in z and θ can be used to fit an arbitrary function $\psi_3(a,\theta,z) = \psi(a,\theta,z)$ on the curved surface of the cylinder, and we obtain for the coefficients

$$E_{mn} = \frac{1}{\pi b\, I_{|m|}\left(\frac{n\pi a}{b}\right)} \int_0^b dz \int_0^{2\pi} d\theta\; \sin\left(\frac{n\pi z}{b}\right)e^{-im\theta}\psi(a,\theta,z). \tag{4.90}$$

We remark in passing that similar techniques would suffice if the boundary condition involved the normal derivatives of the function at the surface instead of the function itself.

4.10   EXTERIOR OF AN INFINITELY LONG CYLINDER

We suppose that $\psi(a,\theta,z)$ is given. In the general solution (4.59) take $\alpha = ik$, with k a continuous variable. Then (4.59) reduces to

$$\psi(r,\theta,z) = \sum_{m=-\infty}^{\infty} \int_{-\infty}^{\infty} dk\; e^{im\theta}e^{ikz}\left[A_m(k)J_{|m|}(ikr) + B_m(k)N_{|m|}(ikr)\right]. \tag{4.91}$$

With the Fourier series in θ and the Fourier integral in k we can fit an arbitrary boundary value function $\psi(a,\theta,z)$.

Before doing that, let us consider the r dependence. It is convenient to rewrite the Bessel and Neumann functions of imaginary argument which appear in (4.91) in terms of I and K functions, so that (4.91) becomes

$$\psi(r,\theta,z) = \sum_{m=-\infty}^{\infty} \int_{-\infty}^{\infty} dk\, e^{im\theta} e^{ikz} \left[C_m(k) I_{|m|}(|k|\,r) + D_m(k) K_{|m|}(|k|\,r) \right]. \quad (4.92)$$

Referring to the formulas (4.56) and (4.57) for their asymptotic forms, we see that the $I_{|m|}(|k|\,r)$ function grows exponentially as $r \to \infty$ while the $K_{|m|}(|k|\,r)$ function decays exponentially. We shall impose the boundary condition that $\psi(r,\theta,z) \to 0$ as $r \to \infty$. This implies $C_m(k) = 0$. Thus we find

$$\psi(r,\theta,z) = \sum_{m=-\infty}^{\infty} \int_{-\infty}^{\infty} dk\, e^{im\theta} e^{ikz} D_m(k) K_{|m|}(|k|\,r). \quad (4.93)$$

Using the inversion formulas for the Fourier series (2.134) and Fourier integral (2.141), we can find $D_m(k)$ such that (4.93) satisfies the boundary condition at $r = a$:

$$D_m(k) = \frac{1}{(2\pi)^2 K_{|m|}(|k|\,a)} \int_0^{2\pi} d\theta \int_{-\infty}^{\infty} dz\, e^{-im\theta} e^{-ikz} \psi(a,\theta,z). \quad (4.94)$$

4.11 CYLINDER IN AN EXTERNAL FIELD

There are cylindrical analogues to the electrostatic and hydrodynamic examples of Sections 3.10 and 3.11. Let us consider here the flow of an incomprehensible fluid transverse to the axis of an infinitely long cylinder as shown in Figure 4.2. The velocity of the fluid is given by

$$\mathbf{v} = -\boldsymbol{\nabla}\psi, \quad (4.95)$$

where ψ satisfies Laplace's equation

$$\boldsymbol{\nabla}^2 \psi = 0. \quad (4.96)$$

At the surface of the cylinder the boundary condition is

$$\left. \frac{\partial \psi}{\partial r} \right|_{r=a} = 0, \quad (4.97)$$

and at $r \to \infty$ the boundary condition is

$$\psi \xrightarrow[r\to\infty]{} -v_0 x = -v_0 r \cos\theta, \quad (4.98)$$

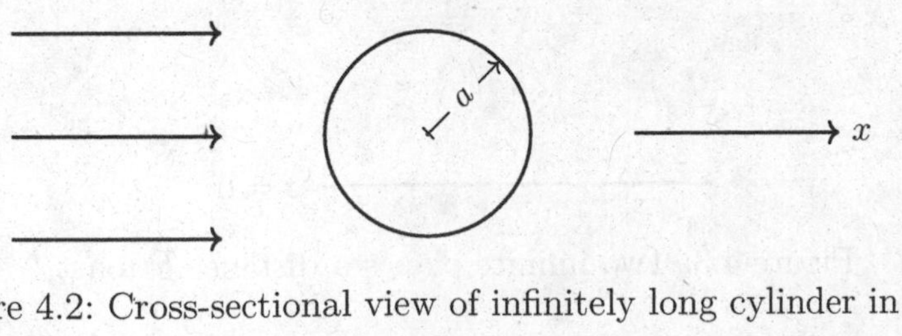

Figure 4.2: Cross-sectional view of infinitely long cylinder in flow

so as to obtain uniform flow in the x direction far from the obstacle:

$$\mathbf{v} \xrightarrow[r \to \infty]{} v_0 \hat{\mathbf{i}}. \tag{4.99}$$

In the problem as posed above there is no z dependence, so we need the solutions (2.31) for the radial equation. The general z independent solution of Laplace's equation then assumes the form

$$\psi(r,\theta) = A_0 + B_0 \ln r + \sum_{m=1}^{\infty} \left(A_m r^m + B_m \frac{1}{r^m} \right) \left(C_m \cos m\theta + D_m \sin m\theta \right). \tag{4.100}$$

This can also be obtained from the more general result (4.59) by taking the limit $\alpha \to 0$ to eliminate the z dependence and using the small argument expansions (4.38) and (4.41) for the Bessel and Neumann functions.

Just as for the spherical examples of Sections 3.10 and 3.11, the angular dependence of the asymptotic boundary condition imposes itself on the whole problem, so that we only need one term from (4.100):

$$\psi(r,\theta) = \left(A_1 r + B_1 \frac{1}{r} \right) \cos \theta. \tag{4.101}$$

We must then have $A_1 = -v_0$ to satisfy (4.98) and $A_1 - B_1/a^2 = 0$ to satisfy (4.97). We finally obtain

$$\psi(r,\theta) = -v_0 \left[r + \frac{a^2}{r} \right] \cos \theta \tag{4.102}$$

$$\mathbf{v} = -\boldsymbol{\nabla}\psi = v_0 \left[1 - \frac{a^2}{r^2} \right] \cos \theta \hat{\mathbf{r}} - v_0 \left[1 + \frac{a^2}{r^2} \right] \sin \theta \hat{\boldsymbol{\theta}}. \tag{4.103}$$

4.12   SPACE BETWEEN TWO INFINITE PLANES

We wish to solve Laplace's equation in the space $0 \leq z \leq b$ between two infinite planes with the boundary condition that ψ is given on the planes $z = 0$ and $z = b$ as shown in Figure 4.3.

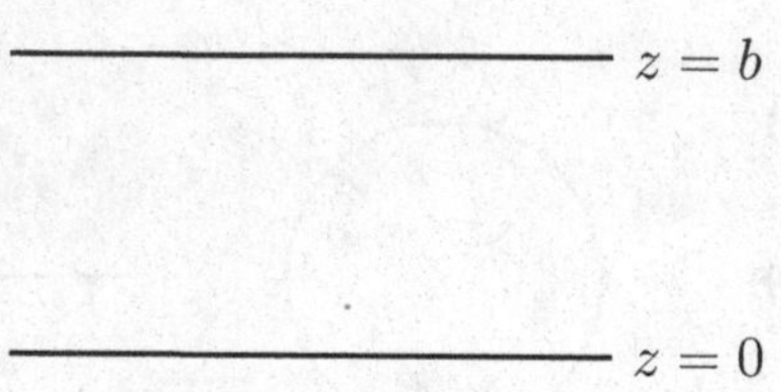

——————————————— $z = b$

——————————————— $z = 0$

Figure 4.3: Two infinite planes a distance b apart

We can work this problem in either cartesian or cylindrical coordinates. In cartesian coordinates the solutions of Laplace's equation are of the form

$$e^{i\alpha x + i\beta y + i\gamma z} \tag{4.104}$$

with

$$\alpha^2 + \beta^2 + \gamma^2 = 0. \tag{4.105}$$

Take $\gamma \pm \sqrt{\alpha^2 + \beta^2}$ and form a superposition of such solutions:

$$\psi(x, y, z) =$$

$$\int_{-\infty}^{\infty} d\alpha \int_{-\infty}^{\infty} d\beta \; e^{i\alpha x + i\beta y} \left[A(\alpha, \beta) e^{\sqrt{\alpha^2 + \beta^2}\, z} + B(\alpha, \beta) e^{-\sqrt{\alpha^2 + \beta^2}\, z} \right]. \tag{4.106}$$

With this double Fourier integral we can satisfy the boundary conditions at $z = 0$ and $z = b$. Inverting the Fourier transforms in (4.106) according to (2.141), we find for $z = 0$

$$A(\alpha, \beta) + B(\alpha, \beta) = \frac{1}{(2\pi)^2} \int_{-\infty}^{\infty} dx \int_{-\infty}^{\infty} dy \; e^{-i\alpha x - i\beta y} \psi(x, y, 0) \tag{4.107}$$

and for $z = b$

$$A(\alpha, \beta) e^{\sqrt{\alpha^2 + \beta^2}\, b} + B(\alpha, \beta) e^{\sqrt{\alpha^2 + \beta^2}\, b}$$

$$= \frac{1}{(2\pi)^2} \int_{-\infty}^{\infty} dx \int_{-\infty}^{\infty} dy \; e^{-i\alpha x - i\beta y} \psi(x, y, b). \tag{4.108}$$

These two equations can be solved for $A(\alpha, \beta)$ and $B(\alpha, \beta)$.

Let us now solve the problem over again in cylindrical coordinates. We can use the general result (4.59). In order to ensure that ψ be finite at $r = 0$, we set $B = 0$. Since there are no boundary conditions on r at finite values of r, α can take on a continuous range of values. Thus (4.59) assumes the form

$$\psi(r, \theta, z) = \sum_m \int_0^{\infty} \alpha \; d\alpha \; J_{|m|}(\alpha r) e^{im\theta} \left[E_m(\alpha) e^{\alpha z} + F_m(\alpha) e^{-\alpha z} \right]. \tag{4.109}$$

We have inserted an extra factor of α here for later convenience.

4.13   FOURIER BESSEL TRANSFORMS

To complete the calculation of (4.109) we need the concept of Fourier Bessel transforms. These are related to the expansion (4.79) in Bessel functions for a finite region $0 \leq r \leq a$ in much the same way as the Fourier integral is related to the Fourier series. Thus we shall imitate here the developments in Section 2.9 by letting $a \to \infty$ in the expansion (4.79), (4.80), (4.82). According to (4.71) the parameter α_{mn} which appears in these equations is related to x_{mn}, the nth zero of $J_m(x)$,

$$\alpha_{mn} = \frac{x_{mn}}{a}. \tag{4.110}$$

For α_{mn} to remain different from 0 as $a \to \infty$ we must have $x_{mn} \to \infty$, i.e., $n \to \infty$, so that we can use the asymptotic expansion (4.42) for $J_m(x)$ to determine these zeros x_{mn}. Thus we see that for large n the distance between successive zeros is

$$x_{m\,n+1} - x_{mn} = \pi,$$

$$\alpha_{m\,n+1} - \alpha_{mn} = d\alpha = \frac{\pi}{a}. \tag{4.111}$$

With $d\alpha \to 0$ we can rewrite the summation in (4.79) as an integral:

$$f(r) = \sum_n c_n J_m(\alpha_{mn} r) \to \int_0^\infty d\alpha \, \frac{a}{\pi} c_n J_m(\alpha r). \tag{4.112}$$

Similarly in the normalization integral (4.82) we can see the asymptotic form (4.42) for $J_m(x)$:

$$J_m'(\alpha_{mn} a) = -\sqrt{\frac{2}{\pi \alpha_{mn} a}} \, \sin\left(\alpha_{mn} a - \frac{m\pi}{2} - \frac{\pi}{4}\right)$$

$$= \pm\sqrt{\frac{2}{\pi \alpha_{mn} a}}, \tag{4.113}$$

since at the zeros of $\cos x$, $\sin x = \pm 1$. Thus (4.82) becomes

$$\frac{a^2}{2} \left[J_m'(\alpha_{mn} a)\right]^2 \to \frac{a}{\pi \alpha} \tag{4.114}$$

so that (4.80) becomes

$$c_n \to \frac{\pi \alpha}{a} \int_0^\infty r dr \, J_m(\alpha r) f(r). \tag{4.115}$$

Finally (4.112) and (4.115) can be rewritten as

$$\boxed{f(r) = \int_0^\infty \alpha \, d\alpha \, g(\alpha) J_m(\alpha r),} \tag{4.116}$$

$$\boxed{g(\alpha) = \int_0^\infty r \, dr \, J_m(\alpha r) f(r).} \tag{4.117}$$

This pair of Fourier Bessel integral transforms provides an expansion for an arbitrary function $f(r)$ in a way quite analogous to the Fourier integral theorem of (2.140), (2.141).

4.14   SPACE BETWEEN TWO INFINITE PLANES - CONTINUED

Armed with the transforms (4.34), (4.35) we can complete the solution of the boundary value problem above, which we had carried to the stage given in (4.109). Setting

$z = 0$ and $z = b$ in that result and inverting the transforms we find

$$E_m(\alpha) + F_m(\alpha) = \frac{1}{2\pi} \int_0^{2\pi} d\theta \, e^{-im\theta} \int_0^{\infty} r \, dr \, J_{|m|}(\alpha r)\psi(r, \theta, 0), \quad (4.118)$$

$$E_m(\alpha)e^{\alpha b} + F_m(\alpha)e^{-\alpha b} = \frac{1}{2\pi} \int_0^{2\pi} d\theta \, e^{-im\theta} \int_0^{\infty} r \, dr \, J_{|m|}(\alpha r)\psi(r, \theta, b). \quad (4.119)$$

These two equations can be solved for $E_m(\alpha)$ and $F_m(\alpha)$.

PROBLEMS

4.1 Find the Wronskian for the following pairs of functions:

    a) $J_m(x)$ and $J_{-m}(x)$,

    b) $J_m(x)$ and $N_m(x)$,

    c) $P_l(x)$ and $Q_l(x)$.

[Use the formula (2.65) to determine the functional form; find the unknown multiplicative constant by using the asymptotic or small argument expansions.]

4.2 By expanding the exponential, rearranging the series and using the series expansions (4.13), (4.14), show that

$$e^{\frac{x}{2}\left(h-\frac{1}{h}\right)} = \sum_{n=-\infty}^{\infty} h^n J_n(x).$$

The quantity on the left is thus a generating function for the Bessel functions of integral order. Use this generating function to derive the following identities:

a)

$$\frac{dJ_n(x)}{dx} = \frac{1}{2}\left[J_{n-1}(x) - J_{n+1}(x)\right],$$

b)

$$\frac{n}{x}J_n(x) = \frac{1}{2}\left[J_{n-1}(x) + J_{n+1}(x)\right],$$

c)

$$J_n(x+y) = \sum_{m=-\infty}^{\infty} J_m(x)J_{n-m}(y),$$

d)

$$e^{ikr\cos\theta} = \sum_{m=-\infty}^{\infty} i^m\, e^{im\theta} J_m(kr).$$

The last expression here is the expansion of a plane wave e^{ikx} in cylindrical waves.

4.3 The ends of a cylinder of length b and radius a are kept at 0 potential. The curved side of the cylinder is held at constant potential V.

    a) Find the series for the electrostatic potential on the inside of the cylinder.

    b) Find the simplified series which gives the potential along the axis of the cylinder.

    c) Evaluate numerically the potential at the center of the cylinder for the special case $a/b = 1/2$.

(Tables of Bessel functions of imaginary argument, i.e., I functions, are given by Jahnke and Emde as well as Abramowitz and Stegun.)

4.4 The part of the x, y plane outside of a circle of radius R centered on the origin is held at zero potential. The inside of the circle is held at constant potential V. Show that the electrostatic potential in the region $z > 0$ above the plane is given by the expression

$$\psi(r, z) = VR \int_0^\infty d\alpha \; J_1(\alpha R) J_0(\alpha r) e^{-\alpha z}.$$

Show that for $r = 0$, i.e., above the center of the circle, the potential is given by

$$\psi(0, z) = V \left[1 - \frac{z}{\sqrt{z^2 + R^2}} \right].$$

(The necessary integral involving a Bessel function is given, for example, by Magnus, Oberhettinger, and Soni, p. 91.)

CHAPTER 5

Normal Mode Eigenvalue Problems

5.1 INTRODUCTION

In Chapters 3 and 4 we have discussed the solutions of Laplace's equation. These solutions are useful for certain electrostatics problems, steady state temperature distributions, and steady flow of an incompressible fluid.

The time dependence in the heat conduction equation and the wave equation adds another dimension compared to Laplace's equation, and after separation of the time dependence, leads to the Helmholtz equation

$$\boxed{\boldsymbol{\nabla}^2 u(\mathbf{r}) + k^2 u(\mathbf{r}) = 0.} \tag{5.1}$$

With suitable boundary conditions the Helmholtz equation becomes an eigenvalue problem with a complete set of eigenfunctions $u_n(\mathbf{r})$ and eigenvalues k_n^2. These are the normal modes of the problem. We can form superpositions of the normal modes to satisfy initial conditions imposed at some initial time, say $t = 0$.

5.2 REDUCTION OF THE DIFFUSION EQUATION AND WAVE EQUATION TO AN EIGENVALUE PROBLEM

Consider first the diffusion equation

$$\boldsymbol{\nabla}^2 \psi(\mathbf{r}, t) = \frac{1}{\kappa} \frac{\partial \psi(\mathbf{r}, t)}{\partial t}. \tag{5.2}$$

We wish to solve this equation in a region V, as per Figure 5.1, subject to a boundary condition and an initial condition of the form

- boundary condition:

$$\psi(\mathbf{r}, t) \text{ on } S \text{ given and independent of time}, \tag{5.3}$$

- initial condition:

$$\psi(\mathbf{r}, t) \text{ given throughout } V. \tag{5.4}$$

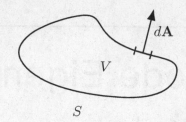

Figure 5.1: Closed volume and associated surface for boundary conditions

In Chapters 3 and 4 we learned how to solve, at least for simple surfaces S, the problem

$$\nabla^2 \psi'(\mathbf{r}) = 0, \quad \psi'(\mathbf{r}) \text{ on } S \text{ given and independent of } t. \tag{5.5}$$

If we now let

$$\psi(\mathbf{r}, t) = \psi'(\mathbf{r}) + \psi''(\mathbf{r}, t), \tag{5.6}$$

we see that $\psi'(\mathbf{r}, t)$ is determined by

$$\nabla^2 \psi''(\mathbf{r}, t) = \frac{1}{\kappa} \frac{\partial \psi''(\mathbf{r}, t)}{\partial t} \tag{5.7}$$

with a homogeneous boundary condition:

$$\psi''(\mathbf{r}, t) = 0 \quad \text{for } \mathbf{r} \text{ on S}. \tag{5.8}$$

The initial condition on $\psi''(\mathbf{r}, t)$ is determined by (5.4), (5.6):

$$\psi''(\mathbf{r}, 0) \quad \text{given throughout } V. \tag{5.9}$$

To proceed further we separate the time dependence as in (2.4)-(2.7) to obtain

$$\psi''(\mathbf{r}, t) = e^{-\kappa k^2 t} u(\mathbf{r}), \tag{5.10}$$

$$\nabla^2 u(\mathbf{r}) + k^2 u(\mathbf{r}) = 0. \tag{5.11}$$

The homogeneous boundary condition (5.8) can be applied to (5.10) and becomes

$$u(\mathbf{r}) = 0 \quad \text{for } \mathbf{r} \text{ on } S. \tag{5.12}$$

The Helmholtz equation (5.11) plus the boundary condition (5.12) form an eigenvalue problem, a three-dimensional Sturm-Liouville problem. There is an infinite set of solutions, i.e., eigenfunctions $u_n(\mathbf{r})$ and eigenvalues k_n^2 satisfying

$$\nabla^2 u_n(\mathbf{r}) + k_n^2 u_n(\mathbf{r}) = 0, \tag{5.13}$$

$$u_n(\mathbf{r}) = 0 \quad \text{for } \mathbf{r} \text{ on } S. \tag{5.14}$$

We can use the identity known as Green's theorem,

$$\boxed{\int_V d^3x \, (u\nabla^2 v - v\nabla^2 u) = \int_S d\mathbf{A} \cdot (u\nabla v - v\nabla u),} \tag{5.15}$$

to prove that eigenfunctions belonging to different eigenvalues are orthogonal.

Proof. For two eigenfunction, eigenvalue pairs $u_n(\mathbf{r})$, k_n^2 and $u_m(\mathbf{r})$, k_m^2 we find

$$\int_V d^3x \left[u_n \boldsymbol{\nabla}^2 u_m - u_m \boldsymbol{\nabla}^2 u_n \right] = \int_S d\mathbf{A} \cdot \left[u_n \boldsymbol{\nabla} u_m - u_m \boldsymbol{\nabla} u_n \right]. \tag{5.16}$$

The right side of (5.16) vanishes in virtue of the boundary condition $(5.14)^1$; using (5.13) to evaluate the left side of (5.16) we find

$$\left(k_n^2 - k_m^2 \right) \int_V d^3x \, u_n(\mathbf{r}) u_m(\mathbf{r}) = 0, \tag{5.17}$$

i.e.

$$\int_V d^3x \, u_n(\mathbf{r}) u_m(\mathbf{r}) = 0, \quad k_n^2 \neq k_m^2. \tag{5.18}$$

□

As usual the eigenfunctions form a complete set and we can expand an arbitrary function $f(\mathbf{r})$:

$$f(\mathbf{r}) = \sum_n c_n u_n(\mathbf{r}). \tag{5.19}$$

Using the orthogonality relation (5.18) we can evaluate the c_n:

$$c_n = \frac{\int_V d^3x \, u_n(\mathbf{r}) f(\mathbf{r})}{\int_V d^3x [u_n(\mathbf{r})]^2}. \tag{5.20}$$

Returning now to the diffusion equation we see that we have produced an infinite set of solutions satisfying the boundary condition (5.8):

$$\psi''(\mathbf{r}, t) = \sum_n C_n e^{-\kappa k_n^2 t} u_n(\mathbf{r}). \tag{5.21}$$

The C_n can now be chosen to satisfy also the initial condition (5.9). Using the expansion (5.19), (5.20) we find

$$C_n = \frac{\int_V d^3x \, u_n(\mathbf{r}) \psi''(\mathbf{r}, 0)}{\int_V d^3x \, [u_n(\mathbf{r})]^2}. \tag{5.22}$$

For the wave equation there is a similar story. Because the equation is second-order in the time, the initial conditions consist of a specification of both ψ and $\frac{\partial \psi}{\partial t}$ at the initial time $t = 0$. The problem thus assumes the form

$$\boldsymbol{\nabla}^2 \psi(\mathbf{r}, t) = \frac{1}{c^2} \frac{\partial^2 \psi(\mathbf{r}, t)}{\partial t^2}, \tag{5.23}$$

- boundary condition:

$$\psi(\mathbf{r}, t) \text{ on } S \text{ given and independent of } t, \tag{5.24}$$

---

[1]Note that the right side of (5.16) vanishes also for the boundary condition $\frac{\partial u}{\partial n}$ = normal derivative =0 on surface S.

- initial conditions:

$$\psi(\mathbf{r}, 0) \text{ given throughout } V, \tag{5.25}$$

$$\left.\frac{\partial \psi}{\partial t}\right|_{t=0} \text{ given throughout } V. \tag{5.26}$$

Subtracting an appropriate solution of Laplace's equation (5.5) we obtain a homogeneous boundary condition:

$$\psi(\mathbf{r}, t) = \psi'(\mathbf{r}) + \psi''(\mathbf{r}, t), \tag{5.27}$$

$$\boldsymbol{\nabla}^2 \psi''(\mathbf{r}, t) = \frac{1}{c^2} \frac{\partial^2 \psi''(\mathbf{r}, t)}{\partial t^2}, \tag{5.28}$$

with boundary conditions

$$\psi''(\mathbf{r}, t) = 0 \quad \text{for } \mathbf{r} \text{ on } S, \tag{5.29}$$

$$\psi''(\mathbf{r}, 0) \quad \text{given throughout } V, \tag{5.30}$$

$$\left.\frac{\partial \psi''}{\partial t}\right|_{t=0} \text{ given throughout } V. \tag{5.31}$$

The separation of the time variable leading to the Helmholtz equation is given by (2.8), (2.9). The eigenvalue problem is exactly the same as for the diffusion equation, (5.13), (5.14). A general solution satisfying the homogeneous boundary condition (5.29) is

$$\psi''(\mathbf{r}, t) = \sum_n \left[A_n \cos k_n ct + B_n \sin k_n ct\right] u_n(\mathbf{r}). \tag{5.32}$$

The constants A_n and B_n can be chosen to satisfy the initial conditions (5.30), (5.31):

$$A_n = \frac{\int_V d^3x \, u_n(\mathbf{r}) \psi''(\mathbf{r}, 0)}{\int_V d^3x \, [u_n(\mathbf{r})]^2}, \tag{5.33}$$

$$B_n = \frac{1}{k_n c} \frac{\int_V d^3x \, u_n(\mathbf{r}) \left.\frac{\partial \psi}{\partial t}\right|_{t=0}}{\int_V d^3x \, [u_n(\mathbf{r})]^2}. \tag{5.34}$$

Finally we point out the similarity to Schrödinger's equation:

$$-\frac{\hbar^2}{2m} \boldsymbol{\nabla}^2 \psi(\mathbf{r}, t) + V(\mathbf{r}) \psi(\mathbf{r}, t) = i\hbar \frac{\partial \psi(\mathbf{r}, t)}{\partial t}. \tag{5.35}$$

If the energy eigenfunctions are $u_n(\mathbf{r})$,

$$-\frac{\hbar^2}{2m} \boldsymbol{\nabla}^2 u_n(\mathbf{r}) + V(\mathbf{r}) u_n(\mathbf{r}) = E_n u_n(\mathbf{r}), \tag{5.36}$$

the general time-dependent solution is

$$\psi(\mathbf{r}, t) = \sum_n c_n e^{-i\frac{E_n}{\hbar}t} u_n(\mathbf{r}). \tag{5.37}$$

The c_n can be used to satisfy an initial condition on ψ at $t = 0$. For $V(\mathbf{r}) = 0$ quantum mechanics is diffusion theory with an imaginary diffusion constant. The time dependence is thus oscillatory rather than damped.

The remainder of this chapter is devoted to examples.

5.3 THE VIBRATING STRING

This system is described by the one-dimensional wave equation (1.17) with the wave velocity given by (1.18). The solution obtained by separating variables is

$$u(x, t) = \sum_n u_n(x)(A_n \cos k_n ct + B_n \sin k_n ct), \tag{5.38}$$

where the u_n are the solution of the Sturm-Liouville problem

$$\frac{d^2 u_n(x)}{dx^2} + k_n^2 u_n(x) = 0, \tag{5.39}$$

$$u_n(0) = u_n(a) = 0. \tag{5.40}$$

In (5.40) it is assumed that the ends of the string are fixed at $x = 0$ and $x = a$. The solutions of (5.39), (5.40) are [compare (2.97), (2.98)]

$$u_n(x) = \sin k_n x, \tag{5.41}$$

$$k_n = \frac{n\pi}{a}. \tag{5.42}$$

These are the normal modes of oscillation of the string with the first three listed in Figure 5.2.

Frequency

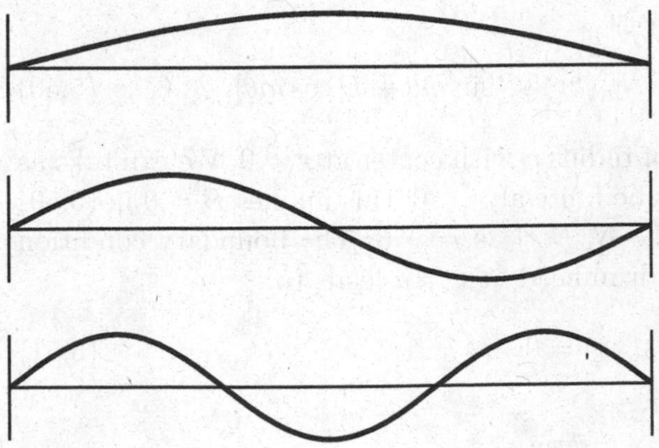

$$n = 1 \qquad \omega_1 = k_1 c = \frac{\pi c}{a}$$

$$n = 2 \qquad \omega_2 = k_2 c = 2\frac{\pi c}{a}$$

$$n = 3 \qquad \omega_3 = k_3 c = 3\frac{\pi c}{a}$$

Figure 5.2: Normal modes of oscillating string

We can satisfy initial conditions on $u(x, 0)$ and $\frac{\partial u}{\partial t}\big|_{t=0}$ with suitable superpositions of these modes. Evaluating (5.38) and its time derivative at $t = 0$, we find

$$u(x, 0) = \sum_n A_n \sin k_n x, \tag{5.43}$$

$$\frac{\partial u}{\partial t}\bigg|_{t=0} = \sum_n k_n c B_n \sin k_n x. \tag{5.44}$$

Using the orthogonality of the $u_n(x)$ to solve for A_n and B_n, we obtain

$$A_n = \frac{2}{a} \int_0^a dx \; \sin k_n x \; u(x, 0), \quad k_n = \frac{n\pi}{a}, \tag{5.45}$$

$$B_n = \frac{2}{k_n c a} \int_0^x dx \; \sin k_n x \; \frac{\partial u}{\partial t}\bigg|_{t=0}. \tag{5.46}$$

5.4   THE VIBRATING DRUMHEAD

For this case the wave equation and wave velocity are given by (1.20), (1.21). We can apply our general result (5.32):

$$u(\mathbf{r}, t) = \sum_n [A_n \cos k_n c t + B_n \sin k_n c t] u_n(\mathbf{r}), \tag{5.47}$$

$$\boldsymbol{\nabla}^2 u_n(\mathbf{r}) + k_n^2 u_n(\mathbf{r}) = 0, \tag{5.48}$$

$$u_n(\mathbf{r}) = 0 \quad \text{on the edge of drumhead.} \tag{5.49}$$

We shall consider a circular drumhead for which it is appropriate to use the solutions of the Helmholtz equation in cylindrical coordinates. These were obtained in Section 2.4. To reduce the three-dimensional case studied there to the two-dimensional case of interest for the drumhead, we eliminate the z dependence by taking $\alpha = k$ in (2.27). Thus we find as solutions of the two-dimensional Helmholtz equation in cylindrical coordinates

$$u(r, \theta) = [A J_m(kr) + B N_m(kr)][C \sin m\theta + D \cos m\theta]. \tag{5.50}$$

We consider a circular drumhead of radius a with center at $r = 0$. We can take as boundary condition that our solutions be finite at $r = 0$; this implies $B = 0$ in (5.50) in view of the singular character of the $N_m(kr)$ as $r \to 0$. The boundary condition that $u(r, \theta)$ vanish on the edge of the drumhead at $r = a$ leads to

$$J_m(ka) = 0, \tag{5.51}$$

which has solutions

$$k_{mn} = \frac{x_{mn}}{a}, \tag{5.52}$$

where x_{mn} is the nth zero of $J_m(x)$. Some of the low order zeros of the low order Bessel functions are[2]

$$x_{01} = 2.405 \qquad x_{11} = 3.832 \qquad x_{21} = 5.136$$
$$x_{02} = 5.520 \qquad x_{12} = 7.016 \qquad x_{22} = 8.417$$
$$x_{03} = 8.654 \qquad x_{13} = 10.173 \qquad x_{23} = 11.620 \qquad (5.53)$$

Thus we obtain as eigenfunctions of the Helmholtz equation the normal modes (see Figure 5.3)

$$u_{mn}(r,\theta) = \begin{cases} J_m(k_{mn}r)\sin m\theta \\ J_m(k_{mn}r)\cos m\theta \end{cases} . \qquad (5.54)$$

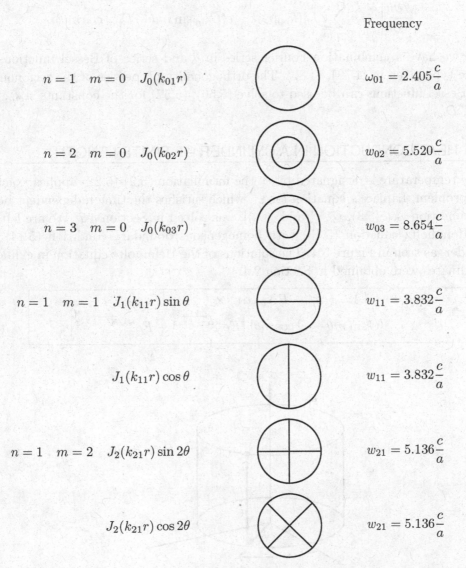

Frequency

$n=1 \quad m=0 \quad J_0(k_{01}r)$ $\qquad \omega_{01} = 2.405\dfrac{c}{a}$

$n=2 \quad m=0 \quad J_0(k_{02}r)$ $\qquad w_{02} = 5.520\dfrac{c}{a}$

$n=3 \quad m=0 \quad J_0(k_{03}r)$ $\qquad w_{03} = 8.654\dfrac{c}{a}$

$n=1 \quad m=1 \quad J_1(k_{11}r)\sin\theta$ $\qquad w_{11} = 3.832\dfrac{c}{a}$

$J_1(k_{11}r)\cos\theta$ $\qquad w_{11} = 3.832\dfrac{c}{a}$

$n=1 \quad m=2 \quad J_2(k_{21}r)\sin 2\theta$ $\qquad w_{21} = 5.136\dfrac{c}{a}$

$J_2(k_{21}r)\cos 2\theta$ $\qquad w_{21} = 5.136\dfrac{c}{a}$

Figure 5.3: Lowest order normal modes and eigenfunctions for drumhead

---

[2]A complete table is given by M. Abramowitz, and I. Stegun op. cit., p. 409.

Superposing the normal modes enables us to satisfy initial conditions on the displacement and velocity of the drumhead. Our general solution is

$$u(r, \theta, t) = \sum_{n,m} J_m(k_{mn}r)[(A_{mn} \sin m\theta + B_{mn} \cos m\theta) \cos k_{mn}ct$$

$$+ (C_{mn} \sin m\theta + D_{mn} \cos m\theta) \sin k_{mn}ct]. \tag{5.55}$$

At $t = 0$ this gives

$$u(r, \theta, 0) = \sum_{n,m} J_m(k_{mn}r)[A_{mn} \sin m\theta + B_{mn} \cos m\theta], \tag{5.56}$$

$$\left.\frac{\partial u}{\partial t}\right|_{t=0} = \sum_{n,m} J_m(k_{mn}r)[k_{mn}c(C_{mn} \sin m\theta + D_{mn} \cos m\theta)]. \tag{5.57}$$

Here we have a combination Fourier series in θ and series of Bessel functions in r of the type given in (4.79), (4.80). The orthogonality properties of the trigonometric and Bessel functions can be used to solve (5.56), (5.57) for the constants A_{mn}, B_{mn}, C_{mn}, D_{mn}.

5.5   HEAT CONDUCTION IN A CYLINDER OF FINITE LENGTH

If the temperature is designated by ψ, the formulation (5.2)-(5.22) applies exactly to this problem. Laplace's equation for ψ', which satisfies the time independent boundary condition [see (5.3), (5.5), and (5.6)], was solved in Section 4.7. We are left with the Helmholtz equation (5.13) with homogeneous boundary condition (5.14) for a cylinder, as seen in Figure 5.4. The solutions of the Helmholtz equation in cylindrical coordinates were obtained in Section 2.4:

$$u(r, \theta, z) = [AJ_m(\alpha r) + BN_m(\alpha r)] \times$$

$$[C \sin m\theta + D \cos m\theta]\left[Ee^{\sqrt{\alpha^2-k^2}\,z} + Fe^{-\sqrt{\alpha^2-k^2}\,z}\right]. \tag{5.58}$$

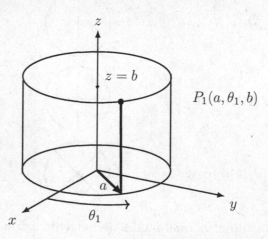

Figure 5.4: Finite length cylinder for heat conduction problem

We can now apply the boundary conditions that $u(r, \theta, z)$ be finite throughout the interior of the cylinder and vanish on the surfaces of the cylinder. In order for $u(r, \theta, z)$ to be finite at $r = 0$, we must have $B = 0$. In order that $u(r, \theta, z)$ vanish for $r = a$, we take

$$\alpha = \alpha_{mn},$$
$$\alpha_{mn}a = x_{mn} = n^{th} \text{ zero of } J_m(x). \tag{5.59}$$

In order that $u(r, \theta, z)$ vanish for $z = 0$ and $z = b$, we must have

$$F = -E,$$
$$\sqrt{\alpha^2 - k^2} = i\beta,$$
$$\beta b = p\pi, \quad p = 1, 2, 3, \dots, \tag{5.60}$$

so that

$$Ee^{\sqrt{\alpha^2 - k^2}\, z} + Fe^{-\sqrt{\alpha^2 - k^2}\, z} = 2iE \sin \frac{p\pi}{b} z. \tag{5.61}$$

Combining (5.59) and (5.60) we obtain the eigenvalues k^2:

$$k^2 = k^2_{mnp} = \alpha^2 + \beta^2 = \alpha^2_{mn} + \frac{p^2\pi^2}{b^2} = \frac{x^2_{mn}}{a^2} + \frac{p^2\pi^2}{b^2}. \tag{5.62}$$

The eigenfunctions corresponding to these eigenvalues are then

$$u_{mnp}(r, \theta, z) = J_m(\alpha_{mn}r) \sin \left(\frac{p\pi z}{b} \right) \begin{cases} \sin m\theta \\ \cos m\theta \end{cases}. \tag{5.63}$$

These eigenfunctions can be used in the general formulas (5.21), (5.22) to satisfy the initial conditions on the temperature.

It is important to appreciate the distinction between solving Laplace's equation and solving the Helmholtz equation as an eigenvalue problem. We have just obtained the eigenfunctions (5.63) for the interior of a cylinder with the boundary condition that u vanish on the surface. On the other hand, the only solution of Laplace's equation for the interior of the cylinder satisfying the boundary condition that $u = 0$ on the surface is the trivial solution $u = 0$ everywhere. This is equivalent to the statement that $k^2 = 0$ is not one of the eigenvalues (5.62) of the Helmholtz equation.

5.6 PARTICLE IN A CYLINDRICAL BOX (QUANTUM MECHANICS)

The energy eigenfunctions for this problem satisfy

$$-\frac{\hbar^2}{2m} \nabla^2 u = Eu \tag{5.64}$$

with the boundary condition $u = 0$ on the walls of the cylindrical box. This is the problem we have just solved in (5.62), (5.63). The energy eigenfunctions are given by (5.63) and the energy eigenvalues are obtained from (5.62):

$$E_{mnp} = \frac{\hbar^2}{2m} \left[\frac{x^2_{mn}}{a^2} + \frac{p^2\pi^2}{b^2} \right]. \tag{5.65}$$

5.7 NORMAL MODES OF AN ACOUSTIC RESONANT CAVITY

We consider sound waves in a gas inside a cylindrical box as shown in Figure 5.5. In Section 1.6 we showed that the density and pressure fluctuations ρ' and p' satisfy the wave equation with the sound velocity given by (1.33). The appropriate boundary condition at the walls of the box is

$$\frac{\partial \rho'}{\partial n} = \frac{\partial p'}{\partial n} = 0, \tag{5.66}$$

where $\partial/\partial n$ means derivative normal to the wall. This boundary condition follows from Newton's second law in the form (1.29),

$$\rho_0 \frac{\partial \mathbf{v}}{\partial t} = -\boldsymbol{\nabla} p', \tag{5.67}$$

and the assumption that there is no motion normal to the walls.

Separating the time dependence in the wave equation we find

$$\rho'(\mathbf{r}, t) = \sum_n u_n(\mathbf{r})[A_n \cos k_n ct + B_n \sin k_n ct], \tag{5.68}$$

where the normal modes $u_n(\mathbf{r})$ are determined by the eigenvalue problem

$$\boldsymbol{\nabla}^2 u_n(\mathbf{r}) + k_n^2 u_n(\mathbf{r}) = 0 \tag{5.69}$$

$$(\hat{\mathbf{n}} \cdot \boldsymbol{\nabla}) u_n(\mathbf{r}) = 0, \quad \mathbf{r} \text{ on surface.} \tag{5.70}$$

The eigenvalue problem here is similar to that solved in (5.62), (5.63). The difference is due to the boundary condition (5.70). This boundary condition that the normal derivative vanish on the surface leads to eigenfunctions

$$u_{mnp}(r, \theta, z) = J_m \left[\frac{y_{mn}}{a} r \right] \cos \left(\frac{p\pi z}{b} \right) \begin{cases} \sin m\theta \\ \cos m\theta \end{cases} \tag{5.71}$$

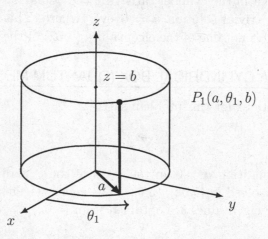

Figure 5.5: Cylindrical box for sound waves

and eigenvalues

$$k_{mnp}^2 = \frac{y_{mn}^2}{a^2} + \frac{p^2\pi^2}{b^2},$$ (5.72)

where y_{mn} is the nth zero of $J_m'(x)$,

$$J_m'(y_{mn}) = 0.$$ (5.73)

Some of the low order solutions of this equation are[3]

$$
\begin{array}{lll}
y_{01} = 0 & y_{11} = 1.841 & y_{21} = 3.054 \\
y_{02} = 3.832 & y_{12} = 5.331 & y_{22} = 6.706 \\
y_{03} = 7.016 & y_{13} = 8.536 & y_{23} = 9.969
\end{array}
$$ (5.74)

Initial values $\rho'(r,0)$ and $\left.\frac{\partial\rho'}{\partial t}\right|_{t=0}$ can be fit by suitable superposition (5.68) of the normal modes.

5.8 ACOUSTIC WAVE GUIDE

Consider the propagation of acoustic waves down a circular pipe of radius a as illustrated in Figure 5.6. The density fluctuations satisfy the wave equation

$$\nabla^2\rho' = \frac{1}{c^2}\frac{\partial^2\rho'}{\partial t^2}.$$ (5.75)

Assume a sinusoidal disturbance of the form

$$\rho'(r,\theta,z,t) = F(r,\theta,z)e^{-i\omega t}.$$ (5.76)

Substituting in the wave equation (5.75) we find the Helmholtz equation

$$\left[\nabla^2 + \frac{\omega^2}{c^2}\right]F = 0,$$ (5.77)

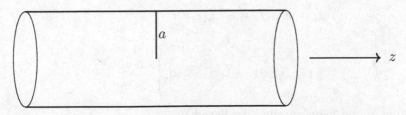

Figure 5.6: Cylindrical acoustic wave guide

---

[3]A complete table is given by M. Abramowitz and I. Stegun, op. cit., p. 411.

with solutions in cylindrical coordinates

$$F(r, \theta, z) = [AJ_m(\alpha r) + BN_m(\alpha r)] \times$$
$$[C \sin m\theta + D \cos m\theta] \left[Ee^{\sqrt{\alpha^2 - \omega^2/c^2}\, z} + Fe^{\sqrt{\alpha^2 - \omega^2/c^2}\, z} \right].$$

(5.78)

In order to obtain from (5.76), (5.78) a wave propagating down the pipe in the z direction we take

$$\sqrt{\alpha^2 - \frac{\omega^2}{c^2}} = ik,$$

$$\frac{\omega^2}{c^2} = \alpha^2 + k^2.$$

(5.79)

Then (5.76) becomes

$$\rho'(r, \theta, z, t) = G(r, \theta)e^{i(kz - \omega t)},$$

(5.80)

where

$$G(r, \theta) = [AJ_m(\alpha r) + BN_m(\alpha r)][C \sin m\theta + D \cos m\theta].$$

(5.81)

We can impose the boundary condition that $G(r, \theta)$ be finite at $r = 0$, which leads to $B = 0$. The boundary condition on the side of the pipe is

$$\left. \frac{\partial G(r, \theta)}{\partial r} \right|_{r=a} = 0.$$

(5.82)

This gives

$$\alpha = \alpha_{mn} = \frac{y_{mn}}{a},$$

(5.83)

with y_{mn} given by (5.73), (5.74).

In summary the allowed modes for propagation down the pipe are

$$\rho'_{mn}(r, \theta, z, t) = G_{mn}(r, \theta)e^{i(kz - \omega t)},$$

(5.84)

$$G_{mn}(r, \theta) = J_m \left[\frac{y_{mn}}{a} r \right] \begin{cases} \sin m\theta \\ \cos m\theta \end{cases},$$

(5.85)

with frequency ω and wave number k related by

$$\frac{\omega^2}{c^2} = k^2 + \frac{y_{mn}^2}{a^2}.$$

(5.86)

We can single out of these modes a particularly simple one corresponding to the zero $y_{01} = 0$ of $J_0'(x)$ at $x = 0$. This mode is independent of r and θ and has the form

$$\rho'_{01} = Ce^{i(kz - \omega t)},$$

(5.87)

with C a constant and ω and k related by $\omega = kc$. This is an evident solution of the wave equation and the boundary condition (5.82).

The other modes have a more complicated form. The wave guide propagates waves of all wavelength λ and wave number $k = 2\pi/\lambda$,

$$0 < k < \infty. \tag{5.88}$$

According to (5.86) the corresponding frequencies lie in the range

$$\omega_{\min}(m, n) < \omega < \infty, \tag{5.89}$$

where

$$\omega_{\min}(m, n) = \frac{c}{a} y_{mn} \tag{5.90}$$

is the minimum frequency which can be propagated in the mode mn. If one attempts to propagate at frequency $\omega < \omega_{\min}(m, n)$ in the mode mn, equation (5.86) tells us that

$$k^2 = \frac{\omega^2 - \omega_{\min}^2(m, n)}{c^2} = -\beta^2 < 0, \tag{5.91}$$

so that $k = \pm i\beta$ is purely imaginary. Thus the wave does not propagate but is in fact exponentially damped:

$$\rho'(r, \theta, z, t) = G_{mn}(r, \theta) e^{-\beta z - i\omega t}. \tag{5.92}$$

PROBLEMS

5.1 A vibrating string of mass per unit length σ and length l, under tension T, is clamped at one end and attached at the other end to a massless ring which slides freely on a fixed rod as shown in Figure 5.7. Find the normal modes and their frequencies. Draw some pictures to illustrate the shapes of the three modes with the lowest frequencies.

Figure 5.7: String with massless ring on rod for Problem 5.1

5.2 a) Find the normal modes and normal frequencies for a square drumhead of side a.

b) By suitable superposition of degenerate modes, i.e., different modes with the same frequency, find the normal modes and normal frequencies for a drumhead having the shape of an isosceles right triangle with dimension shown in Figure 5.8.

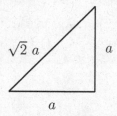

Figure 5.8: Isosceles triangle drum head for Problem 5.2

5.3 A drumhead has the shape of a sector with opening angle α of a circle of radius R as shown in Figure 5.9. The drumhead is clamped down on the sides and has mass σ per unit area and is under tension T.

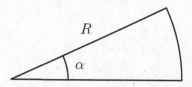

Figure 5.9: Pie slice shaped drumhead for Problem 5.3

a) Find the normal modes of vibration for this drumhead and their frequencies.

b) For which values of the opening angle α are all the normal frequencies for the full circular drumhead? Which subset?

c) Show that for $\alpha = 2\pi/(2l + 1)$, $l = 0, 1, 2, \ldots$, the modes with no angular nodes (except the edges of the drum) are expressible in terms of spherical Bessel functions - see (6.6). Study in particular the case $l = 0$. Find the normal modes and normal frequencies exactly. Is such a drumhead realizable physically?

5.4 A circular drumhead of radius a, tension T, mass per unit area σ, is struck at the center at $t = 0$ by a drumstick so that the displacement and velocity at $t = 0$ are given by

$$u(r, \theta, 0) = 0$$

$$\left.\frac{\partial u(r, \theta, t)}{\partial t}\right|_{t=0} = V\frac{\delta(r)}{2\pi r}.$$

Find the subsequent motion of the drumhead.

5.5 A cylindrical object of radius a, length b, is made of a thermally conducting material with $\kappa = K/c\rho$. At $t = 0$ the object has a uniform temperature T_0 throughout. At this time it is immersed in a heat bath held at a different uniform temperature T_1.

a) Find the temperature throughout the interior of the object at later times.

b) Find a simple expression for the temperature at large times. What defines large times?

5.6 Work Problem 5.5 for an object having the shape of a cylindrical wedge of opening angle α as shown in Figure 5.10 .

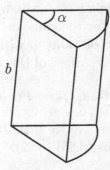

Figure 5.10: Cylindrical wedge for Problem 5.6

5.7 A thermally conducting object with arbitrary shape is kept in a heat bath at fixed temperature T_0 (see Figure 5.11). At time $t = 0$ it is heated locally at the point $\mathbf{r}'$ so that at this instant the temperature is given by

$$T(\mathbf{r}, 0) = T_0 + T_1\delta(\mathbf{r} - \mathbf{r}').$$

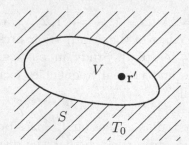

Figure 5.11: Arbitrary volume in heat bath for Problem 5.7

a) Find the temperature at later times in terms of the eigenfunctions and eigenvalues of the Helmholtz equation:

$$\nabla^2 u_n(\mathbf{r}) + k_n^2 u_n(\mathbf{r}) = 0,$$
$$u_n(\mathbf{r}) = 0 \quad \text{on surface } S.$$

Assume the eigenfunctions are normalized

$$\int_V d^3x \, [u_n(\mathbf{r})]^2 = 1.$$

b) Find the dominant term in the series at large times. What determines large times?

c) For short times assume the effect of the boundary is negligible and so use the eigenfunctions for an infinite region in your series derived in part (a). Find an explicit formula for $T(\mathbf{r}, t)$. Compare with Problem 4 from Chapter 1. At what time will the approximation fail?

5.8 Work out a theory for the emission of cylindrical waves analogous to the theory presented in Section 6.5 for the emission of spherical waves. Suppose the radiating source is a cylinder of radius a with axis along the z axis and take the boundary condition at the surface of this cylinder to be

$$\psi(a, \theta, z, t) = F(\theta)e^{-i\omega t},$$

where $F(\theta)$ is a given function. Notice that the source is z-independent. Work out the analogues of equations (6.85)-(6.92).

5.9 Work out a theory for the scattering of a plane wave by a cylindrical obstacle analogous to the theory presented in Section 6.6 for the scattering by a spherical obstacle. Take the incoming wave to be along the x axis and use the result in Problem 2 from Chapter 4 for the expansion of a plane wave in cylindrical waves. Assume the cylindrical obstacle is of radius a with its axis along the z axis, and take the boundary condition to be $\psi = 0$ at the surface of the cylinder. Work out the analogues of (6.105), (6.110), (6.99).

CHAPTER 6

Spherical Bessel Functions and Applications

6.1 INTRODUCTION

The solutions we obtained in Section 2.5 for the Helmholtz equation

$$\boldsymbol{\nabla}^2 u(\mathbf{r}) + k^2 u(\mathbf{r}) = 0 \tag{6.1}$$

in spherical coordinates were of the form

$$u = \left\{ A \frac{1}{\sqrt{kr}} J_\beta(kr) + B \frac{1}{\sqrt{kr}} N_\beta(kr) \right\} Y(\theta, \varphi), \quad \beta = \sqrt{\lambda + \frac{1}{4}}. \tag{6.2}$$

The boundary condition that the angular function $Y(\theta, \varphi)$ be everywhere single values and finite leads to

$$\lambda = l(l+1),$$
$$Y = Y_l^m(\theta, \varphi), \tag{6.3}$$

which implies that the order of the Bessel functions in (6.2) is

$$\beta = \sqrt{\lambda + \frac{1}{4}} = l + \frac{1}{2}, \quad l = 0, 1, 2, \dots . \tag{6.4}$$

Thus the solutions (6.2) of the Helmholtz equation in spherical coordinates become

$$u = \{ A j_l(kr) + B n_l(kr) \} Y_l^m(\theta, \varphi), \tag{6.5}$$

where the so-called spherical Bessel and Neumann functions are defined by

$$j_l(x) = \sqrt{\frac{\pi}{2x}} J_{l+\frac{1}{2}}(x), \tag{6.6}$$

$$n_l(x) = \sqrt{\frac{\pi}{2x}} N_{l+\frac{1}{2}}(x). \tag{6.7}$$

These functions give the radial dependence of monochromatic spherical waves. They are important in all problems having to do with the emission or scattering of waves by systems of finite extent such as atoms, nuclei, or elementary particles.

6.2 FORMULAS FOR SPHERICAL BESSEL FUNCTIONS IN TERMS OF ELEMENTARY FUNCTIONS

Although expressed in terms of Bessel and Neumann functions in (6.6), (6.7) the spherical Bessel and Neumann functions can also be expressed in terms of sines, cosines, and inverse powers of x. Consider first $j_0(x)$ and refer to the series (4.13) for the Bessel function. We have

$$j_0(x) = \sqrt{\frac{\pi}{2x}} J_{\frac{1}{2}}(x) = \frac{\sqrt{\pi}}{2} \sum_{r=0}^{\infty} \frac{(-1)^r}{r!\, \Gamma\left(\frac{1}{2}+r+1\right)} \left(\frac{x}{2}\right)^{2r}. \tag{6.8}$$

Using the recursion relation of the gamma function and the value of $\Gamma(1/2)$ we find

$$\Gamma\left(\frac{1}{2}+r+1\right) = \left(r+\frac{1}{2}\right)\left(r+\frac{1}{2}-1\right)\cdot\ldots\cdot\frac{1}{2}\Gamma\left(\frac{1}{2}\right)$$

$$= \frac{(2r+1)(2r-1)\cdot\ldots\cdot 1}{2^{r+1}}\Gamma\left(\frac{1}{2}\right)$$

$$= \frac{(2r+1)!}{2^{2r+1}\, r!}\sqrt{\pi}. \tag{6.9}$$

Substituting this in (6.8) we obtain

$$\boxed{j_0(x) = \sum_{r=0}^{\infty} \frac{(-1)^r}{(2r+1)!} x^{2r} = \frac{\sin x}{x}} \tag{6.10}$$

by recognizing the well-known Taylor series for $\sin x$.

By using a recursion relation for the Bessel function we can find $j_l(x)$. The series (4.13) for the Bessel function can be written in the form

$$\frac{J_p(x)}{\left(\frac{x}{2}\right)^p} = \sum_{r=0}^{\infty} \frac{(-1)^r}{r!\, \Gamma(r+p+1)} \left(\frac{x}{2}\right)^{2r}$$

$$= \sum_{r=0}^{\infty} \frac{(-1)^r}{r!\, \Gamma(r+p+1)} \frac{\left(\frac{x^2}{2}\right)^r}{2^r}. \tag{6.11}$$

If we now apply the operator

$$\frac{1}{x}\frac{d}{dx} = \frac{d}{d\left(\frac{x^2}{2}\right)} \tag{6.12}$$

l times to (6.11), we find

$$\left[\frac{1}{x}\frac{d}{dx}\right]^l \frac{J_p(x)}{\left(\frac{x}{2}\right)^p} = \sum_{r=l}^{\infty} \frac{(-1)^r}{(r-l)!\, \Gamma(r+p+1)} \frac{\left(\frac{x^2}{2}\right)^{r-l}}{2^r}$$

$$= \sum_{s=0}^{\infty} \frac{(-1)^{l+s}}{s! \; \Gamma(s+l+p+1)} \frac{\left(\frac{x^2}{2}\right)^s}{2^{l+s}}$$

$$= \frac{(-1)^l}{2^l} \frac{J_{p+l}(x)}{\left(\frac{x}{2}\right)^{p+l}}, \tag{6.13}$$

i.e.,

$$\frac{J_{p+l}(x)}{x^p} = (-x)^l \left[\frac{1}{x}\frac{d}{dx}\right]^l \frac{J_p(x)}{x^p}. \tag{6.14}$$

Setting $p = \frac{1}{2}$ in this general result and using (6.6) and (6.10) we find

$$\boxed{j_l(x) = (-x)^l \left[\frac{1}{x}\frac{d}{dx}\right]^l \left[\frac{\sin x}{x}\right].} \tag{6.15}$$

As for $n_l(x)$, we find, using the definition (4.19) for the Neumann function and $\cos(l+\frac{1}{2})\pi = 0$ for l integral,

$$n_l(x) = \sqrt{\frac{\pi}{2x}} N_{l+\frac{1}{2}}(x)$$

$$= \sqrt{\frac{\pi}{2x}} \frac{J_{l+\frac{1}{1}}(x)\cos(l+\frac{1}{2})\pi - J_{-l-\frac{1}{2}}(x)}{\sin(l+\frac{1}{2})\pi}$$

$$= (-1)^{l+1} \sqrt{\frac{\pi}{2x}} J_{-l-\frac{1}{2}}(x). \tag{6.16}$$

For $l = 0$ this gives

$$n_0(x) = -\sqrt{\frac{\pi}{2x}} J_{-\frac{1}{2}}(x)$$

$$= -\sqrt{\frac{\pi}{2x}} \left(\frac{x}{2}\right)^{-\frac{1}{2}} \sum_{r=0}^{\infty} \frac{(-1)^r}{r! \; \Gamma\left(r-\frac{1}{2}+1\right)} \left(\frac{x}{2}\right)^{2r}. \tag{6.17}$$

From (6.9) we have

$$\Gamma\left(r-\frac{1}{2}+1\right) = \frac{(2r-1)!}{2^{2r-1}(r-1)!}\sqrt{\pi} = \frac{(2r)!}{2^{2r}\, r!}\sqrt{\pi}. \tag{6.18}$$

So we find

$$\boxed{n_0(x) = -\frac{1}{x}\sum_{r=0}^{\infty}\frac{(-1)^r}{(2r)!}x^{2r} = -\frac{\cos x}{x}.} \tag{6.19}$$

To find $n_l(x)$ we need a recursion relation similar to but different from (6.14):

$$\frac{J_{-p}(x)}{(x/2)^p} = \sum_{r=0}^{\infty} \frac{(-1)^r}{r!\,\Gamma(r-p+1)} \frac{\left(\frac{x^2}{2}\right)^{r-p}}{2^{r-p}}, \tag{6.20}$$

$$\left[\frac{1}{x}\frac{d}{dx}\right]^l \frac{J_{-p}(x)}{(x/2)^p} = \sum_{r=0}^{\infty} \frac{(-1)^r}{r!\,\Gamma(r-p-l+1)} \frac{\left(\frac{x^2}{2}\right)^{r-p-l}}{2^{r-p}}$$

$$= \frac{x^{-p-l}}{2^{-p}} \sum_{r=0}^{\infty} \frac{(-1)^r}{r!\,\Gamma(r-p-l+1)} \left(\frac{x}{2}\right)^{2r-p-l}$$

$$= \frac{x^{-p-l}}{2^{-p}} J_{-p-l}(x), \tag{6.21}$$

or

$$J_{-p-l}(x) = x^{p+l} \left[\frac{1}{x}\frac{d}{dx}\right]^l \frac{J_{-p}(x)}{x^p}. \tag{6.22}$$

Setting $p = 1/2$ and using (6.16) and (6.19) we find

$$\boxed{n_l(x) = -(-x)^l \left[\frac{1}{x}\frac{d}{dx}\right]^l \left[\frac{\cos x}{x}\right].} \tag{6.23}$$

We can easily calculate $j_l(x)$ and $n_l(x)$ for small l from (6.15) and (6.23):

$$j_0(x) = \frac{\sin x}{x},$$

$$j_1(x) = \frac{\sin x}{x^2} - \frac{\cos x}{x}, \tag{6.24}$$

$$j_2(x) = \left[\frac{3}{x^3} - \frac{1}{x}\right]\sin x - \frac{3}{x^2}\cos x,$$
$$\vdots$$

and

$$n_0(x) = -\frac{\cos x}{x},$$

$$n_1(x) = -\frac{\sin x}{x} - \frac{\cos x}{x^2}, \tag{6.25}$$

$$n_2(x) = -\frac{3}{x^2}\sin x - \left[\frac{3}{x^3} - \frac{1}{x}\right]\cos x.$$
$$\vdots$$

The spherical Hankel functions are defined by the same formulas as used for the ordinary Hankel functions, (4.44), (4.45):

$$h_l^{(1)}(x) = j_l(x) + in_l(x) = (-x)^l \left[\frac{1}{x}\frac{d}{dx}\right]^l \left[\frac{e^{ix}}{ix}\right],$$ (6.26)

$$h_l^{(2)}(x) = j_l(x) - in_l(x) = (-x)^l \left[\frac{1}{x}\frac{d}{dx}\right]^l \left[\frac{e^{-ix}}{-ix}\right].$$ (6.27)

The small argument expansions of these functions are easily obtained from the power series:

$$j_l(x) = \sqrt{\frac{\pi}{2x}} \left(\frac{x}{2}\right)^{l+\frac{1}{2}} \sum_{r=0}^{\infty} \frac{(-1)^r}{r!\,\Gamma\left(r+l+\frac{1}{2}+1\right)} \left(\frac{x}{2}\right)^{2r}$$

$$\xrightarrow[x\to 0]{} \sqrt{\frac{\pi}{2x}} \frac{1}{\Gamma\left(l+1+\frac{1}{2}\right)} \left(\frac{x}{2}\right)^{l+\frac{1}{2}}$$

$$= \frac{2^l\,l!}{(2l+1)!}\,x^l,$$ (6.28)

$$n_l(x) = (-1)^{l+1}\sqrt{\frac{\pi}{2x}}\,J_{-l-\frac{1}{2}}(x)$$

$$= (-1)^{l+1}\sqrt{\frac{\pi}{2x}} \left(\frac{x}{2}\right)^{-l-\frac{1}{2}} \sum_{r=0}^{\infty} \frac{(-1)^r}{r!\,\Gamma\left(r-l-\frac{1}{2}+1\right)}\,x^{2r}$$

$$n_l(x) \xrightarrow[x\to 0]{} (-1)^{l+1}\sqrt{\frac{\pi}{2x}} \frac{1}{\Gamma\left(-l-\frac{1}{2}+1\right)} \left(\frac{x}{2}\right)^{-l-\frac{1}{2}}$$

$$= -\frac{(2l)!}{2^l\,l!} \frac{1}{x^{l+1}}.$$ (6.29)

To obtain the asymptotic expansions we could use the results quoted in Section 4.2 for Bessel and Neumann functions – (4.42), (4.43), (4.46), (4.47). We can also proceed directly from the formulas (6.26), (6.27) just derived. For $x \to \infty$ the largest contribution in these formulas is obtained by applying all derivatives to e^{ix} rather than the inverse powers of x. Thus we find

$$h_l^{(1)}(x) \xrightarrow[x \to \infty]{} \frac{e^{i\left[x - (l+1)\frac{\pi}{2}\right]}}{x}, \tag{6.30}$$

$$h_l^{(2)}(x) \xrightarrow[x \to \infty]{} \frac{e^{-i\left[x - (l+1)\frac{\pi}{2}\right]}}{x}. \tag{6.31}$$

Combining these, we obtain

$$j_l(x) \xrightarrow[x \to \infty]{} \frac{1}{x} \cos\left[x - (l+1)\frac{\pi}{2}\right], \tag{6.32}$$

$$n_l(x) \xrightarrow[x \to \infty]{} \frac{1}{x} \sin\left[x - (l+1)\frac{\pi}{2}\right]. \tag{6.33}$$

6.3 EIGENVALUE PROBLEM AND APPLICATION OF THE EXPANSION THEOREM

The functions $j_l(kr)$ satisfy the differential equation [see (2.36)]

$$\frac{1}{r} \frac{d^2}{dr^2}\left[r j_l(kr)\right] - \frac{l(l+1)}{r^2} j_l(kr) = -k^2 j_l(kr). \tag{6.34}$$

This can be put in a slightly simpler form by defining

$$u(kr) = r j_l(kr), \tag{6.35}$$

which satisfies

$$\mathcal{L}u = \frac{d^2}{dr^2} u(kr) - \frac{l(l+1)}{r^2} u(kr) = -k^2 u(kr). \tag{6.36}$$

By imposing suitable boundary conditions we obtain a Sturm-Liouville eigenvalue problem for the spherical Bessel functions analogous to the eigenvalue problem for ordinary Bessel functions discussed in Section 4.8 – see (4.74), (4.75). Thus we take as boundary conditions

$$u(kr) = 0 \quad \text{at} \quad r = 0 \quad \text{and} \quad r = a. \tag{6.37}$$

The Sturm-Liouville eigenvalue problem defined by (6.36) and (6.37) has as solutions

$$u_n(k_{ln}r) = r j_l(k_{ln}r), \tag{6.38}$$

where

$$k_{ln}a = z_{ln} = \text{n}^{th} \text{ zero of } j_l(x)$$

$$= x_{l+\frac{1}{2}n} = \text{n}^{th} \text{ zero of } j_{l+\frac{1}{2}}(x). \tag{6.39}$$

We note that the functions $rn_l(kr)$ are not allowed by the boundary condition of $r = 0$.

It is easy to check that the differential operator on the left of (6.36) satisfies the condition

$$\int_0^a dr \, u\mathcal{L}v = \int_0^a dr \, v\mathcal{L}u \tag{6.40}$$

for functions u and v satisfying the boundary conditions (6.37) so that by the usual argument, (2.106)-(2.110), eigenfunctions belonging to different eigenvalues are orthogonal:

$$\int_0^a dr \, rj_l(k_{ln}r) \, rj_l(k_{lm}r) = 0, \quad n \neq m. \tag{6.41}$$

We can obtain the normalization integral by using the corresponding result (4.82) for the ordinary Bessel functions and the definition (6.6):

$$\int_0^a r^2 \, dr \, [j_l(k_{ln}r)]^2 = \frac{\pi}{2k_{ln}} \int_0^a r \, dr \, \left[J_{l+\frac{1}{2}}(k_{ln}r) \right]^2$$

$$= \frac{\pi}{2k_{ln}} \frac{a^2}{2} \left[J'_{l+\frac{1}{2}}(k_{ln}a) \right]^2$$

$$= \frac{1}{2}a^3 \left[j'_l(k_{ln}a) \right]^2. \tag{6.42}$$

The functions (6.38) form a complete set and we can expand an arbitrary function $rf(r)$ in terms of them, or after canceling a factor r,

$$f(r) = \sum_{n=1}^{\infty} c_n j_l(k_{ln}r). \tag{6.43}$$

Using (6.41), (6.42) we obtain the expansion coefficients

$$c_n = \frac{\int_0^a r^2 \, dr \, j_l(k_{ln}r)f(r)}{\int_0^a r^2 \, dr \, [j_l(k_{ln}r)]^2}. \tag{6.44}$$

Finally by letting $a \to \infty$ we obtain a Fourier Bessel transform in terms of spherical Bessel functions analogous to (4.116), (4.117) in terms of ordinary Bessel functions. Rather than go through the limiting process all over again we can apply the results (4.116), (4.117) with $m = l + \frac{1}{2}$ to a function $\sqrt{r}\, f(r)$ and use the definition (6.6) of the spherical Bessel functions to obtain

$$\boxed{f(r) = \int_0^{\infty} k^2 \, dk \, h(k)j_l(kr),} \tag{6.45}$$

$$\boxed{h(k) = \frac{2}{\pi} \int_0^{\infty} r^2 \, dr \, j_l(kr)f(r).} \tag{6.46}$$

A completeness relation can be obtained in the usual way by substituting (6.46) for $h(k)$ in (6.45) and noting that since $f(r)$ is arbitrary we must have

$$\int_0^\infty k^2 \, dk \, j_l(kr)j_l(kr') = \frac{\pi}{2r^2}\delta(r-r').$$

(6.47)

6.4 EXPANSION OF PLANE AND SPHERICAL WAVES IN SPHERICAL COORDINATES

Consider the spatial part

$$u(\mathbf{r}) = e^{i\mathbf{k}\cdot\mathbf{r}} = e^{ikr\cos\theta}$$

(6.48)

or a plane wave $e^{i(\mathbf{k}\cdot\mathbf{r}-\omega t)}$ propagating in the z direction. This is a solution of the Helmholtz equation

$$(\boldsymbol{\nabla}^2 + k^2)u(\mathbf{r}) = 0$$

(6.49)

and so can be expanded in terms of the solutions (6.5) of the Helmholtz equation in spherical coordinates. For propagation in the z direction there is no dependence on φ, so only contributions (6.5) with $m = 0$, $Y_l^0(\theta) \propto P_l(\cos\theta)$, will enter. Since the plane wave (6.48) is finite at $r = 0$ there will be no contribution from the singular terms $n_l(kr)$. Thus we find

$$e^{ikr\cos\theta} = \sum_{l=0}^\infty c_l \, j_l(kr)P_l(\cos\theta).$$

(6.50)

To find the coefficients, c_l we use the orthogonality properties (3.35) of the $P_l(\cos\theta)$:

$$c_l \, j_l(kr) = \frac{2l+1}{2} \int_{-1}^1 dx \, e^{ikrx}P_l(x).$$

(6.51)

To calculate the integral, use Rodrigues's formula (3.23) for $P_l(x)$ and integrate by parts:

$$c_l \, j_l(kr) = \frac{2l+1}{2^{l+1}\,l!} \int_{-1}^1 dx \, e^{ikrx} \frac{d^l}{dx^l}(x^2-1)^l$$

$$= \frac{2l+1}{2^{l+1}\,l!}(-ikr)^l \int_{-1}^1 dx \, e^{ikrx}(x^2-1)^l.$$

(6.52)

The integral in (6.52) can be further reduced by integration by parts:

$$\int_{-1}^1 dx \, e^{ikrx}(x^2-1)^l = \int_{-1}^1 dx \left[\frac{d}{dx}\frac{e^{ikrx}}{ikr}\right](x^2-1)^l$$

$$= -2l \int_{-1}^1 dx \, \frac{e^{ikrx}}{ikr}x(x^2-1)^{l-1}$$

$$= \frac{2l}{kr} \frac{d}{d(kr)} \int_{-1}^{1} dx \, e^{ikrx} (x^2 - 1)^{l-1}$$

$$= [2l][2(l-1)] \frac{1}{kr} \frac{d}{d(kr)} \frac{1}{kr} \frac{d}{d(kr)} \int_{-1}^{1} dx \, e^{ikrx} (x^2 - 1)^{l-2}$$

$$= 2^l \, l! \left[\frac{1}{kr} \frac{d}{d(kr)} \right]^l \int_{-1}^{1} dx \, e^{ikrx}$$

$$= 2^{l+1} \, l! \left[\frac{1}{kr} \frac{d}{d(kr)} \right]^l \frac{\sin kr}{kr}. \tag{6.53}$$

Substituting this result in (6.52) we find

$$c_l \, j_l(kr) = (2l+1)i^l(-kr)^l \left[\frac{1}{kr} \frac{d}{d(kr)} \right]^l \frac{\sin kr}{kr}. \tag{6.54}$$

Comparing with the formula (6.15) for $j_l(kr)$, we see that (6.54) is consistent, and we find

$$c_l = (2l+1)i^l. \tag{6.55}$$

Substituting (6.55) in (6.50) leads to the desired expansion

$$\boxed{ e^{ikr\cos\theta} = \sum_{l=0}^{\infty} (2l+1)i^l \, j_l(kr)P_l(\cos\theta). } \tag{6.56}$$

A slightly more general result for propagation in an arbitrary direction $\mathbf{k}$ can be obtained by using the spherical harmonics addition theorem (3.114) for $P_l(\cos\Theta)$ with $\cos\Theta = \frac{\mathbf{k}\cdot\mathbf{r}}{kr}$:

$$\boxed{ e^{i\mathbf{k}\cdot\mathbf{r}} = 4\pi \sum_{l=0}^{\infty} \sum_{m=-l}^{l} i^l \, j_l(kr)Y_l^{m*}(\theta_{\mathbf{k}}, \varphi_{\mathbf{k}})Y_l^m(\theta_{\mathbf{r}}, \varphi_{\mathbf{r}}). } \tag{6.57}$$

These expansions of a plane wave in spherical waves are essential in the theory of scattering – see Section 6.6.

A related expansion can be obtained for a displaced spherical wave. From (6.5) for $l = 0$, or by elementary calculations, we see that

$$h_0^{(1)}(kr) = \frac{e^{ikr}}{ikr} \tag{6.58}$$

is a solution of the Helmholtz equation. If we displace the origin of this wave from $\mathbf{r} = 0$ to $\mathbf{r} = \mathbf{r}'$ we obtain, since the Laplacian is invariant with respect to translations, another solution of the Helmholtz equation:

$$v(\mathbf{r}, \mathbf{r}') = \frac{e^{ik|\mathbf{r}-\mathbf{r}'|}}{ik \, |\mathbf{r} - \mathbf{r}'|} = \frac{e^{ik\sqrt{r^2 + r'^2 - 2rr'\cos\Theta}}}{ik\sqrt{r^2 + r'^2 - 2rr'\cos\Theta}}, \tag{6.59}$$

with $\cos\Theta = \frac{\mathbf{r}\cdot\mathbf{r}'}{rr'}$.

We can expand (6.59) in Legendre polynomials,

$$v(\mathbf{r},\mathbf{r}') = \sum_{l=0}^{\infty} f_l(r,r') P_l(\cos\Theta), \tag{6.60}$$

and since v is a solution of the Helmholtz equation, we see by comparison with (6.5) that $f_l(r,r')$ must be a combination of spherical Bessel and Neumann functions. The particular combination needed can be inferred from the form and properties of the function $v(\mathbf{r},\mathbf{r}')$, (6.59):

a) $v(\mathbf{r},\mathbf{r}') = v(\mathbf{r}',\mathbf{r})$ is the same function of $\mathbf{r}'$ as it is of $\mathbf{r}$.

b) As $r \to 0$, $v(\mathbf{r},\mathbf{r}')$ remains finite so that in this limit only $j_l(kr)$ is allowed, not $n_l(kr)$.

c) As $r \to \infty$, $v(\mathbf{r},\mathbf{r}') \to \frac{e^{ikr}}{r} F(\frac{r'}{r}, \cos\Theta)$, so that in this limit only the spherical Hankel function $h_l^{(1)}(kr)$ is allowed, not $h_l^{(2)}(kr)$ – see (6.26), (6.27).

d) The discontinuous jump from behavior (b) to behavior (c) occurs at $r = r'$ where the function (6.59) is infinite for $\cos\Theta = 1$.

From these properties we infer

$$f_l(r,r') = \begin{cases} a_l\, j_l(kr')\, h_l^{(1)}(kr), & r > r' \\[2mm] a_l\, h_l^{(1)}(kr')\, j_l(kr), & r < r' \end{cases}$$

$$= a_l\, j_l(kr_<)\, h_l^{(1)}(kr_>). \tag{6.61}$$

To obtain the coefficients, a_l we consider the limit $r' \to \infty$. According to (6.61) we find for $r' > r$

$$v(\mathbf{r},\mathbf{r}') = \sum_{l=0}^{\infty} a_l\, h_l^{(1)}(kr')\, j_l(kr)\, P_l(\cos\Theta). \tag{6.62}$$

In the limit $r' \to \infty$ we have

$$\sqrt{r^2 + r'^2 - rr'\cos\Theta} \xrightarrow[r'\to\infty]{} r' - r\cos\Theta,$$

$$v(\mathbf{r},\mathbf{r}') \xrightarrow[r'\to\infty]{} \frac{e^{ikr' - ikr\cos\Theta}}{ikr'},$$

$$h_l^{(1)}(kr') \xrightarrow[r'\to\infty]{} \frac{e^{i[kr' - (l+1)\frac{\pi}{2}]}}{kr'}, \tag{6.63}$$

so that (6.62) becomes

$$e^{-ikr\cos\Theta} = \sum_{l=0}^{\infty} a_l(-i)^l\, j_l(kr)P_l(\cos\Theta). \tag{6.64}$$

Comparing with complex conjugate of (6.56) we find

$$a_l = 2l + 1. \tag{6.65}$$

Combining our results (6.60), (6.61), (6.65) and the spherical harmonics addition theorem (3.114) we finally obtain

$$\frac{e^{ik|\mathbf{r}-\mathbf{r}'|}}{ik\,|\mathbf{r}-\mathbf{r}'|} = \frac{e^{ik\sqrt{r^2+r'^2-2rr'\cos\Theta}}}{ik\sqrt{r^2+r'^2-2rr'\cos\Theta}}$$

$$= \sum_{l=0}^{\infty}(2l+1)j_l(kr_<)h_l^{(1)}(kr_>)P_l(\cos\Theta)$$

$$\boxed{\frac{e^{ik|\mathbf{r}-\mathbf{r}'|}}{ik\,|\mathbf{r}-\mathbf{r}'|} = 4\pi\sum_{l=0}^{\infty}\sum_{m=-l}^{l} j_l(kr_<)h_l^{(1)}(kr_>)Y_l^m(\theta_\mathbf{r},\varphi_\mathbf{r})Y_l^{m*}(\theta_{\mathbf{r}'},\varphi_{\mathbf{r}'}).} \tag{6.66}$$

This formula, which is a generalization of (3.122), is needed for the multipole expansion of a system radiating waves.

6.5   THE EMISSION OF SPHERICAL WAVES

As we discussed in Chapter 1, the propagation of many types of waves is described by the wave equation

$$\boldsymbol{\nabla}^2\psi = \frac{1}{c^2}\frac{\partial^2\psi}{\partial t^2}, \tag{6.67}$$

where c is the wave velocity. Simple plane wave solutions of this equation are

$$\psi = Ae^{\pm i\mathbf{k}\cdot\mathbf{r}\pm i\omega t} \tag{6.68}$$

with

$$\mathbf{k}^2 = \frac{\omega^2}{c^2}, \quad \omega = kc. \tag{6.69}$$

By superposition of the solutions (6.68) we obtain other solutions. Thus

$$\psi = \begin{cases} e^{i(\mathbf{k}\cdot\mathbf{r}-\omega t)} \\ \cos(\mathbf{k}\cdot\mathbf{r}-\omega t) \end{cases} \tag{6.70}$$

are traveling waves moving in the direction $\mathbf{k}$,

$$\psi = \begin{cases} e^{i(\mathbf{k}\cdot\mathbf{r}+\omega t)} \\ \cos(\mathbf{k}\cdot\mathbf{r}+\omega t) \end{cases} \tag{6.71}$$

are traveling waves moving in the direction $(-\mathbf{k})$, and

$$\psi = \begin{cases} \cos \mathbf{k} \cdot \mathbf{r} \; e^{-i\omega t} \\[1em] \sin \mathbf{k} \cdot \mathbf{r} \; e^{-i\omega t} \\[1em] \cos \mathbf{k} \cdot \mathbf{r} \; \cos \omega t \end{cases} \tag{6.72}$$

are standing waves.

More generally if we separate off the time dependence

$$\psi(\mathbf{r}, t) = u(\mathbf{r})e^{-i\omega t}, \tag{6.73}$$

the wave equation (6.67) leads to the Helmholtz equation

$$\boldsymbol{\nabla}^2 u(\mathbf{r}) + k^2 u(\mathbf{r}) = 0, \quad k = \frac{\omega}{c}, \tag{6.74}$$

for $u(\mathbf{r})$. The spherical wave solutions of this are of the form

$$u(\mathbf{r}) = \begin{cases} j_l(kr)Y_l^m(\theta, \varphi) \\[1em] n_l(kr)Y_l^m(\theta, \varphi), \end{cases} \tag{6.75}$$

yielding

$$\boxed{\psi(\mathbf{r}, t) = [Aj_l(kr) + Bn_l(kr)]Y_l^m(\theta, \varphi)e^{-i\omega t}.} \tag{6.76}$$

The waves (6.76) are spherical standing waves. Like the plane standing waves (6.72) they have nodal surfaces where $\psi = 0$ for all t. Such standing waves are useful for solving normal mode problems.

Thus, for example, the normal modes for sound waves inside a spherical enclosure of radius a are

$$\psi_{lmn}(r, t) = j_l(k_{ln}r)Y_l^m(\theta, \varphi)e^{-i\omega_{ln}t} \tag{6.77}$$

with

$$\begin{aligned} j_l'(k_{ln}a) &= 0, \\ k_{ln}a &= \; n^{th} \text{ zero of } j_l'(x). \end{aligned} \tag{6.78}$$

To obtain traveling waves instead of the standing waves of (6.76) we use spherical Hankel functions in place of the spherical Bessel and Neumann functions:

$$\psi(r, t) = \begin{cases} h_l^{(1)}(kr)Y_l^m(\theta, \varphi)e^{-i\omega t} & \tag{6.79} \\[1em] h_l^{(2)}(kr)Y_l^m(\theta, \varphi)e^{-i\omega t} & \tag{6.80} \end{cases}$$

with $\omega = kc$. For large r we can use the asymptotic forms (6.30), (6.31), and (6.79), (6.80) reduce to

$$\psi(r,t) \xrightarrow[r\to\infty]{} \begin{cases} (-i)^{l+1} \dfrac{e^{i(kr-\omega t)}}{kr} Y_l^m(\theta,\varphi) & (6.81) \\[3mm] i^{l+1} \dfrac{e^{-i(kr-\omega t)}}{kr} Y_l^m(\theta,\varphi). & (6.82) \end{cases}$$

For any value of r we see from (6.26), (6.27) that

$$h_l^{(1)}(kr) \propto \frac{e^{ikr}}{r}, \tag{6.83}$$

$$h_l^{(2)}(kr) \propto \frac{e^{-ikr}}{r}. \tag{6.84}$$

Thus with the conventional time dependence $e^{-i\omega t}$, $\omega > 0$, (6.79) involving $h_l^{(1)}(kr)$ is a traveling <u>outgoing spherical wave</u> and (6.80) involving $h_l^{(2)}(kr)$ is a traveling <u>ingoing spherical wave</u>. We emphasize that this depends on the convention for the time dependence; with time dependence $e^{+i\omega t}$ the roles of $h_l^{(1)}(kr)$ and $h_l^{(2)}(kr)$ would be reversed.

Consider now the radiation of waves by a source of some type, a loud speaker or antenna for example. When we discuss Green's functions we will be able to express the radiation in terms of a source density function. For the present we specify the source by a boundary condition at some radius $r = a$ within which the source is located. For a monochromatic source we then take

$$\psi(a,\theta,\varphi,t) = F(\theta,\varphi)e^{-i\omega t} \tag{6.85}$$

with $F(\theta,\varphi)$ a given function determined by the source.

For $r > a$ the radiation is a superposition of waves of the type (6.79):

$$\psi(r,\theta,\varphi,t) = \sum_{l,m} a_{lm} h_l^{(1)}(kr) Y_l^m(\theta,\varphi) e^{-i\omega t}. \tag{6.86}$$

Here we include only outgoing waves. Ingoing waves (6.80) are omitted for a source emitting radiation, not absorbing it. The coefficients a_{lm} are determined by the boundary condition (6.85). Evaluating (6.86) at $r = a$ and using the orthogonality properties of spherical harmonics, we find

$$F(\theta,\varphi) = \sum_{l,m} a_{lm} h_l^{(1)}(ka) Y_l^m(\theta,\varphi), \tag{6.87}$$

$$a_{lm} = \frac{1}{h_l^{(1)}(ka)} \int d\Omega \, Y_l^{m*}(\theta,\varphi) F(\theta,\varphi). \tag{6.88}$$

For $r \to \infty$, far from the source, we can use the asymptotic form (6.81) for the spherical waves in (6.86) to obtain

$$\psi(r,\theta,\varphi,t) \xrightarrow[r\to\infty]{} f(\theta,\varphi)\frac{e^{i(kr-\omega t)}}{r} \tag{6.89}$$

with

$$f(\theta,\varphi) = \frac{1}{k}\sum_{l,m}(-i)^{l+1}a_{lm}Y_l^m(\theta,\varphi). \tag{6.90}$$

The exact expressions for energy density and energy flux depend on the kind of waves under consideration. However, it always turns out that for large r, with a simple monochromatic wave like (6.89), the flux of energy is in the direction $\mathbf{r}$ with a magnitude

$$\text{Energy flux} = c\times \text{Energy density} \propto c\,|\psi(r,\theta,\varphi,t)|^2$$

$$= c\,\frac{|f(\theta,\varphi)|^2}{r^2}. \tag{6.91}$$

The rate of emission of energy through solid angle $d\Omega$ is then obtained by multiplying by the area $r^2 d\Omega$ subtended by $d\Omega$. Finally, the rate of emission of energy per unit solid angle, i.e., power per unit solid angle, is

$$\frac{dP}{d\Omega} \propto c\,|f(\theta,\varphi)|^2. \tag{6.92}$$

We see how the factors of r have cancelled to give a sensible result.

6.6   SCATTERING OF WAVES BY A SPHERE

Suppose a monochromatic plane wave of a unit amplitude moving in the z direction,

$$\psi = e^{i(kz-\omega t)} = e^{ikr\cos\theta - i\omega t}, \tag{6.93}$$

impinges on a spherical obstacle of radius a centered at $r = 0$ as seen in Figure 6.1.

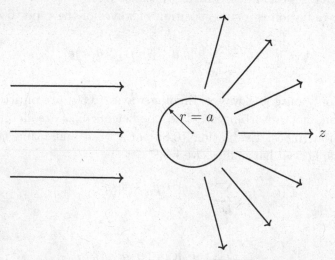

Figure 6.1: Waves scattered by a sphere

The result of the interaction of the incoming plane wave with the spherical obstacle will be the production of an outgoing spherical wave, called the scattered wave, in addition to the incoming plane wave. With a monochromatic incoming wave such as (6.93) we can expect a monochromatic scattered wave with the same frequency as response. Thus the total disturbance will be of the form

$$\psi = e^{i(kz-\omega t)} + \psi_{sc}(\mathbf{r}, t), \tag{6.94}$$

where ψ_{sc} is a superposition of outgoing spherical waves of the type (6.79):

$$\psi_{sc}(\mathbf{r}, t) = \sum_{l,m} b_{lm} h_l^{(1)}(kr) Y_l^m(\theta, \varphi) e^{-i\omega t}, \quad \omega = kc. \tag{6.95}$$

The coefficients b_{lm} are determined by the boundary conditions, whatever they happen to be, at the surface $r = a$ of the spherical obstacle. We discuss some examples below. Before doing that, let us consider the asymptotic form of (6.95) for $r \to \infty$. Using the asymptotic form (6.81) for the spherical waves we find

$$\psi_{sc}(\mathbf{r}, t) \xrightarrow[r\to\infty]{} f(\theta, \varphi) \frac{e^{i(kr-\omega t)}}{r}, \tag{6.96}$$

with the so-called scattering amplitude $f(\theta, \varphi)$ given by

$$f(\theta, \varphi) = \frac{1}{k} \sum_{l,m} b_{lm} (-i)^{l+1} Y_l^m(\theta, \varphi). \tag{6.97}$$

Using the same kind of expressions for energy flux alluded to at the end of Section 6.5, we can calculate the so-called cross section:

$$d\sigma = \frac{\text{Flux through } d\Omega \text{ of scattered wave}}{\text{Flux per unit area of incident wave}}$$

$$= \frac{c |f(\theta, \varphi)|^2 \, d\Omega}{c \left| e^{i(kz-\omega t)} \right|^2}, \tag{6.98}$$

or

$$\frac{d\sigma}{d\Omega} = |f(\theta, \varphi)|^2 . \tag{6.99}$$

As a first example consider the scattering of sound waves by a spherical obstacle. For this case the boundary condition is

$$\left. \frac{\partial \psi}{\partial r} \right|_{r=a} = 0. \tag{6.100}$$

Because of the symmetry about the z axis the solution of this problem must be independent of φ. Thus we need only the terms with $m = 0$ in (6.95) and can write

$$\psi_{sc}(\mathbf{r}, t) = \sum_l a_l h_l^{(1)}(kr) P_l(\cos\theta) e^{-i\omega t}. \tag{6.101}$$

The expansion of the incoming plane wave is given by (6.56):

$$e^{i(kr\cos\theta - \omega t)} = \sum_{l=0}^{\infty}(2l+1)i^l\, j_l(kr)\, P_l(\cos\theta)e^{-i\omega t}. \tag{6.102}$$

Adding (6.101) and (6.102) to obtain the total field ψ [see (6.94)], and using the boundary condition (6.100), we find

$$\sum_l \left[a_l h_l^{(1)\prime}(ka) + (2l+1)i^l\, j_l'(ka) \right] P_l(\cos\theta) = 0, \tag{6.103}$$

which implies

$$a_l = -(2l+1)i^l\, \frac{j_l'(ka)}{h_l^{(1)\prime}(ka)}. \tag{6.104}$$

This determines the scattered wave (6.101) for all $r > a$. The scattering amplitude is determined from the asymptotic form of (6.101):

$$\psi_{sc}(\mathbf{r}, t) \xrightarrow[r\to\infty]{} f(\theta)\frac{e^{i(kr-\omega t)}}{r}, \tag{6.105}$$

$$f(\theta) = \frac{1}{k}\sum_l a_l(-i)^{l+1} P_l(\cos\theta)$$

$$= \frac{i}{k}\sum_l (2l+1)\, \frac{j_l'(ka)}{h_l^{(1)\prime}(ka)}\, P_l(\cos\theta). \tag{6.106}$$

This can be simplified after some manipulations:

$$\frac{i\, j_l'(ka)}{h_l^{(1)\prime}(ka)} = \frac{i\, j_l'(ka)}{j_l'(ka) + in_l'(ka)} = \frac{1}{\frac{n_l'(ka)}{j_l'(ka)} - i}. \tag{6.107}$$

Define the so-called phase shift δ_l by

$$\tan\delta_l(k) = \frac{j_l'(ka)}{n_l'(ka)}, \tag{6.108}$$

and use

$$\frac{1}{\cot\delta_l(k) - i} = e^{i\delta_l(k)}\sin\delta_l(k). \tag{6.109}$$

Substituting in (6.106) we then find for the scattering amplitude

$$f(\theta) = \frac{1}{k}\sum_{l=0}^{\infty}(2l+1)e^{i\delta_l(k)}\sin\delta_l(k)\, P_l(\cos\theta). \tag{6.110}$$

To see the origin of the name phase shift we can rewrite the asymptotic form of the wave:

$$\psi(\mathbf{r}, t) = e^{i(kr\cos\theta - \omega t)} + \psi_{sc}(\mathbf{r}, t)$$

$$\xrightarrow[r\to\infty]{} \sum_l (2l+1)\left[i^l \frac{\cos\{kr - (l+1)\frac{\pi}{2}\}}{kr} + \frac{e^{i\delta_l}\sin\delta_l}{k}\frac{e^{ikr}}{r}\right] P_l(\cos\theta)e^{-i\omega t}$$

$$= \frac{1}{2ik}\sum_l (2l+1)\left[\frac{e^{ikr}}{r} + (-1)^{l+1}\frac{e^{-ikr}}{r} + (e^{2i\delta_l} - 1)\frac{e^{ikr}}{r}\right] P_l(\cos\theta)e^{-i\omega t}$$

$$= \frac{1}{2ik}\sum_l (2l+1)\left[e^{2i\delta_l}\frac{e^{ikr}}{r} + (-1)^{l+1}\frac{e^{-ikr}}{r}\right] P_l(\cos\theta)e^{-i\omega t}. \tag{6.111}$$

The effect of the sphere has been to shift the phase of the outgoing wave by $2\delta_l$, i.e., $1 \to e^{2i\delta_l}$.

The formulas (6.108), (6.110) simplify somewhat in the limit of large wavelengths or small wave numbers. For

$$ka \ll 1, \quad \lambda \gg a, \tag{6.112}$$

we can use the small argument expansions (6.28), (6.29) for the spherical Bessel and Neumann functions to evaluate (6.108):

$$\tan\delta_l(k) \xrightarrow[ka\ll 1]{} \frac{l}{l+1}\frac{(2^l\, l!)^2}{(2l)!\,(2l+1)!}(ka)^{2l+1}. \tag{6.113}$$

In this limit we then have $\delta_l \ll 1$, so that

$$e^{i\delta_l}\sin\delta_l \simeq \delta_l \simeq \tan\delta_l. \tag{6.114}$$

The phase shifts fall off rapidly with increasing l so that we only need to keep the first few terms in the series (6.110). For this particular problem the term $l = 0$ is a special case because of the factor l in the numerator of (6.113). For $l = 0$ we must keep the next higher term in the expansion. Using the explicit formulas (6.24) and (6.25) for $j_0(ka)$ and $n_0(ka)$ we find

$$\tan\delta_0(k) = \frac{j_0'(ka)}{n_0'(ka)} = ka\,\frac{1 - \dfrac{\tan ka}{ka}}{1 + ka\tan ka} \xrightarrow[ka\to 0]{} -\frac{1}{3}(ka)^3. \tag{6.115}$$

From (6.113) we have

$$\delta_1(k) = \frac{1}{6}(ka)^3. \tag{6.116}$$

Keeping just the first two terms in (6.110), we thus find, for $ka \ll 1$,

$$f(\theta) = \frac{1}{k} \left[-\frac{1}{3}(ka)^3 P_0(\cos\theta) + 3\left(\frac{1}{6}\right)(ka)^3 P_1(\cos\theta) \right]$$

$$= -\frac{1}{3}k^2 a^3 \left[1 - \frac{3}{2}\cos\theta \right]. \tag{6.117}$$

In the above example we have used the boundary condition (6.100) appropriate for sound waves. In quantum mechanics one frequently studies so-called hard sphere scattering, which is described by the formulation above with the one change that the boundary condition (6.100) is replaced by

$$\psi(a, \theta, \varphi) = 0. \tag{6.118}$$

Proceeding as above, we find a scattering amplitude given by the formula (6.110) with (6.108) replaced by

$$\tan\delta_l(k) = \frac{j_l(ka)}{n_l(ka)}. \tag{6.119}$$

PROBLEMS

6.1 Find the frequencies of the lowest three modes, i.e., the three modes with the lowest frequencies, for an acoustic resonant cavity consisting of a sphere of radius R.

6.2 Find the radiation by a spherical source of radius a. The disturbance satisfies the wave equation

$$\nabla^2 \psi - \frac{1}{c^2} \frac{\partial^2 \psi}{\partial t^2} = 0,$$

and the value of the disturbance at the surface of the source is given by

$$\psi(r = a, \theta, \varphi, t) = F(\theta) \cos \omega t$$

with

$$F(\theta) = \begin{cases} +V, & 0 < \theta < \frac{\pi}{2} \\ \\ -V, & \frac{\pi}{2} < 0 < \pi \end{cases}$$

and $V = $ constant. Write your result in the form

$$\psi(r, \theta, \varphi, t) \xrightarrow[r \to \infty]{} \operatorname{Re} f(\theta) \frac{e^{i(kr - \omega t)}}{r},$$

and evaluate $f(\theta)$ numerically for the case $ka = \frac{\omega}{c} a = 0.20$.

6.3 Consider the propagation of sinusoidal surface temperature variations into a semi-infinite thermally conducting solid with $\kappa = \frac{K}{\rho c}$ as shown in Figure 6.2. Suppose the surface temperature is given by

$$T(x = 0, t) = T_0 \cos \omega t = \operatorname{Re} T_o e^{i \omega t}.$$

Assume a complex solution of the heat conduction equation of the form

$$f(x) e^{i \omega t},$$

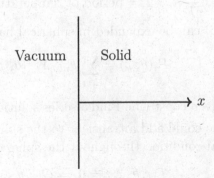

Figure 6.2: Semi-infinite thermally conducting solid for Problem 6.3

substitute in the heat conduction equation, solve for $f(x)$, impose an appropriate boundary condition at $x \to \infty$, and obtain

$$T(x,t) = \operatorname{Re} f(x)e^{i\omega t} = T_0 e^{-\beta x}\cos(\omega t - \beta x),$$

with

$$\beta = \sqrt{\frac{\omega}{2\kappa}}.$$

Apply this formula to study the penetration of surface temperature variations into the earth. Take $\kappa = 0.01 \text{ cm}^2/\text{sec}$ for rock. At what depth have daily and yearly temperature fluctuations fallen off by $1/e$? At what depth are daily and yearly temperature fluctuations $180°$ out of phase with the variations at the surface?

6.4 Study the propagation of sinusoidal surface temperature variations into a thermally conducting sphere of radius R with $\kappa = \frac{K}{\rho c}$. Suppose the surface temperature is given by

$$T(r = R, t) = T_0 \cos \omega t = \operatorname{Re} T_0 e^{i\omega t}.$$

Obtain a solution as the real part of a complex solution of the heat conduction equation:

$$T(r,t) = \operatorname{Re} f(r)e^{i\omega t}.$$

You need to impose an appropriate boundary condition at $r = 0$.

6.5 a) Work Problem 6.4 for a general boundary condition. Suppose the temperature at the surface of the sphere is given by

$$T(r = R, \theta, \varphi, t) = F(\theta, \varphi, t),$$

where $F(\theta, \varphi, t)$ is a given periodic function of time. $F(\theta, \varphi, t)$ can be expanded in a Fourier series

$$F(\theta, \varphi, t) = \sum_{n=-\infty}^{\infty} F_n(\theta, \varphi)e^{i\omega_n t}$$

with

$$\omega_n = \frac{2\pi}{T}n, \quad T = \text{period of temperature variation},$$

and the $F_n(\theta, \varphi)$ can be expanded in spherical harmonics

$$F_n(\theta, \varphi) = \sum_{l,m} F_{nlm} Y_l^m(\theta, \varphi).$$

The F_{nlm} are given constants. Find a series solution for $T(r, \theta, \varphi, t)$, $r \le a$.

b) Discuss how one could add a transient to the solution of part (a) so as to satisfy an initial condition throughout the sphere:

$$T(r, \theta, \varphi, t = 0) = G(r, \theta, \varphi).$$

Find the appropriate series for the transient and use it to satisfy the initial condition.

6.6 a) Work Problem 6.5 for the wave equation. Assume that $\psi(r, \theta, \varphi, t)$ satisfies the wave equation

$$\nabla^2 \psi - \frac{1}{c^2} \frac{\partial^2 \psi}{\partial t^2} = 0$$

inside a sphere of radius R and a boundary condition

$$\psi(r = R, \theta, \varphi, t) = F(\theta, \varphi, t),$$

where $F(\theta, \varphi, t)$ is a given periodic function of time. Find $\psi(r, \theta, \varphi, t)$, $r < a$.

b) Discuss the analogue of the transient of Problem 6.5 (b). Note that it is no longer transient, i.e., does not decay in time. Show how it can be used to satisfy initial conditions on ψ and $\frac{\partial \psi}{\partial t}$ throughout the interior of the sphere.

Summary of Part I

SUMMARY OF RESULTS FROM SEPARATING VARIABLES IN LAPLACE'S EQUATION AND HELMHOLTZ'S EQUATION

Cartesian Coordinates

Helmholtz's Equation:

$$(\boldsymbol{\nabla}^2 + k^2)\psi(\mathbf{r}) = 0,$$

$$\psi(\mathbf{r}) = \sum A(\mathbf{k})e^{i\mathbf{k}\cdot\mathbf{r}} = \sum A(\mathbf{k})e^{i(k_1 x + k_2 y + k_3 z)},$$

$$\text{with } \mathbf{k}^2 = k_1^2 + k_2^2 + k_3^2 = k^2;$$

alternatively

$$\psi = \sum (A\sin k_1 x + B\cos k_1 x)(C\sin k_2 y + D\cos k_2 y)(E\sin k_3 z + F\cos k_3 z).$$

Laplace's Equation:

$$\boldsymbol{\nabla}^2\psi = 0,$$

$$\psi = \sum A e^{i(k_1 x + k_2 y) - \alpha z}, \quad \alpha^2 = k_1^2 + k_2^2,$$
$$\psi = \sum (A\sin k_1 x + B\cos k_1 x)(C\sin k_2 y + D\cos k_2 y)(Ee^{\alpha z} + Fe^{-\alpha z}).$$

The sums above and below can be taken over all allowed values of the separation constants subject to the restrictions such as $k_1^2 + k_2^2 + k_3^2 = k^2$, etc. Depending on the problem, the sums may be integrals over continuous ranges of parameters k_1, k_2, etc.

Cylindrical Coordinates

Helmholtz's Equation:

$$(\boldsymbol{\nabla}^2 + k^2)\psi = 0,$$

$$\psi = \sum [AJ_m(\alpha r) + BN_m(\alpha r)]\left[Ce^{im\theta} + De^{-im\theta}\right]\left[Ee^{\sqrt{\alpha^2 - k^2}\, z} + Fe^{-\sqrt{\alpha^2 - k^2}\, z}\right]$$

Laplace's Equation:

$$\nabla^2 \psi = 0,$$

$$\psi = \sum \left[A J_m(\alpha r) + B N_m(\alpha r) \right] \left[C e^{im\theta} + D e^{-im\theta} \right] \left[E e^{\alpha z} + F e^{-\alpha z} \right]$$

Spherical Coordinates

Helmholtz's Equation:

$$(\nabla^2 + k^2)\psi = 0,$$

$$\psi = \sum \left[A j_l(kr) + B n_l(kr) \right] Y_l^m(\theta, \varphi)$$

Laplace's Equation:

$$\nabla^2 \psi = 0,$$

$$\psi = \sum \left[A r^l + \frac{B}{r^{l+1}} \right] Y_l^m(\theta, \varphi)$$

THE WAVE EQUATION, THE HEAT CONDUCTION EQUATION, EIGENVALUE PROBLEMS, AND LAPLACE'S EQUATION

The Helmholtz equation arises from separating the time in the heat conduction equation or the wave equation:

$$\frac{1}{\kappa} \frac{\partial u}{\partial t} = \nabla^2 u,$$

$$u(\mathbf{r}, t) = \sum A \psi(\mathbf{r}) e^{-k^2 \kappa t},$$

$$(\nabla^2 + k^2)\psi(\mathbf{r}) = 0.$$

and

$$\frac{1}{c^2} \frac{\partial^2 v}{\partial t^2} = \nabla^2 v,$$

$$v(\mathbf{r}, t) = \sum \psi(\mathbf{r})[A \sin kct + B \cos kct],$$

$$(\nabla^2 + k^2)\psi(\mathbf{r}) = 0.$$

The separation constants k^2, m, l, α, etc. are determined by the boundary conditions. For the case of the Helmholtz equation (i.e., as it arises from the heat conduction or wave equations) one can impose homogeneous boundary conditions:

$$\psi(\mathbf{r}) = 0 \quad \text{on a closed surface}$$

or

$$\frac{\partial \psi}{\partial n} = 0 \quad \text{on a closed surface}.$$

Boundary conditions of this type together with the Helmholtz equation form an eigenvalue problem:

$$(\nabla^2 + k_n^2)\psi_n(\mathbf{r}) = 0.$$

The infinite set of solutions of this problem form a complete orthogonal set:

$$\int_V d^3r \; \psi_n^*(\mathbf{r})\psi_m(\mathbf{r}) = 0, \quad n \neq m.$$

The $\psi_n(\mathbf{r})$ are the normal modes of the system. The solution of the heat conduction equation is then

$$u(\mathbf{r}, t) = \sum_n A_n \psi_n(\mathbf{r}) e^{-k_n^2 \kappa t}.$$

The initial value of u, $u(\mathbf{r}, 0)$, can be specified arbitrarily; the A_n are determined to fit this value:

$$u(\mathbf{r}, 0) = \sum_n A_n \psi_n(\mathbf{r}), \quad A_n = \frac{\int_V d^3x \; \psi_n^*(\mathbf{r})u(\mathbf{r}, 0)}{\int_V d^3x \; \psi_n^*(\mathbf{r})\psi_n(\mathbf{r})}.$$

The solution of the wave equation is similarly

$$v(\mathbf{r}, t) = \sum_n \psi_n(\mathbf{r})[A_n \sin k_n ct + B_n \cos k_n ct].$$

The two sets of constants A_n, B_n can be chosen to fit arbitrary values for v and $\partial v/\partial t$ at $t = 0$:

$$v(\mathbf{r}, 0) = \sum_n \psi_n(\mathbf{r})B_n, \qquad B_n = \frac{\int_V d^3x \; \psi_n^*(\mathbf{r})v(\mathbf{r}, 0)}{\int_V d^3x \; \psi_n^*(\mathbf{r})\psi_n(\mathbf{r})}$$

$$\left.\frac{\partial v}{\partial t}\right|_{t=0} = \sum_n \psi_n(\mathbf{r})k_n cA_n, \qquad k_n cA_n = \frac{\int_V d^3x \; \psi_n^*(\mathbf{r})\left[\frac{\partial v}{\partial t}\right]_{t=0}}{\int_V d^3x \; \psi_n^*(\mathbf{r})\psi_n(\mathbf{r})}.$$

One can solve the wave equation or the heat conduction equation with time independent but nonhomogeneous (nonzero) boundary conditions on the surface by adding an appropriate time independent solution of Laplace's equation which satisfies the new boundary conditions.

Laplace's equation is the special case of the Helmholtz equation with $k^2 = 0$. However, the problem to be solved is somewhat different. Homogeneous boundary conditions (of the type applied above to the Helmholtz equation) lead to a trivial solution of Laplace's equation (either 0 or a constant depending on the boundary conditions). This is because $k^2 = 0$ is not an eigenvalue of the Helmholtz equation (or is a trivial one).

For Laplace's equation the problem to be solved is an inhomogeneous boundary value problem:

$$\psi(\mathbf{r}) = \text{given function not everywhere 0 on surface}$$

or

$$\frac{\partial \psi}{\partial n} = \text{given function not everywhere 0 on surface.}$$

Sometimes it is convenient to obtain a solution as the sum of several parts; for each part ψ (or $\partial \psi / \partial n$) is 0 on all surfaces except one. There will be an infinite series of solutions of one of these subproblems. The solutions form a complete set. The complete set of solutions can be used to fit the nonzero boundary condition on the one remaining surface.

EXPANSION IN COMPLETE SETS

Cartesian Coordinates

$$0 < x < a, \quad f(x) = \sum_{n=-\infty}^{\infty} c_n e^{i \frac{2\pi n x}{a}},$$

$$c_n = \frac{1}{a} \int_0^a dx \; e^{-i \frac{2\pi n x}{a}} f(x).$$

Also

$$0 < x < a \quad f(x) = \sum_{n=1}^{\infty} b_n \sin\left(\frac{n\pi x}{a}\right),$$

$$b_n = \frac{2}{a} \int_0^a dx \; \sin\left(\frac{n\pi x}{a}\right) f(x).$$

For an infinite region we obtain Fourier integrals:

$$-\infty < x < \infty, \quad f(x) = \int_{-\infty}^{\infty} dk \; g(k) e^{ikx},$$

$$g(k) = \frac{1}{2\pi} \int_{-\infty}^{\infty} dx \; e^{-ikx} f(x).$$

Completeness relation: $\delta(x - x') = \frac{1}{2\pi} \int_{-\infty}^{\infty} dk \; e^{ik(x-x')}.$

Cylindrical Coordinates

$$0 < r < a, \quad f(r) = \sum_n c_n J_m \left(\frac{x_{mn} r}{a} \right),$$

$$x_{mn} = \text{n}^{th} \text{ root of } J_m(x) = 0 \,,$$

$$c_n = \frac{\int_0^a r \, dr \, J_m \left(\frac{x_{mn} r}{a} \right) f(r)}{\frac{a^2}{2} \left[J_m'(x_{mn}) \right]^2}.$$

$$0 < r < \infty, \quad f(r) = \int_0^\infty k \, dk \, g(k) J_m(kr),$$

$$g(k) = \int_0^\infty r \, dr \, J_m(kr) f(r).$$

Completeness relation: $\dfrac{\delta(r - r')}{r} = \displaystyle\int_0^\infty k \, dk \, J_m(kr) J_m(kr').$

Spherical Coordinates

$$0 \leq \theta \leq \pi$$

$$0 \leq \varphi \leq 2\pi, \quad f(\theta, \varphi) = \sum_{lm} c_{lm} Y_l^m(\theta, \varphi),$$

$$c_{lm} = \int_0^\pi \sin\theta d\theta \int_0^{2\pi} d\varphi \, Y_l^{m*}(\theta, \varphi) f(\theta, \varphi).$$

$$0 < r < a, \quad f(r) = \sum_n c_n j_l \left[\frac{z_{ln} r}{a} \right], \quad z_{ln} = \text{n}^{\text{th}} \text{ root of } j_l(x) = 0,$$

$$c_n = \frac{\int_0^a r^2 \, dr \, j_l \left[\frac{z_{ln} r}{a} \right] f(r)}{\frac{a^3}{2} \left[j_l'(z_{ln}) \right]^2}.$$

$$0 < r < \infty, \quad f(r) = \int_0^\infty k^2 \, dk \, g(k) j_l(kr),$$

$$g(k) = \frac{2}{\pi} \int_0^\infty r^2 dr \, j_l(kr) f(r).$$

Completeness relation: $\dfrac{\pi}{2r^2} \delta(r - r') = \displaystyle\int_0^\infty k^2 \, dk \, j_l(kr) j_l(kr').$

SOME PROPERTIES OF SPECIAL FUNCTIONS

Bessel Functions

The four functions

$$Z_m(x) = \begin{cases} J_m(x) \\[1.5em] N_m(x) \\[1.5em] H_m^{(1)}(x) \\[1.5em] H_m^{(2)}(x) \end{cases}$$

with

$$H_m^{(1)}(x) = J_m(x) + iN_m(x)$$
$$H_m^{(2)}(x) = J_m(x) - iN_m(x)$$

are solutions of Bessel's equation

$$\frac{d^2 Z_m(x)}{dx^2} + \frac{1}{x}\frac{dZ_m(x)}{dx} + \left[1 - \frac{m^2}{x^2}\right] Z_m(x) = 0.$$

For small x

$$J_m(x) \xrightarrow[x\to 0]{} \frac{1}{\Gamma(m+1)} \left(\frac{x}{2}\right)^m,$$

$$N_m(x) \xrightarrow[x\to 0]{} \begin{cases} -\dfrac{\Gamma(m)}{\pi} \left(\dfrac{x}{2}\right)^{-m}, & m \neq 0 \\[1.5em] \dfrac{2}{\pi}\left[\ln\dfrac{x}{2} + \gamma\right], & \gamma = 0.5772\dots, \quad m = 0. \end{cases}$$

For large x

$$J_m(x) \xrightarrow[x\to\infty]{} \sqrt{\frac{2}{\pi x}} \, \cos\left[x - \frac{m\pi}{2} - \frac{\pi}{4}\right],$$

$$N_m(x) \xrightarrow[x\to\infty]{} \sqrt{\frac{2}{\pi x}} \, \sin\left[x - \frac{m\pi}{2} - \frac{\pi}{4}\right].$$

Usually the Hankel functions $H_m^{(1)}(x)$ and $H_m^{(2)}(x)$ are most useful for dealing with the behavior at $x \to \infty$.

$$H_m^{(1)}(x) \xrightarrow[x\to\infty]{} \sqrt{\frac{2}{\pi x}} \, e^{i\left(x - \frac{m\pi}{2} - \frac{\pi}{4}\right)},$$

$$H_m^{(2)}(x) \xrightarrow[x\to\infty]{} \sqrt{\frac{2}{\pi x}}\, e^{-i\left(x-\frac{m\pi}{2}-\frac{\pi}{4}\right)}$$

Note that both $H_m^{(1)}(x)$ and $H_m^{(2)}(x)$ are singular at $x \to 0$.

Spherical Bessel Functions

$$j_l(x) = \sqrt{\frac{\pi}{2x}} J_{l+\frac{1}{2}}(x)$$

$$= (-x)^l \left[\frac{1}{x}\frac{d}{dx}\right]^l \left[\frac{\sin x}{x}\right]$$

$$n_l(x) = \sqrt{\frac{\pi}{2x}} N_{l+\frac{1}{2}}(x)$$

$$= -(-x)^l \left[\frac{1}{x}\frac{d}{dx}\right]^l \left[\frac{\cos x}{x}\right]$$

$$h_l^{(1)}(x) = j_l(x) + i n_l(x)$$

$$= (-x)^l \left[\frac{1}{x}\frac{d}{dx}\right]^l \left[\frac{e^{ix}}{ix}\right]$$

$$h_l^{(2)}(x) = j_l(x) - i n_l(x)$$

$$= (-x)^l \left[\frac{1}{x}\frac{d}{dx}\right]^l \left[\frac{e^{-ix}}{-ix}\right]$$

These four functions are the solutions of the equation

$$\frac{d^2 z_l(x)}{dx^2} + \frac{2}{x}\frac{dz_l(x)}{dx} + \left[1 - \frac{l(l+1)}{x^2}\right] z_l(x) = 0.$$

The small and large argument expansions are

$$j_l(x) \xrightarrow[x\to 0]{} \frac{2^l\, l!}{(2l+1)!} x^l,$$

$$n_l(x) \xrightarrow[x \to 0]{} \frac{-(2l)!}{2^l \, l!} \frac{1}{x^{l+1}},$$

$$j_l(x) \xrightarrow[x \to \infty]{} \frac{1}{x} \cos\left[x - (l+1)\frac{\pi}{2}\right],$$

$$n_l(x) \xrightarrow[x \to \infty]{} \frac{1}{x} \sin\left[x - (l+1)\frac{\pi}{2}\right],$$

$$h_l^{(1)}(x) \xrightarrow[x \to \infty]{} \frac{1}{x} e^{i\left[x - (l+1)\frac{\pi}{2}\right]},$$

$$h_l^{(2)}(x) \xrightarrow[x \to \infty]{} \frac{1}{x} e^{-i\left[x - (l+1)\frac{\pi}{2}\right]}.$$

Spherical Harmonics

These are the eigenfunctions of the angular part of the Laplacian with the boundary condition that Y_l^m be single valued and finite:

$$\frac{1}{\sin\theta} \frac{\partial}{\partial\theta} \left[\sin\theta \frac{\partial Y_l^m}{\partial\theta}\right] + \frac{1}{\sin^2\theta} \frac{\partial^2 Y_l^m}{\partial\varphi^2} = -l(l+1)Y_l^m,$$

$$Y_l^m(\theta, \varphi) = (-1)^m \sqrt{\frac{2l+1}{2} \frac{(l-m)!}{(l+m)!}} \, P_l^m(\cos\theta) \frac{e^{im\varphi}}{\sqrt{2\pi}}, \quad m > 0,$$

$$Y_l^{-m} = (-1)^m Y_l^{m*},$$

$$P_l^m(x) = \frac{(1-x^2)^{\frac{m}{2}}}{2^l \, l!} \frac{d^{l+m}}{dx^{l+m}} (x^2 - 1)^l$$

$$= \frac{(-1)^m (l+m)!}{2^l \, l! \, (l-m)!} \frac{1}{(1-x^2)^{\frac{m}{2}}} \frac{d^{l-m}}{dx^{l-m}} (x^2 - 1)^l,$$

with

$$\int d\Omega \, Y_l^{m*}(\theta, \varphi) Y_{l'}^{m'}(\theta, \varphi) = \delta_{ll'} \delta_{mm'}.$$

Useful Relations

$$\frac{1}{\sqrt{1 - 2ux + u^2}} = \sum_{l=0}^{\infty} P_l(x)u^l, \quad |u| < 1,$$

$$P_l(\cos \Theta) = \frac{4\pi}{2l + 1} \sum_{m=-l}^{l} Y_l^m(\theta, \varphi) Y_l^{m*}(\theta', \varphi'),$$

$$\cos \Theta = \cos \theta \cos \theta' + \sin \theta \sin \theta' \cos(\varphi - \varphi'),$$

$$e^{ikr \cos \theta} = \sum_{l=0}^{\infty} (2l + 1)i^l \, j_l(kr) P_l(\cos \theta),$$

$$\frac{e^{ik|\mathbf{r} - \mathbf{r}'|}}{ik \, |\mathbf{r} - \mathbf{r}'|} = \sum_{l=0}^{\infty} (2l + 1) j_l(kr_<) h_l^{(1)}(kr_>) P_l(\cos \Theta),$$

$$\Theta = \text{ angle between } \mathbf{r} \text{ and } \mathbf{r}'.$$

II

Inhomogeneous Problems,
Green's Functions,
and
Integral Equations

CHAPTER 7

Dielectric and Magnetic Media

7.1 INTRODUCTION

In addition to some mathematical apparatus, we have so far discussed the solutions of homogeneous partial differential equations representing the behavior of various systems in regions free from the sources (of the fields, waves, etc.). In this second part of the book we study the production of the fields, waves, etc., by their sources. Green's functions, the study of which is begun in the following chapter, provide a general method for dealing with such problems. In the present chapter we discuss the methods for dealing with a special class of source terms, the charges and currents due to polarization and magnetization of a medium. As we show below, for these cases the introduction of auxiliary fields enables us to "hide" the source terms so that the equations appear homogeneous.

7.2 MACROSCOPIC ELECTROSTATICS IN THE PRESENCE OF DIELECTRICS

In the interior of a piece of matter the electric field is determined by

$$\nabla \cdot E = 4\pi\rho \tag{7.1}$$
$$\nabla \times E = 0 \tag{7.2}$$

the solution of which is

$$\mathbf{E} = -\nabla\phi, \tag{7.3}$$

$$\phi(\mathbf{r}) = \int d^3x' \, \frac{\rho(\mathbf{r}')}{|\mathbf{r} - \mathbf{r}'|}. \tag{7.4}$$

These fields are very complicated functions with singularities at the position of each electron and nucleus. For most macroscopic problems we are not interested in these detailed microscopic fields, but rather in average fields in which the singularities due to individual charges have been smoothed out.

The quantities of interest for a macroscopic description are given by averaging over a small neighborhood of the point in question:

$$\langle E(\mathbf{r}) \rangle = \frac{1}{\Delta V} \int_{\Delta V} d^3 x \; \mathbf{E}(\mathbf{r} + \mathbf{x}), \tag{7.5}$$

$$\langle \rho(\mathbf{r}) \rangle = \frac{1}{\Delta V} \int_{\Delta V} d^3 x \; \rho(\mathbf{r} + \mathbf{x}). \tag{7.6}$$

Here ΔV is a volume containing many atoms, a volume large enough to smooth out local variations but small enough so that $\langle \mathbf{E}(\mathbf{r}) \rangle$ still contains the gross macroscopic variations. Thus we take

$$d_a \ll (\Delta V)^{\frac{1}{3}} \ll d_m, \tag{7.7}$$

where d_a is an atomic dimension and d_m a length characterizing a macroscopic variation. From the form of (7.5), (7.6) we see that

$$\boldsymbol{\nabla} \cdot \langle \mathbf{E}(\mathbf{r}) \rangle = \langle \boldsymbol{\nabla} \cdot \mathbf{E}(\mathbf{r}) \rangle, \tag{7.8}$$

etc., so that when we average (7.1), (7.2) we obtain

$$\boldsymbol{\nabla} \cdot \langle \mathbf{E}(\mathbf{r}) \rangle = 4\pi \langle \rho(\mathbf{r}) \rangle, \tag{7.9}$$

$$\boldsymbol{\nabla} \times \langle \mathbf{E}(\mathbf{r}) \rangle = 0. \tag{7.10}$$

Because of the linear character of all the relations involved, the averaged equations (7.9), (7.10) have the same form as the original equations (7.1), (7.2). Thus from now on we drop the bracket notation and use (7.1), (7.2) with the understanding that $\mathbf{E}$ and ρ stand for the averaged quantities (7.5), (7.6).

The source of the electric field is the charge density ρ. For a dielectric this is the sum of two parts,

$$\rho = \rho_F + \rho_P. \tag{7.11}$$

The free charge ρ_F is external charge placed on the dielectric. The polarization charge ρ_P arises from polarizing the atoms of the medium, i.e., separating the positive charge of the nucleus from the negative charge of the electrons.

The polarization charge ρ_P can be calculated in terms of the properties of the dielectric. Suppose the charge density in an atom is given by $\rho(\mathbf{r}, \mathbf{x})$. Here $\mathbf{r}$ locates the center of mass of the atom and $\mathbf{x}$ the position inside the atom relative to the center of mass as illustrated in Figure 7.1. The charge density $\rho(\mathbf{r}, \mathbf{x})$ depends on both variables

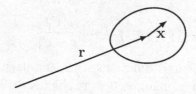

Figure 7.1: Conceptual diagram of charge density distribution of an atom

x and **r** since atoms at different positions **r** will in general be polarized differently. In fact this variation of polarization with **r** just gives rise to ρ_P. Let us calculate the charge in a fixed volume due to polarization. If all the charge in each atom were located at the center of mass of that atom, the charge in this volume would be zero since each atom is neutral. But if the charge is spread out according to $\rho(\mathbf{r}, \mathbf{x})$ for each atom, some charge crosses the boundary, and the volume may be left charged as illustrated in Figure 7.2.

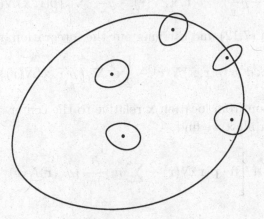

Figure 7.2: Conceptual diagram of atomic charge near boundary in dielectric

Consider a surface element dA on the side of the volume. How much charge crosses dA when we allow the charge in each atom to spread out from the center of mass of the atom to the distribution $\rho(\mathbf{r}, \mathbf{x})$? We first calculate the contribution of $\rho(\mathbf{r}, \mathbf{x})d^3x$, the charge at the same position **x** relative to the center of mass of the atoms, in those atoms at positions **r** near the surface. The charge $\rho(\mathbf{r}, \mathbf{x})d^3x$ from all atoms the centers of mass of which lie in a cylinder with side **x** will move across the boundary surface.

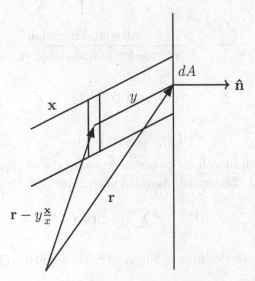

Figure 7.3: Vectors for density of charge at boundary

Thus if $N(\mathbf{r})$ is the density of atoms at position $\mathbf{r}$, we find for the charge crossing the boundary

$$dq = -dA \frac{\hat{\mathbf{n}} \cdot \mathbf{x}}{x} d^3x \int_0^x dy\, \rho(\mathbf{r} - y\frac{\mathbf{x}}{x}, \mathbf{x}) N(\mathbf{r} - y\frac{\mathbf{x}}{x}), \qquad (7.12)$$

where variables from Figure 7.3 have been leveraged. We now expand in a Taylor series

$$\rho(\mathbf{r} - y\frac{\mathbf{x}}{x}, \mathbf{x}) N(\mathbf{r} - y\frac{\mathbf{x}}{x}) = \rho(\mathbf{r}, \mathbf{x}) N(\mathbf{r}) - y\frac{\mathbf{x}}{x} \cdot \boldsymbol{\nabla}_\mathbf{r} \{\rho(\mathbf{r}, \mathbf{x}) N(\mathbf{r})\} + \dots \;. \qquad (7.13)$$

Substituting (7.13) in (7.12) and carrying out the integration over y we find

$$dq = -dA\, \hat{\mathbf{n}} \cdot \mathbf{x}\, d^3x \left[\rho(\mathbf{r}, \mathbf{x}) N(\mathbf{r}) - \frac{1}{2} \mathbf{x} \cdot \boldsymbol{\nabla}_\mathbf{r} \{\rho(\mathbf{r}, \mathbf{x}) N(\mathbf{r})\} + \dots \right]. \qquad (7.14)$$

This is the contribution from location $\mathbf{x}$ relative to the center of mass of the atoms. Integrating $\int d^3x$ over all $\mathbf{x}$ we find

$$dq = -dA \left[\hat{\mathbf{n}} \cdot \mathbf{p}(\mathbf{r}) N(\mathbf{r}) - \sum_{i,j} n_i \frac{\partial}{\partial r_j} \{q_{ij}(\mathbf{r}) N(\mathbf{r})\} + \dots \right], \qquad (7.15)$$

where

$$\mathbf{p}(\mathbf{r}) = \int d^3x\, \mathbf{x} \rho(\mathbf{r}, \mathbf{x}), \qquad (7.16)$$

$$q_{ij}(\mathbf{r}) = \frac{1}{2} \int d^3x\, x_i x_j \rho(\mathbf{r}, \mathbf{x}). \qquad (7.17)$$

These are the dipole and quadrupole moments of an atom at position $\mathbf{r}$ – see (3.128), (3.130).

Henceforth we shall keep only the dipole moment term in (7.15). This is justified when

$$\frac{\dfrac{\partial}{\partial r_j} \{q_{ij}(\mathbf{r}) N(\mathbf{r})\}}{p_i(\mathbf{r}) N(\mathbf{r})} \sim \frac{\text{atomic dimension}}{\text{distance for macroscopic variation}} \ll 1. \qquad (7.18)$$

Then we have

$$dq = -dA\, \hat{\mathbf{n}} \cdot \mathbf{P}(\mathbf{r}) \qquad (7.19)$$

with

$$\mathbf{P}(\mathbf{r}) = N(\mathbf{r})\mathbf{p}(\mathbf{r}) \qquad (7.20)$$

the dipole moment per unit volume or polarization of the dielectric. If we have several kinds of atoms, we add the contributions of each type:

$$\mathbf{P}(\mathbf{r}) = \sum_i N_i(\mathbf{r})\mathbf{p}_i(\mathbf{r}). \qquad (7.21)$$

The total charge in the volume is obtained by integrating (7.19) over the surface:

$$q = \int dq = -\int dA\, \mathbf{n} \cdot \hat{\mathbf{P}}(\mathbf{r}) = -\int d^3x\, \boldsymbol{\nabla} \cdot \mathbf{P}(\mathbf{r}). \qquad (7.22)$$

Here we have used Gauss's theorem to transform the surface integral into a volume integral. The result (7.22) implies a density of polarization charge given by

$$\rho_P(\mathbf{r}) = -\boldsymbol{\nabla} \cdot \mathbf{P}(\mathbf{r}). \tag{7.23}$$

With the formula (7.23) for the polarization charge contribution to the total charge (7.11) we can rewrite the electrostatics equations (7.1), (7.2) or (7.9), (7.10) as

$$\boldsymbol{\nabla} \cdot \mathbf{E} = 4\pi\rho_F - 4\pi\boldsymbol{\nabla} \cdot \mathbf{P}, \tag{7.24}$$

$$\boldsymbol{\nabla} \times \mathbf{E} = 0. \tag{7.25}$$

At this point it is convenient to introduce a new field $\mathbf{D}$ called the displacement,

$$\mathbf{D} = \mathbf{E} + 4\pi\mathbf{P}, \tag{7.26}$$

in terms of which (7.24) becomes

$$\boldsymbol{\nabla} \cdot \mathbf{D} = 4\pi\rho_F. \tag{7.27}$$

By this procedure we succeed in "hiding" part of the source of the electric field.

To proceed, we assume that $\mathbf{P}$ is a function of $\mathbf{E}$. Expanding the function in a Taylor series we find

$$P_i = a_i + \sum_j b_{ij}E_j + \sum_{jk} c_{ijk}E_jE_k + \ldots . \tag{7.28}$$

We assume there is no permanent polarization so that $\mathbf{P} \to 0$ as $\mathbf{E} \to 0$, i.e., $a_i = 0$. We shall make the approximation of ignoring the nonlinear terms in (7.28), so that we obtain

$$P_i = \sum_j b_{ij}E_j. \tag{7.29}$$

This is a valid approximation for sufficiently small fields. Since electric fields obtainable in the laboratory are usually small compared to the fields which hold an atom together,

$$E_{\text{atomic}} \sim \frac{e}{r^2} \sim \frac{5 \times 10^{-10} \text{ esu}}{(10^{-8} \text{ cm})^2} = 5 \times 10^6 \text{ esu}$$

$$= 1.5 \times 10^9 \text{ volts/cm}, \tag{7.30}$$

the small field approximation (7.29) is usually valid.

For an anisotropic substance b_{ij} will be a tensor. We consider here only an isotropic substance such that $\mathbf{E}$ is parallel to $\mathbf{P}$ and the ratio P/E is independent of the direction of $\mathbf{E}$. In this case we have

$$\mathbf{P} = \chi\mathbf{E}. \tag{7.31}$$

The actual evaluation of the electric susceptibility χ (or b_{ij}) is a problem in solid state physics. We do not enter into that here at all. We take χ as a given constant. Substituting (7.31) in (7.26) we find

$$\mathbf{D} = \mathbf{E} + 4\pi\mathbf{P} = \varepsilon\mathbf{E}, \tag{7.32}$$

where the dielectric constant ε is given by

$$\varepsilon = 1 + 4\pi\chi. \tag{7.33}$$

We can now summarize the electrostatic equations for a dielectric medium

$$\boldsymbol{\nabla} \cdot \mathbf{D} = 4\pi\rho_F, \tag{7.34}$$
$$\boldsymbol{\nabla} \times \mathbf{E} = 0, \tag{7.35}$$

with constitutive relations

$$\rho_P = -\boldsymbol{\nabla} \cdot \mathbf{P}, \tag{7.36}$$
$$\mathbf{D} = \mathbf{E} + 4\pi\mathbf{P} = \varepsilon\mathbf{E}, \tag{7.37}$$
$$\mathbf{P} = \chi\mathbf{E}, \tag{7.38}$$
$$\varepsilon = 1 + 4\pi\chi. \tag{7.39}$$

In addition to the equation (7.34)-(7.39) we need the boundary conditions satisfied at the interface between two dielectrics. These can be obtained by applying (7.34) and (7.35) to the rapidly changing region which forms the interface. Applying Gauss's theorem to (7.34) in a pillbox-shaped region of vanishing thickness straddling the interface, as shown in Figure 7.4, we obtain

$$\mathbf{D}_1 \cdot \hat{\mathbf{n}}_1 + \mathbf{D}_2 \cdot \hat{\mathbf{n}}_2 = 4\pi\sigma_F, \tag{7.40}$$

where σ_F is the surface density of free charge on the interface.

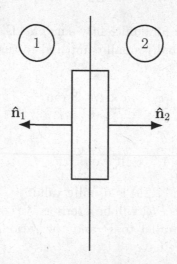

Figure 7.4: Pillbox region across an interface for dielectric

If $\sigma_F = 0$, which is usually the case, we find

$$\boxed{\text{Normal component of } \mathbf{D} \text{ continuous.}} \tag{7.41}$$

Applying stokes theorem to a loop of vanishing width across the surface, as illustrated in Figure 7.5, we find

$$\mathbf{E}_1 \cdot \hat{\mathbf{n}}_1 \times \hat{\mathbf{l}} + \mathbf{E}_2 \cdot \hat{\mathbf{n}}_2 \times \hat{\mathbf{l}} = 0, \tag{7.42}$$

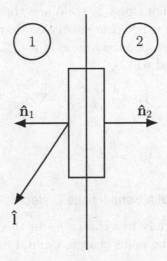

Figure 7.5: Loop crossing an interface for dielectric

or since $\hat{\mathbf{l}}$ is an arbitrary vector parallel to the surface,

$$\mathbf{E}_1 \times \hat{\mathbf{n}}_1 + \mathbf{E}_2 \times \hat{\mathbf{n}}_2 = 0, \tag{7.43}$$

which is equivalent to

$$\boxed{\text{Tangential component of } \mathbf{E} \text{ continuous.}} \tag{7.44}$$

With $\mathbf{E} = -\boldsymbol{\nabla}\phi$, (7.44) implies that the discontinuity in ϕ across the surface is independent of position on the surface. Since additive constants are immaterial in the definition of ϕ, we may take

$$\boxed{\phi \text{ continuous across surface.}} \tag{7.45}$$

7.3   BOUNDARY VALUE PROBLEMS IN DIELECTRICS

7.3.1   Free Charge Distribution ρ_F Embedded in an Infinite Uniform Dielectric with a Constant Dielectric Constant ε

Substituting (7.37) in (7.34) and dividing by ε, we find

$$\boldsymbol{\nabla} \cdot \mathbf{E} = 4\pi \frac{\rho_F}{\varepsilon}, \tag{7.46}$$

$$\boldsymbol{\nabla} \times \mathbf{E} = 0. \tag{7.47}$$

Comparing (7.46), (7.47) with (7.1), (7.2) or (7.9), (7.10) and (7.11) we find

$$\rho = \rho_F + \rho_P = \frac{\rho_F}{\varepsilon}, \tag{7.48}$$

so that

$$\rho_P = -\left(1 - \frac{1}{\varepsilon}\right)\rho_F. \tag{7.49}$$

The polarization of the medium produces a polarization charge proportional to the free charge and of opposite sign (for $\varepsilon > 1$). When the polarization charge is added to the free charge, we obtain a net charge smaller than the free charge by a factor $1/\varepsilon$. In particular if we place a point charge q in a dielectric, the net charge will be q/ε and the field and potential will be

$$\mathbf{E} = \frac{1}{\varepsilon}\frac{q}{r^2}\frac{\mathbf{r}}{r} = -\boldsymbol{\nabla}\phi, \tag{7.50}$$

$$\phi = \frac{1}{\varepsilon}\frac{q}{r}. \tag{7.51}$$

7.3.2 Point Charge in Front of a Semi-infinite Dielectric

As shown in Figure 7.6, the only free charge ρ_F in this problem is the point charge q. Except at the position of the point charge, we find from (7.34), (7.37), (7.35)

$$\boldsymbol{\nabla}\cdot\mathbf{E} = 0, \tag{7.52}$$

$$\boldsymbol{\nabla}\times\mathbf{E} = 0, \tag{7.53}$$

which imply

$$\mathbf{E} = -\boldsymbol{\nabla}\phi, \tag{7.54}$$

$$\nabla^2\phi = 0. \tag{7.55}$$

Near q, $\phi \to q/R$, the singular potential of the point charge.

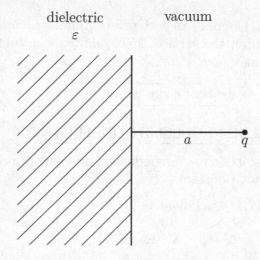

Figure 7.6: Point charge in vacuum a distance a from dielectric

This problem can be worked with the image method familiar from elementary electricity and magnetism courses. Assume the potential in the vacuum is given by the potential of q plus the potential of an "image charge" q', as in Figure 7.7:

$$\phi_{\text{vac}} = -\frac{q}{R_1} + \frac{q'}{R_2}. \tag{7.56}$$

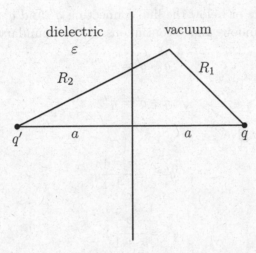

Figure 7.7: Potential in vacuum from charge, q, and image charge, q'

Notice that this has the correct singularity near q. In the dielectric assume the potential is due to an effective charge q'' located at the position of q, as in Figure 7.8:

$$\phi_{\text{diel}} = \frac{q''}{R_1}. \tag{7.57}$$

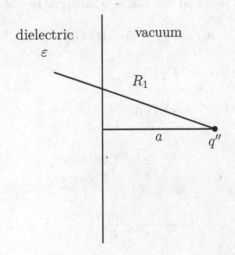

Figure 7.8: Potential in dielectric from effective charge, q''

The charges q' and q'' are now determined to satisfy the boundary conditions along the interface, where

$$R_1 = R_2 = \sqrt{r^2 + a^2},$$

$$\frac{\partial}{\partial n}\frac{1}{R_1} = -\frac{\partial}{\partial n}\frac{1}{R_2} = \frac{a}{R^3} \qquad (7.58)$$

with r a distance above or below the line connecting q'' and q'. We find from (7.41), (7.45), normal $\mathbf{D}$ continuous and ϕ continuous at the boundary, respectively,

$$q - q' = \varepsilon q'' \qquad (7.59)$$

$$q + q' = q'', \qquad (7.60)$$

which can be solved to give

$$q' = -\frac{\varepsilon - 1}{\varepsilon + 1}q, \qquad (7.61)$$

$$q'' = \frac{2}{\varepsilon + 1}q. \qquad (7.62)$$

The polarization charge is given by (7.36), (7.38):

$$\rho_P = -\boldsymbol{\nabla} \cdot \mathbf{P} = -\chi\boldsymbol{\nabla} \cdot \mathbf{E}. \qquad (7.63)$$

According to (7.52) this vanishes except on the surface of the dielectric where the discontinuity in $\mathbf{P}$ leads to a surface charge. [This is a general result for problems with constant ε.] To obtain the surface charge density, see Figure 7.9, integrate (7.63) over a pillbox on the surface and use Gauss's theorem to obtain

$$\sigma_P = -(\mathbf{P}_1 \cdot \hat{\mathbf{n}} - \mathbf{P}_2 \cdot \hat{\mathbf{n}})$$

$$= \mathbf{P}_2 \cdot \hat{\mathbf{n}}$$

$$= \chi\mathbf{E}_2 \cdot \hat{\mathbf{n}}$$

$$= -\frac{\varepsilon - 1}{4\pi}\left.\frac{\partial\phi_{\text{diel}}}{\partial n}\right|_{\text{surface}}$$

$$= -\frac{\varepsilon - 1}{4\pi}q''\frac{a}{R^3}$$

$$= -\frac{1}{2\pi}\frac{\varepsilon - 1}{\varepsilon + 1}q\frac{a}{(r^2 + a^2)^{\frac{3}{2}}}. \qquad (7.64)$$

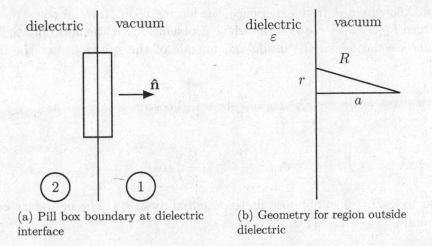

(a) Pill box boundary at dielectric interface

(b) Geometry for region outside dielectric

Figure 7.9: Boundary and geometry for surface charge density on dielectric

7.3.3 Dielectric Sphere in a Uniform External Electric Field

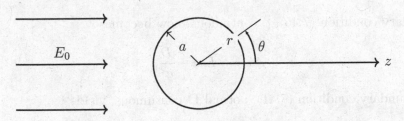

Figure 7.10: Dielectric sphere in uniform field

This problem is similar to Section 3.9.4 with Figure 7.10 as reference. With the dielectric constant ε a constant throughout the sphere we find from (7.34), (7.35), (7.37)

$$\nabla \cdot \mathbf{E} = 0,$$
$$\nabla \times \mathbf{E} = 0, \tag{7.65}$$

so that

$$\mathbf{E} = -\nabla \phi,$$
$$\nabla^2 \phi = 0, \tag{7.66}$$

both inside and outside the sphere. The external field determines the boundary condition far from the sphere:

$$\phi \xrightarrow[r \to \infty]{} -E_0 z = -E_0 r \cos \theta. \tag{7.67}$$

At the surface of the sphere we must satisfy the boundary conditions (7.41) and (7.44) or (7.45).

As we know by now from experience, the forcing term (7.67) imposes the angular dependence $P_L(\cos\theta) = \cos\theta$ on the whole problem so that the appropriate solutions of Laplace's equation for the inside and outside of the sphere are of the form [see (3.142)]

$$\phi_{\text{in}} = \left(Ar + \frac{B}{r^2}\right)\cos\theta, \tag{7.68}$$

$$\phi_{\text{out}} = \left(Cr + \frac{D}{r^2}\right)\cos\theta. \tag{7.69}$$

We must have $B = 0$ to obtain a finite potential at $r = 0$. The boundary condition (7.67) implies $C = -E_0$. Thus we have

$$\phi_{\text{in}} = Ar\cos\theta, \tag{7.70}$$

$$\phi_{\text{out}} = \left(-E_0 r + \frac{D}{r^2}\right)\cos\theta. \tag{7.71}$$

The boundary condition (7.45), ϕ continuous, now becomes

$$Aa = -E_0 a + \frac{D}{a^2}, \tag{7.72}$$

and the boundary condition (7.41), normal $\mathbf{D}$ continuous, yields

$$\varepsilon A = -E_0 - \frac{2D}{a^3}. \tag{7.73}$$

These two equations can be solved for A and D:

$$A = \frac{-3}{\varepsilon + 2}E_0, \tag{7.74}$$

$$D = \frac{\varepsilon - 1}{\varepsilon + 2}E_0 a^3. \tag{7.75}$$

Substituting the value of D in (7.71) and comparing with (3.132) we see that the dielectric sphere has acquired an induced dipole moment

$$\mathbf{P} = \frac{\varepsilon - 1}{\varepsilon + 2}a^3 \mathbf{E}_0. \tag{7.76}$$

The electric field inside the sphere is obtained from (7.70) with the value (7.74) for A:

$$\mathbf{E}_{\text{in}} = \frac{3}{\varepsilon + 2}\mathbf{E}_0. \tag{7.77}$$

For $\varepsilon > 1$, $E_{in} < E_0$. The polarization charge on the surface of the sphere is given by a calculation similar to (7.64):

$$\sigma_P = \frac{\varepsilon - 1}{4\pi} \mathbf{E}_{in} \cdot \hat{\mathbf{n}}$$

$$= \frac{3}{4\pi} \frac{\varepsilon - 1}{\varepsilon + 2} E_0 \cos\theta. \tag{7.78}$$

The field due to the polarization charge opposes the external field inside the sphere, as illustrated in Figure 7.11, reducing the internal field to the value (7.77).

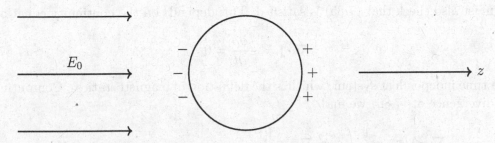

Figure 7.11: Polarization charge on surface of dielectric sphere

7.4   MAGNETOSTATICS AND THE MULTIPOLE EXPANSION FOR THE VECTOR POTENTIAL

The basic equations

$$\boldsymbol{\nabla} \cdot \mathbf{B} = 0, \tag{7.79}$$

$$\boldsymbol{\nabla} \times \mathbf{B} = \frac{4\pi}{c}\mathbf{j} \tag{7.80}$$

have already been introduced in Chapter 1 [equations (1.60)-(1.69)].

Before proceeding with that discussion we note that if $\mathbf{j} = 0$, it is possible to introduce a scalar potential ϕ as in electrostatics. From (7.79) and (7.80) with $\mathbf{j} = 0$ we find

$$\mathbf{B} = \boldsymbol{\nabla}\phi, \tag{7.81}$$

$$\nabla^2 \phi = 0. \tag{7.82}$$

This is sometimes convenient for finding the field around a magnet. The method is however not general; it can be used only in regions with $\mathbf{j} = 0$.

The more general procedure, as outlined in equations (1.60)-(1.69), is to introduce a vector potential $\mathbf{A}$:

$$\mathbf{B} = \boldsymbol{\nabla} \times \mathbf{A}. \tag{7.83}$$

With a suitable choice of gauge we have as equations to determine $\mathbf{A}$

$$\nabla^2 \mathbf{A} = -\frac{4\pi}{c}\mathbf{j}, \tag{7.84}$$

$$\nabla \cdot \mathbf{A} = 0. \tag{7.85}$$

By analogy with the solution (7.4) of Poisson's equation we can infer the solution of (7.84):

$$\mathbf{A}(\mathbf{r}) = \frac{1}{c}\int d^3x' \, \frac{\mathbf{j}(\mathbf{r}')}{|\mathbf{r} - \mathbf{r}'|}. \tag{7.86}$$

We must also check that (7.85) is satisfied. This depends on the continuity equation

$$\nabla \cdot \mathbf{j} = -\frac{\partial \rho}{\partial t} = 0 \tag{7.87}$$

for a time independent system (which is the definition of magnetostatics). Computing the divergence of (7.86) we find

$$\nabla \cdot \mathbf{A}(\mathbf{r}) = \frac{1}{c}\int d^3x' \, \nabla_{\mathbf{r}} \cdot \frac{\mathbf{j}(\mathbf{r}')}{|\mathbf{r} - \mathbf{r}'|}$$

$$= \frac{1}{c}\int d^3x' \, \mathbf{j}(\mathbf{r}') \cdot \nabla_{\mathbf{r}} \frac{1}{|\mathbf{r} - \mathbf{r}'|}$$

$$= -\frac{1}{c}\int d^3x' \, \mathbf{j}(\mathbf{r}') \cdot \nabla_{\mathbf{r}'} \frac{1}{|\mathbf{r} - \mathbf{r}'|}$$

$$= -\frac{1}{c}\int d^3x' \left[\nabla_{\mathbf{r}'} \cdot \left[\frac{\mathbf{j}(\mathbf{r}')}{|\mathbf{r} - \mathbf{r}'|} \right] - \frac{1}{|\mathbf{r} - \mathbf{r}'|}\nabla_{\mathbf{r}'} \cdot \mathbf{j}(\mathbf{r}') \right]$$

$$= -\frac{1}{c}\int_S d\mathbf{A}' \cdot \frac{\mathbf{j}(\mathbf{r}')}{|\mathbf{r} - \mathbf{r}'|} + \frac{1}{c}\int d^3x' \, \frac{\nabla_{\mathbf{r}'} \cdot \mathbf{j}(\mathbf{r}')}{|\mathbf{r} - \mathbf{r}'|}, \tag{7.88}$$

where we have used Gauss's theorem to transform a volume integral to a surface integral over a distant surface. If the currents are confined to a finite region, the first term on the right side of the last line of (7.88) vanishes. The second term vanishes in view of (7.87). Thus we verify that the solution (7.86) satisfies the condition (7.85).

We would now like a multipole expansion for the vector potential analogous to that which we found in Section 3.8 for the scalar potential of electrostatics. To do this in complete generality for a vector quantity such as $\mathbf{A}$, we would need the so-called vector spherical harmonics.[1] To avoid this we keep only the first few terms, in which

[1]See J. M. Blatt and V. F. Weisskopf, Theoretical Nuclear Physics (Wiley, New York, 1952) Appendix B; M. E. Rose, Multipole Fields (Wiley, New York, 1955); J. D. Jackson, Classical Electrodynamics (Wiley, New York, 1975) Chapter 16.

case we can use a Taylor expansion. We assume the currents which contribute to (7.86) are confined to a finite region and study the vector potential $\mathbf{A}(\mathbf{r})$ far outside this region as shown in Figure 7.12. Expanding in powers of r'/r we find

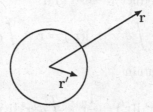

Figure 7.12: Vectors for region of charge distribution, $\mathbf{r}'$, and point of observation, $\mathbf{r}$

$$\frac{1}{|\mathbf{r} - \mathbf{r}'|} = \frac{1}{r} + \frac{1}{r^3}\mathbf{r} \cdot \mathbf{r}' + \dots \, , \tag{7.89}$$

$$\mathbf{A}(\mathbf{r}) = \frac{1}{cr} \int d^3 x' \, \mathbf{j}(\mathbf{r}') + \frac{\mathbf{r}}{cr^3} \cdot \int d^3 x' \, \mathbf{r}' \mathbf{j}(\mathbf{r}) + \dots \, . \tag{7.90}$$

We can show that the monopole contribution to (7.90) vanishes. We have[2]

$$\boldsymbol{\nabla} \cdot (\mathbf{j}\mathbf{r}) = (\boldsymbol{\nabla} \cdot \mathbf{j})\mathbf{r} + (\mathbf{j} \cdot \boldsymbol{\nabla})\mathbf{r} = (\boldsymbol{\nabla} \cdot \mathbf{j})\mathbf{r} + \mathbf{j}, \tag{7.91}$$

so that

$$\int d^3 x' \, \mathbf{j}(\mathbf{r}') = \int d^3 x' \left[\boldsymbol{\nabla}' \cdot (\mathbf{j}\mathbf{r}') - (\boldsymbol{\nabla} \cdot \mathbf{j})\mathbf{r}' \right]$$

$$= \int_S d\mathbf{A}' \, \mathbf{j}\mathbf{r}' - \int d^3 x' \, (\boldsymbol{\nabla} \cdot \mathbf{j})\mathbf{r}'$$

$$= 0, \tag{7.92}$$

since $\boldsymbol{\nabla} \cdot \mathbf{j} = 0$ and the currents are confined to a finite region. The dipole contribution to (7.90) can be rewritten. We have

$$\boldsymbol{\nabla} \cdot (\mathbf{j}\,\mathbf{r}\,\mathbf{r}) = (\boldsymbol{\nabla} \cdot \mathbf{j})\mathbf{r}\,\mathbf{r} + (\mathbf{j} \cdot \boldsymbol{\nabla}\mathbf{r})\mathbf{r} + \mathbf{r}(\mathbf{j} \cdot \boldsymbol{\nabla}\mathbf{r})$$

$$= (\boldsymbol{\nabla} \cdot \mathbf{j})\mathbf{r}\,\mathbf{r} + \mathbf{j}\mathbf{r} + \mathbf{r}\mathbf{j}, \tag{7.93}$$

so that for the same reasons that led to (7.92) we find

$$\int d^3 x' \, [\mathbf{j}(\mathbf{r}')\,\mathbf{r}' + \mathbf{r}'\,\mathbf{j}(\mathbf{r}')] = 0 \tag{7.94}$$

---

[2]In the dyadic expressions which follow, the order of the factors is important. Vector operations are carried out between adjacent factors separated by dots or crosses. Derivatives operate on factors to the right inside parentheses.

and

$$\mathbf{r} \cdot \int d^3x' \, \mathbf{j}(\mathbf{r}') \, \mathbf{r}' + \mathbf{r} \cdot \int d^3x' \, \mathbf{r}' \mathbf{j}(\mathbf{r}') = 0. \tag{7.95}$$

By expanding a triple cross product we also find

$$-\mathbf{r} \cdot \int d^3x' \, \mathbf{j}(\mathbf{r}') \, \mathbf{r}' + \mathbf{r} \cdot \int d^3x' \, \mathbf{r}' \mathbf{j}(\mathbf{r}') = \mathbf{r} \times \int d^3x' \, [\mathbf{j}(\mathbf{r}') \times \mathbf{r}'] . \tag{7.96}$$

Adding (7.95) and (7.96) we obtain

$$\mathbf{r} \cdot \int d^3x' \, \mathbf{r}' \mathbf{j}(\mathbf{r}') = \frac{1}{2}\mathbf{r} \times \int d^3x' \, [\mathbf{j}(\mathbf{r}') \times \mathbf{r}'] . \tag{7.97}$$

Substituting (7.97) in (7.90) we obtain

$$\mathbf{A}(\mathbf{r}) = \frac{\mathbf{m} \times \mathbf{r}}{r^3} + \dots , \tag{7.98}$$

where

$$\mathbf{m} = \frac{1}{2c} \int d^3x' \, \mathbf{r}' \times \mathbf{j}(\mathbf{r}') \tag{7.99}$$

is the magnetic dipole moment. Higher terms in the series (7.89), (7.90), (7.98) would include the contributions of magnetic quadrupoles, etc.

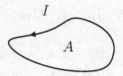

Figure 7.13: Planar current loop

The simplest application of (7.99) is to a planar current loop, see Figure 7.13. The direction of $\mathbf{m}$ is normal to the plane of the loop in the sense of the right screw rule and the magnitude m is given by

$$m = \frac{IA}{c}, \tag{7.100}$$

where I is the current in the loop.

7.5 MAGNETIC MEDIA

By analogy with the electrostatic case discussed in Section 7.2 the macroscopic magnetic field and current in a piece of matter are defined as averages:

$$\langle \mathbf{B}(\mathbf{r}) \rangle = \frac{1}{\Delta V} \int_{\Delta V} d^3x \, \mathbf{B}(\mathbf{r} + \mathbf{x}), \tag{7.101}$$

$$\langle \mathbf{j}(\mathbf{r}) \rangle = \frac{1}{\Delta V} \int_{\Delta V} d^3x \, \mathbf{j}(\mathbf{r} + \mathbf{x}). \tag{7.102}$$

Since the magnetostatic equations (7.79), (7.80) are linear, we immediately obtain

$$\nabla \cdot \langle \mathbf{B}(\mathbf{r}) \rangle = 0, \tag{7.103}$$

$$\nabla \times \langle \mathbf{B}(\mathbf{r}) \rangle = \frac{4\pi}{c} \langle \mathbf{j}(\mathbf{r}) \rangle. \tag{7.104}$$

Henceforth we drop the bracket notation but keep in mind that we are dealing with the average fields.

By analogy with the split of ρ into two contributions for the electrostatic case we shall here split $\mathbf{j}$ into two contributions:

$$\mathbf{j} = \mathbf{j}_c + \mathbf{j}_m. \tag{7.105}$$

In this equation $\mathbf{j}_c$ is the conduction current which arises from the transport of charged particles through the medium. The magnetization current $\mathbf{j}_m$ arises when the magnetization due to intra-atomic currents varies with position in the medium; for example there is a net current downward through the square in the diagram of Figure 7.14.

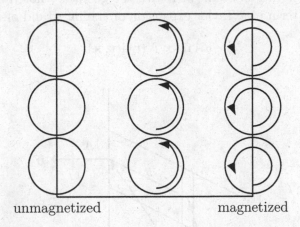

unmagnetized   magnetized

Figure 7.14: Current loops at boundary of material

We can obtain the magnetization current from a calculation similar to that used to obtain the polarization charge in Section 7.2. Suppose that $\mathbf{i}(\mathbf{r}, \mathbf{x})$ is the current density at the position $\mathbf{x}$ relative to the center of mass of an atom whose center of mass is located at $\mathbf{r}$ as shown in Figure 7.15. The average of these atomic currents in a volume ΔV is

$$\mathbf{j}_m = \frac{1}{\Delta V} \int \mathbf{i} \, dV \tag{7.106}$$

according to (7.102).

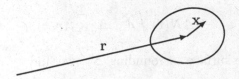

Figure 7.15: Conceptualization of atom position and current density

For an atom completely inside ΔV we find

$$\int \mathbf{i}\, dV = 0 \tag{7.107}$$

according to (7.92). For an atom on the edge we have

$$\int_{\text{inside } \Delta V} \mathbf{i}\, dV = -\int_{\text{outside } \Delta V} \mathbf{i}\, dV. \tag{7.108}$$

Consider the pieces of atoms, at positions $\mathbf{x}$ relative to the centers of mass of the atoms, which cross a little area dA on the side of ΔV when the atoms, originally imagined collapsed on their centers of mass, are expanded according to the distribution $\mathbf{i}(\mathbf{r}, \mathbf{x})$. The contribution to (7.108) from these pieces is

$$-dA \frac{\hat{\mathbf{n}} \cdot \mathbf{x}}{x} \int_0^x dy\, N(\mathbf{r} - y\frac{\mathbf{x}}{x})\, \mathbf{i}(\mathbf{r} - y\frac{\mathbf{x}}{x}, \mathbf{x}), \tag{7.109}$$

where $N(\mathbf{r})$ is the density of atoms at $\mathbf{r}$ with the geometry noted in Figure 7.16. We keep only the first term in a Taylor expansion of the integrand of (7.109) to obtain

$$-dA\, \hat{\mathbf{n}} \cdot \mathbf{x}\, N(\mathbf{r})\, \mathbf{i}(\mathbf{r}, \mathbf{x}). \tag{7.110}$$

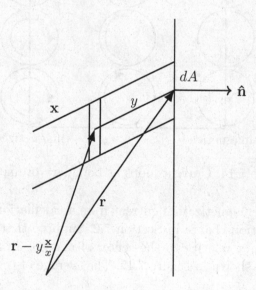

Figure 7.16: Vectors for density of current at boundary

Integrating over positions $\mathbf{x}$ in the atoms we obtain

$$-dA\, N(\mathbf{r}) \int d^3x\, \hat{\mathbf{n}} \cdot \mathbf{x}\, i(r, x), \tag{7.111}$$

and integrating over the surface surrounding ΔV we find

$$\mathbf{j}_m = -\frac{1}{\Delta V} \int dA\, N(\mathbf{r}) \int d^3x\, \hat{\mathbf{n}} \cdot \mathbf{x}\, i(r, x). \tag{7.112}$$

Using the result (7.97) with $\mathbf{r}$ replaced by $\hat{\mathbf{n}}$ we have

$$\hat{\mathbf{n}} \cdot \int d^3x \; \mathbf{x}\, i(\mathbf{r},\mathbf{x}) = -\frac{1}{2}\hat{\mathbf{n}} \times \int d^3x \; \mathbf{x} \times i(\mathbf{r},\mathbf{x})$$

$$= -c\hat{\mathbf{n}} \times \mathbf{m}(\mathbf{r}), \tag{7.113}$$

where

$$\mathbf{m}(\mathbf{r}) = \frac{1}{2c} \int d^3x \; \mathbf{x} \times i(\mathbf{r},\mathbf{x}) \tag{7.114}$$

is the magnetic moment of an atom at the position $\mathbf{r}$. If we introduce the magnetic moment per unit volume or magnetization,

$$\mathbf{M}(\mathbf{r}) = N(\mathbf{r})\mathbf{m}(\mathbf{r}), \tag{7.115}$$

we see that (7.112) can be written as

$$\mathbf{j}_m = \frac{c}{\Delta V} \int dA \; \hat{\mathbf{n}} \times \mathbf{M}(\mathbf{r}). \tag{7.116}$$

The integral in this expression can be converted to a volume integral:

$$\int dA \; (\hat{\mathbf{n}} \times \mathbf{M})_x = \int dA \; (\hat{\mathbf{n}} \times \mathbf{M}) \cdot \hat{\mathbf{x}}$$

$$= \int dA \; \hat{\mathbf{n}} \cdot (\mathbf{M} \times \hat{\mathbf{x}})$$

$$= \int dV \; \boldsymbol{\nabla} \cdot (\mathbf{M} \times \hat{\mathbf{x}})$$

$$= \int dV \left[\frac{\partial M_z}{\partial y} - \frac{\partial M_y}{\partial z} \right]$$

$$= \int dV \; (\boldsymbol{\nabla} \times \mathbf{M})_x. \tag{7.117}$$

Thus (7.116) becomes

$$\mathbf{j}_m = \frac{c}{\Delta V} \int dV \; \boldsymbol{\nabla} \times \mathbf{M} \tag{7.118}$$

or, since by definition the macroscopic quantity $\mathbf{M}$ does not vary appreciably over ΔV,

$$\mathbf{j}_m = c\boldsymbol{\nabla} \times \mathbf{M}. \tag{7.119}$$

Substituting (7.105), (7.119) in the macroscopic equations (7.103), (7.104) we obtain

$$\boldsymbol{\nabla} \cdot \mathbf{B} = 0 \tag{7.120}$$

$$\boldsymbol{\nabla} \times \mathbf{H} = \frac{4\pi}{c}\mathbf{j}_c, \tag{7.121}$$

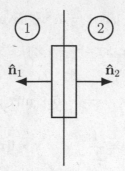

Figure 7.17: Pillbox across the interface for magnetic media

where the field **H** is defined by

$$\mathbf{H} = \mathbf{B} - 4\pi\mathbf{M}. \tag{7.122}$$

For an isotropic paramagnetic or diamagnetic substance **M**, **B**, **H** are related by equations similar to their electrostatic analogues (7.37)-(7.39):

$$\mathbf{M} = \chi_m \mathbf{H}, \tag{7.123}$$

$$\mathbf{B} = \mu \mathbf{H}, \tag{7.124}$$

$$\mu = 1 + 4\pi\chi_m. \tag{7.125}$$

In a ferromagnetic substance the relations between **M**, **B**, **H** are nonlinear and depend on the previous history of the specimen.

As for the electrostatic case the boundary conditions at the interface between two magnetic media are obtained by integrating (7.120), (7.121) over the rapidly changing region in the neighborhood of the interface. Applying Gauss's theorem to (7.120) in a pillbox-shaped region straddling the interface, as illustrated in Figure 7.17, we find

$$\mathbf{B}_1 \cdot \hat{\mathbf{n}}_1 + \mathbf{B}_2 \cdot \hat{\mathbf{n}}_2 = 0 \tag{7.126}$$

or

$$\boxed{\text{normal component of } \mathbf{B} \text{ continuous.}} \tag{7.127}$$

Applying Stoke's theorem to (7.121) on a closed loop across the interface, as illustrated in Figure 7.18 we obtain

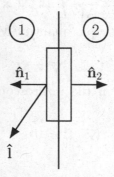

Figure 7.18: Loop at boundary of magnetic media

$$\mathbf{H}_1 \cdot \hat{\mathbf{n}}_1 \times \hat{\mathbf{l}} + \mathbf{H}_2 \cdot \hat{\mathbf{n}}_2 \times \hat{\mathbf{l}} = -\frac{4\pi}{c} \mathbf{K}_c \cdot \hat{\mathbf{l}}, \tag{7.128}$$

where $\mathbf{K}_c$ is the surface conduction current. Since $\hat{\mathbf{l}}$ is an arbitrary vector parallel to the surface, (7.128) becomes

$$\mathbf{H}_1 \times \hat{\mathbf{n}}_1 + \mathbf{H}_2 \times \hat{\mathbf{n}}_2 = \frac{-4\pi}{c} \mathbf{K}_c. \tag{7.129}$$

if $\mathbf{K}_c = 0$, the usual case, (7.129) is equivalent to

$$\boxed{\text{tangential component of } \mathbf{H} \text{ continuous.}} \tag{7.130}$$

7.6 BOUNDARY VALUE PROBLEMS IN MAGNETIC MEDIA

7.6.1 Uniformly Magnetized Sphere, M Given

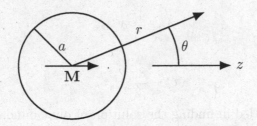

Figure 7.19: Uniformly magnetized sphere

Consider the uniformly magnetized sphere as shown in Figure 7.19, outside the sphere we have $\boldsymbol{\nabla} \cdot \mathbf{B} = \boldsymbol{\nabla} \times \mathbf{B} = 0$ so that

$$\mathbf{B} = -\boldsymbol{\nabla}\phi, \tag{7.131}$$
$$\boldsymbol{\nabla}^2 \phi = 0. \tag{7.132}$$

We shall guess the answer is a dipole field outside,

$$\phi = \frac{A}{r^2} \cos\theta, \tag{7.133}$$

$$B_r^{\text{out}} = H_r^{\text{out}} = \frac{2A}{r^3} \cos\theta, \tag{7.134}$$

$$B_\theta^{\text{out}} = H_\theta^{\text{out}} = \frac{A}{r^3} \sin\theta, \tag{7.135}$$

and a uniform field inside,

$$\mathbf{B}^{\text{in}} = C\,\hat{\mathbf{z}}, \tag{7.136}$$
$$B_r^{\text{in}} = C\cos\theta, \quad B_\theta^{\text{in}} = -C\sin\theta. \tag{7.137}$$

With a uniform magnetization $\mathbf{M} = M\,\hat{\mathbf{z}}$, we have

$$\mathbf{H}^{\text{in}} = \mathbf{B}^{\text{in}} - 4\pi\mathbf{M}, \tag{7.138}$$

$$H_r^{\text{in}} = (C - 4\pi M)\cos\theta, \tag{7.139}$$

$$H_\theta^{\text{in}} = -(C - 4\pi M)\sin\theta. \tag{7.140}$$

These fields certainly satisfy the appropriate equations (7.120), (7.121). It only remains to check the boundary conditions (7.127) and (7.130) that B_r and H_θ be continuous at the surface of the sphere. These become

$$C\cos\theta = \frac{2A}{a^3}\cos\theta, \tag{7.141}$$

$$-(C - 4\pi M)\sin\theta = \frac{A}{a^3}\sin\theta. \tag{7.142}$$

These equations are evidently consistent and have a solution

$$A = \frac{4\pi}{3}a^3 M, \tag{7.143}$$

$$C = \frac{8\pi}{3}M, \tag{7.144}$$

so that we have succeeded in finding the solution of our boundary value problem:

$$\mathbf{B}^{\text{out}} = \mathbf{H}^{\text{out}} = -\boldsymbol{\nabla}\phi, \quad \phi = \frac{4\pi}{3}a^3 M\frac{\cos\theta}{r^2}, \tag{7.145}$$

$$\mathbf{B}^{\text{in}} = \frac{8\pi}{3}M\hat{\mathbf{z}}, \tag{7.146}$$

$$\mathbf{H}^{\text{in}} = \mathbf{B}^{\text{in}} - 4\pi\mathbf{M} = -\frac{4\pi}{3}M\hat{\mathbf{z}}. \tag{7.147}$$

There are no conduction currents $\mathbf{j}_c$ in this problem. The magnetization current

$$\mathbf{j}_m = c\,\boldsymbol{\nabla}\times\mathbf{M} \tag{7.148}$$

vanishes except on the surface of the sphere where the discontinuity in $\mathbf{M}$ leads to a surface magnetization current $\mathbf{K}_m$. To evaluate $\mathbf{K}_m$, use Gauss's theorem in the form (7.117) to integrate $\mathbf{j}_m$ over a small pillbox-shaped volume of vanishingly small thickness on the surface, see Figure 7.20:

Figure 7.20: Pillbox on surface of magnetized sphere

$$\int dV \, \mathbf{j}_m = \int dA \, \mathbf{K}_m = c \int dV \, \boldsymbol{\nabla} \times \mathbf{M}$$

$$= c \int dA \, \hat{\mathbf{n}} \times \mathbf{M}. \qquad (7.149)$$

This implies

$$\mathbf{K}_m = -c \, \hat{\mathbf{n}} \times \mathbf{M}, \qquad (7.150)$$

where $\hat{\mathbf{n}}$ is now the external normal to the sphere, as shown in Figure 7.21. We see that $\mathbf{K}_m$ is in the φ direction and of magnitude

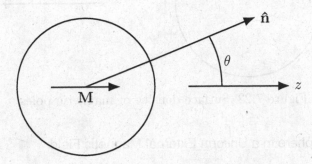

Figure 7.21: Magnetization, $\mathbf{M}$, and surface normal of magnetized sphere

$$(K_m)_\varphi = cM \sin\theta. \qquad (7.151)$$

This surface current is the source of the field $\mathbf{B}$ according to the equations

$$\boldsymbol{\nabla} \cdot \mathbf{B} = 0, \qquad (7.152)$$

$$\boldsymbol{\nabla} \times \mathbf{B} = \frac{4\pi}{c} \mathbf{j}_m. \qquad (7.153)$$

An alternate, somewhat artificial, way to treat the problem is to regard $\mathbf{H}$ as the fundamental field:

$$\boldsymbol{\nabla} \times \mathbf{H} = 0, \qquad (7.154)$$
$$\boldsymbol{\nabla} \cdot \mathbf{H} = \boldsymbol{\nabla} \cdot (\mathbf{B} - 4\pi\mathbf{M}) = -4\pi\boldsymbol{\nabla} \cdot \mathbf{M}. \qquad (7.155)$$

Comparing these with the corresponding electrostatic equations (7.1), (7.2) we see that

$$-\boldsymbol{\nabla} \cdot \mathbf{M} \qquad (7.156)$$

can be regarded as the magnetic pole density source of the $\mathbf{H}$ field. With a uniform $\mathbf{M}$, $\boldsymbol{\nabla} \cdot \mathbf{M}$ is nonzero only on the surface of the magnetized sphere where the discontinuity in $\mathbf{M}$ leads to a surface density σ, as illustrated in Figure 7.22 of magnetic poles:

$$\int \sigma dA = \int (-\boldsymbol{\nabla} \cdot \mathbf{M}) dV = -\int dA \, \mathbf{M} \cdot \mathbf{n} \qquad (7.157)$$

or

$$\sigma = \mathbf{M} \cdot \mathbf{n}, \quad \hat{\mathbf{n}} = \text{outward normal}$$
$$= M \cos\theta. \tag{7.158}$$

It is the demagnetizing effect of these poles which leads to $\mathbf{H}$ being opposite in direction to $\mathbf{M}$ inside the sphere.

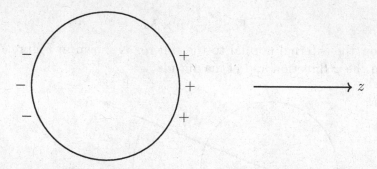

Figure 7.22: Surface density of magnetic poles

7.6.2 Magnetic Sphere in a Uniform External Magnetic Field

Consider the magnetic sphere in a uniform external magnetic field as shown in Figure 7.23, the solution of this problem is obtained by adding the external field $\mathbf{B}_0$ to the solution (7.145)-(7.147) of Section 7.6.1:

$$\mathbf{B}^{\text{out}} = \mathbf{H}^{\text{out}} = \mathbf{B}_0 - \boldsymbol{\nabla}\phi, \quad \phi = \frac{4\pi}{3}a^3 M \frac{\cos\theta}{r^2}, \tag{7.159}$$

$$\mathbf{B}^{\text{in}} = \mathbf{B}_0 + \frac{8\pi}{3}\mathbf{M}, \tag{7.160}$$

$$\mathbf{H}^{\text{in}} = \mathbf{B}^{\text{in}} - 4\pi\mathbf{M} = \mathbf{B}_0 - \frac{4\pi}{3}\mathbf{M}. \tag{7.161}$$

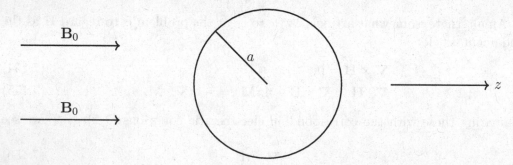

Figure 7.23: Magnetic sphere in uniform magnetic field

It is easy to verify that these fields satisfy the equations (7.120), (7.121) and the boundary conditions (7.127), (7.130) (normal $\mathbf{B}$ and tangential $\mathbf{H}$ continuous, respectively).

To proceed further we need some relation between $\mathbf{M}$ and $\mathbf{B}_0$. For a paramagnetic or diamagnetic substance we can use (7.124):

$$\mathbf{B}^{\text{in}} = \mu \mathbf{H}^{\text{in}}, \tag{7.162}$$

$$\mathbf{B}_0 + \frac{8\pi}{3}\mathbf{M} = \mu\left(\mathbf{B}_0 - \frac{4\pi}{3}\mathbf{M}\right), \tag{7.163}$$

$$\mathbf{M} = \frac{3}{4\pi}\frac{\mu - 1}{\mu + 2}\mathbf{B}_0. \tag{7.164}$$

This is completely analogous to the result we obtained for a dielectric sphere in an external field – see (7.76) and note $\mathbf{m} = (4\pi a^3/3)\mathbf{M}$ is the magnetic moment of the sphere.

For the case of a ferromagnetic substance we need a hysteresis curve.

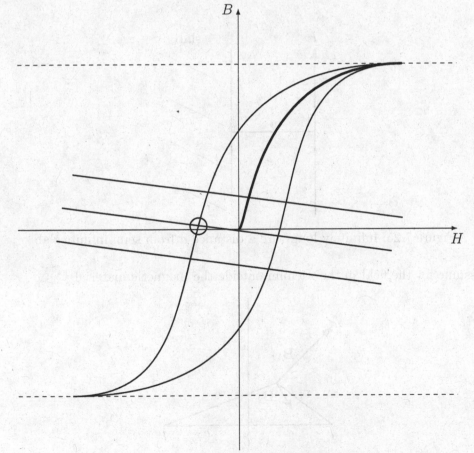

[Note that the B scale is much compressed relative]
to the H scale for a ferromagnetic substance.]

Figure 7.24: Hysteresis curve of B vs. H

This gives us one relation between B^{in} and H^{in} (the analogue of $B^{in} = \mu H^{in}$). Eliminating M from (7.160) and (7.161) we obtain another relation:

$$B^{in} = 3B_0 - 2H^{in}. \tag{7.165}$$

These are straight lines with slope (-2) on the B vs. H plot shown in Figure 7.24. The intersection of the hysteresis curve with one of these lines determines B and H inside the magnet in terms of the external field B_0. For example if we saturate the ferromagnetic substance and then reduce the external field to zero, we end up at the point marked $\odot$ on the diagram.

7.6.3 Long Straight Wire Carrying Current I Parallel to a Semi-infinite Slab of Material of Permeability μ

This problem, as illustrated in Figure 7.25, can be worked by the image method. The magnetic field around an infinitely long straight wire is of the form

$$\mathbf{B} = \frac{2I}{c} \frac{1}{r} \hat{\varphi}. \tag{7.166}$$

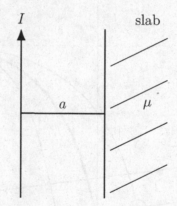

Figure 7.25: Infinitely long wire a distance a from semi-infinite slab

Assume for the field in the vacuum outside the magnetic material

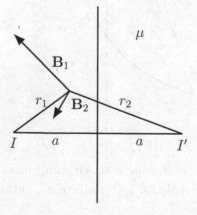

Figure 7.26: Field in vacuum from current, I and image current I'

$$\mathbf{B} = \mathbf{B}_1 + \mathbf{B}_2, \tag{7.167}$$

$$\mathbf{B}_1 = \frac{2I}{c} \frac{1}{r_1} \hat{\boldsymbol{\varphi}}_1, \tag{7.168}$$

$$\mathbf{B}_2 = \frac{2I'}{c} \frac{1}{r_2} \hat{\boldsymbol{\varphi}}_2, \tag{7.169}$$

where I' is an image current and the geometry shown in Figure 7.26, and assume for the field inside the magnetic material

$$\mathbf{B} = \frac{2I''}{c} \frac{1}{r_1} \hat{\boldsymbol{\varphi}}_1. \tag{7.170}$$

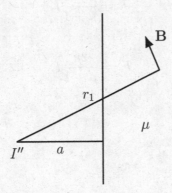

Figure 7.27: Field in slab from effective current, I''

With geometry of the boundary shown in Figure 7.28, the boundary conditions (7.127) and (7.130) that the normal component of $\mathbf{B}$ and the tangential component of $\mathbf{H}$ be continuous at the boundary between the vacuum and the magnetic material assume the form

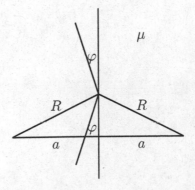

Figure 7.28: Geometry of boundary conditions at slab

$$\frac{2I}{C}\frac{1}{R}\sin\varphi + \frac{2I'}{c}\frac{1}{R}\sin\varphi = \frac{2I''}{c}\frac{1}{R}\sin\varphi, \tag{7.171}$$

$$\frac{2I}{C}\frac{1}{R}\cos\varphi - \frac{2I'}{c}\frac{1}{R}\cos\varphi = \frac{1}{\mu}\frac{2I''}{c}\frac{1}{R}\cos\varphi, \tag{7.172}$$

or

$$I + I' = I'', \tag{7.173}$$

$$I - I' = \frac{1}{\mu}I'', \tag{7.174}$$

which can be solved to give

$$I' = \frac{\mu - 1}{\mu + 1}I, \tag{7.175}$$

$$I'' = \frac{2\mu}{\mu + 1}I. \tag{7.176}$$

PROBLEMS

7.1   a) A potential difference V is maintained across the plates of an infinitely long cylindrical condenser, see Figure 7.29, of inner and outer radii a and b filled with a material of dielectric constant ε. Find the electric field in the condenser, the free charge ρ_F on the condenser plates, and the polarization charge ρ_P on the dielectric.

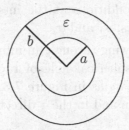

Figure 7.29: Cross section view of cylindrical condenser for Problem 7.1a)

   b) Work the problem for the case that only a sector of the condenser is filled with dielectric (region between dashed lines in Figure 7.30).

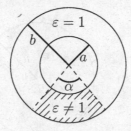

Figure 7.30: Condenser with section filled with dielectric for Problem 7.1b)

7.2 A point charge q is located a distance b from the center of a sphere of material of dielectric constant ε and radius $a < b$ as shown in Figure 7.31. Find the electrostatic potential $V(r, \theta)$ at all points both inside and outside the sphere. Find the polarization charge on the sphere.

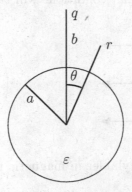

Figure 7.31: Charge q a distance b from center of a dielectric sphere for 7.2

7.3 The steady state flow of current through a conducting medium is described by the equations

$$\nabla \cdot \mathbf{j} = 0,$$
$$\nabla \times \mathbf{E} = 0,$$
$$\mathbf{j} = \sigma \mathbf{E}.$$

a) Find the boundary conditions at the interface between two media with different conductivities, σ_1 and σ_2.

b) Suppose current is flowing through an infinite conducting medium of conductivity σ_1 with a spherical hole of radius a containing a medium of different conductivity σ_2 as in Figure 7.32. Far from the sphere the current density is uniform and in the z direction:

$$\mathbf{j} \xrightarrow[r \to \infty]{} j_0 \, \hat{\mathbf{k}}.$$

Find the current density everywhere. Compare with the analogous hydrodynamic problem for the case $\sigma_2 = 0$.

c) Find the surface charge density on the sphere.

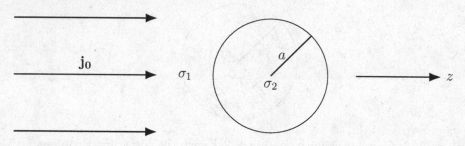

Figure 7.32: Spherical conducting medium within current flow for Problem 7.3

7.4 A magnetic shield consists of a cylindrical layer of a material of permeability μ as in Figure 7.33. The magnetic field $\mathbf{B_0}$ outside the cylinder and far from it is uniform and transverse to the axis of the cylinder. Find the magnetic field $\mathbf{B_1}$ inside the shield. Under what conditions will the shield be a good shield?

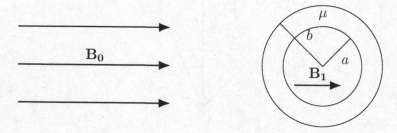

Figure 7.33: Magnetic shield cylinder in magnetic field for Problems 7.4 and 7.5

7.5 Work Problem 7.4 for a spherical shield.

7.6 A magnet has the shape of a cylinder of radius a and length b as seen in Figure 7.34. It is uniformly magnetized with a magnetization M_0 per unit volume along the axis of the cylinder.

a) Find the magnetization current $\mathbf{j_M} = c\,\mathbf{\nabla} \times \mathbf{M}$ and use it to find the fields $\mathbf{B}$ and $\mathbf{H}$ at points along the axis of the cylinder, both inside and outside the magnet.

b) Find the density of magnetic poles $\rho_M = -\mathbf{\nabla} \cdot \mathbf{M}$ and use it to find the fields $\mathbf{B}$ and $\mathbf{H}$ at points along the axis of the cylinder, both inside and outside the magnet.

Do your results from parts (a) and (b) agree?

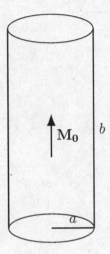

Figure 7.34: Cylindrical magnet for Problem 7.6

CHAPTER 8

Green's Functions: Part One

8.1 INTRODUCTION

Green's functions provide a method for dealing with source terms in differential equations. In the partial differential equations

$$\nabla^2 \psi = -4\pi\rho, \tag{8.1}$$

$$\nabla^2 \psi - \frac{1}{\kappa}\frac{\partial \psi}{\partial t} = -4\pi\rho, \tag{8.2}$$

$$\nabla^2 \psi - \frac{1}{c^2}\frac{\partial^2 \psi}{\partial t^2} = -4\pi\rho, \tag{8.3}$$

the inhomogeneous terms $4\pi\rho$ on the right are the source terms and are regarded as given. The method can also be applied to ordinary differential equations. We shall begin with this case in the following section and gradually introduce the additional complications arising from more than one space dimension and where the response and driving term have the same sinusoidal time dependence. This is addressed first since the time dependence can be easily described and results in the Helmholtz equation for a focus on the geometry bounding the problem. Expanding the problem to infinite regions and general time dependency will be continued in the following chapter.

8.2 ORDINARY DIFFERENTIAL EQUATIONS

We wish to solve the inhomogeneous differential equation

$$\mathcal{L}u(x) \equiv \frac{d}{dx}\left[p(x)\frac{du(x)}{dx}\right] - q(x)u(x) = \phi(x) \tag{8.4}$$

in the interval $a \leq x \leq b$ with $\phi(x)$ a specified known function and with boundary conditions at $x = a$ and $x = b$ to be specified below. As mentioned in Section 2.8 the assumed self-adjoint form of the differential operator $\mathcal{L}$ in (8.4) is rather general since after multiplication by a suitable factor any second-order linear differential operator can be put in this form. For the self-adjoint form of $\mathcal{L}$ we have the Green's theorem identity (2.105):

$$\boxed{\int_a^b dx \ [v\mathcal{L}u - u\mathcal{L}v] = \left[p\left(v\frac{du}{dx} - u\frac{dv}{dx}\right)\right]_a^b.} \tag{8.5}$$

In order to solve (8.4) for a general source $\phi(x)$ we introduce the so-called Green's function $G(x, x')$, which is the solution for a point source at x':

$$\mathcal{L}G(x, x') = \delta(x - x'), \tag{8.6}$$

$\delta(x - x')$ being the Dirac delta function. We then use the linear character of (8.4) to find the solution for a source $\phi(x)$ by superposition:

$$\phi(x) = \int dx' \ \delta(x - x')\phi(x'), \tag{8.7}$$

$$u(x) = \int dx' \ G(x, x')\phi(x'). \tag{8.8}$$

8.3   GENERAL THEORY, VARIOUS BOUNDARY CONDITIONS

In order to elaborate on the remarks of the previous paragraph, with suitable attention to the boundary conditions, let us suppose the solution of (8.6) can be found; we show how to do this in practice below. Apply Green's theorem (8.5) to $u(x)$ and $v(x) = G(x, x')$;

$$\int_a^b dx \ [G(x, x')\mathcal{L}u(x) - u(x)\mathcal{L}G(x, x')]$$

$$= \left[p(x)\left\{G(x, x')\frac{du(x)}{dx} - u(x)\frac{d}{dx}G(x, x')\right\}\right]_{x=a}^b$$

$$= \int_a^b dx \ G(x, x')\phi(x) - u(x'), \tag{8.9}$$

where we have used (8.4), (8.6) and the integral properties of delta functions. Interchanging x and x', (8.9) becomes

$$\boxed{\begin{aligned} u(x) &= \int_a^b dx' \ G(x', x)\phi(x') \\[2mm] &\quad - \left[p(x')\left\{G(x', x)\frac{du(x')}{dx'} - u(x')\frac{d}{dx'}G(x', x)\right\}\right]_{x'=a}^b. \end{aligned}} \tag{8.10}$$

We can now consider several possibilities for the boundary conditions on $u(x)$ at $x = a, b$.

a) u(a) and u(b) given.

For this case choose as boundary conditions on $G(x', x)$

$$G(a, x) = G(b, x) = 0. \tag{8.11}$$

This has the desirable advantage of eliminating from the formula (8.10) the unknown quantities $du(x')/dx'$ at $x' = a, b$. In fact (8.10) reduces to

$$u(x) = \int_a^b dx' \; G(x', x)\phi(x') + \left[p(x')u(x')\frac{d}{dx'}G(x', x)\right]_{x'=a}^b. \tag{8.12}$$

This is the solution of (8.4) in terms of known quantities.

b) $u(a)$ and $\left.\frac{du(x)}{dx}\right|_{x=b}$ given.

For this case choose as boundary conditions on $G(x', x)$

$$G(a, x) = 0, \quad \left.\frac{d}{dx'}G(x', x)\right|_{x'=b} = 0. \tag{8.13}$$

Again the unknown quantities, in this case $u(b)$ and $du(x)/dx$ at $x = a$ are eliminated from the formula (8.10) and we obtain the solution of (8.4) in terms of known quantities:

$$u(x) = \int_a^b dx' \; G(x', x)\phi(x')$$

$$- p(a)u(a)\left.\frac{d}{dx'}G(x', x)\right|_{x'=a} - p(b)G(b, x)\left.\frac{du(x')}{dx'}\right|_{x'=b}. \tag{8.14}$$

c) General case: $Au(a) + Bu'(a) = X$ given and $Cu(b) + Du'(b) = Y$ given.

Choose

$$AG(a, x) + B\left.\frac{d}{dx'}G(x', x)\right|_{x'=a} = 0,$$

$$CG(b, x) + D\left.\frac{d}{dx'}G(x', x)\right|_{x'=b} = 0. \tag{8.15}$$

The quantity needed in (8.10) at $x' = a$ is then

$$G(a, x)\left.\frac{du(x')}{dx'}\right|_{x'=a} - u(a)\left.\frac{d}{dx'}G(x', x)\right|_{x'=a}$$

$$= G(a, x)\left.\frac{du(x')}{dx'}\right|_{x'=a} - u(a)\left\{-\frac{A}{B}G(a, x)\right\}$$

or

$$G(a,x)\left.\frac{du(x')}{dx'}\right|_{x'=a} - u(a)\left.\frac{d}{dx'}G(x',x)\right|_{x'=a} = \frac{G(a,x)}{B}\left[Au(a) + B\left.\frac{du(x')}{dx'}\right|_{x'=a}\right]$$

$$= \frac{G(a,x)}{B}X$$

$$= -\frac{1}{A}\left.\frac{d}{dx'}G(x',x)\right|_{x'=a}X, \qquad (8.16)$$

which is a known quantity. A similar manipulation shows that the quantity needed at $x' = b$ in (8.10) can be expressed in terms of the given quantity Y.

In each case we can choose the boundary conditions on G so as to eliminate unknown values of u and u' at the boundaries. In this way we obtain a solution of the inhomogeneous differential equation (8.4) once we find the Green's function $G(x',x)$.

We can prove that the Green's function is symmetric in its two arguments.

Proof. To see this, apply the Green's theorem (8.5) to $u(x) = G(x,x')$ and $v(x) = G(x,x'')$. Using (8.6) we find

$$G(x',x'') - G(x'',x') = \left[p(x)\left\{G(x,x'')\frac{dG(x,x')}{dx} - G(x,x')\frac{dG(x,x'')}{dx}\right\}\right]_{x=a}^{b}. \quad (8.17)$$

For the general case with boundary conditions (8.15) the right-hand side of (8.17) vanishes and we find

$$\boxed{G(x',x'') = G(x'',x').} \qquad (8.18)$$

□

Let us now return to the differential equation (8.6) for $G(x,x')$:

$$\frac{d}{dx}\left[p(x)\frac{d}{dx}G(x,x')\right] - q(x)G(x,x') = \delta(x - x'). \qquad (8.19)$$

For $x \neq x'$ this reduces to the homogeneous equation

$$\frac{d}{dx}\left[p(x)\frac{d}{dx}G(x,x')\right] - q(x)G(x,x') = 0, \quad x \neq x', \qquad (8.20)$$

so we can use whatever techniques are available for homogeneous equations in solving for $G(x,x')$ in the two regions $x > x'$ and $x < x'$. At $x = x'$ we must match the two solutions so obtained. The delta function $\delta(x - x')$ on the right of (8.19) is to be interpreted as due to a discontinuity in $\frac{dG(x,x')}{dx}$ at $x = x'$. In fact, integrating (8.19) over an interval $x = x' - \epsilon$ to $x = x' + \epsilon$ of infinitesimal length about $x = x'$ and assuming $q(x)G(x,x')$ is finite in this region so that

$$\int_{x'-\epsilon}^{x'+\epsilon} dx\, q(x)G(x,x') \xrightarrow[\epsilon\to 0]{} 0, \qquad (8.21)$$

we find

$$p(x) \frac{d}{dx} G(x, x') \bigg|_{x=x'+\epsilon} - p(x) \frac{d}{dx} G(x, x') \bigg|_{x=x'-\epsilon} = 1, \qquad (8.22)$$

or

$$\frac{d}{dx} G(x, x') \bigg|_{x=x'-\epsilon}^{x'+\epsilon} = \frac{1}{p(x')}, \qquad (8.23)$$

provided $p(x)$ is continuous at $x = x'$. On the other hand, we must take $G(x, x')$ to be continuous at $x = x'$:

$$G(x' + \epsilon, x') = G(x' - \epsilon, x'). \qquad (8.24)$$

If there were a discontinuity in $G(x, x')$ itself, the first derivative $\frac{d}{dx} G(x, x')$ would then contain a term proportional to $\delta(x - x')$ and the second derivative $\frac{d^2}{dx^2} G(x, x')$ would then contain a term proportional to $\frac{d}{dx} \delta(x - x')$, in contradiction to the defining equation (8.19).

The homogeneous equation (8.20) in the two regions $x < x'$ and $x > x'$ plus the two conditions (8.23) and (8.24) at $x = x'$ plus the two boundary conditions at $x = a$ and $x = b$ discussed previously completely determine the Green's function, the four conditions fixing the two pairs of integration constants obtained on solving the differential equation in the two regions.

8.4   THE BOWED STRETCHED STRING

As an example we can consider the bowed stretched string. Suppose the bowing force (assumed transverse to the string) per unit length at position x is proportional to $f(x, t)$ with the correct constant factors so that the differential equation describing the motion of the string is

$$\frac{\partial^2 u(x, t)}{\partial x^2} - \frac{1}{c^2} \frac{\partial^2 u(x, t)}{\partial t^2} = f(x, t). \qquad (8.25)$$

We assume a sinusoidal bowing force

$$f(x, t) = f(x) e^{-i\omega t}. \qquad (8.26)$$

The forced response of the system to the bowing will then have the same time dependence:

$$u(x, t) = u(x) e^{-i\omega t}, \qquad (8.27)$$

where $u(x)$ is determined by

$$\frac{d^2 u(x)}{dx^2} + k^2 u(x) = f(x), \quad k = \frac{\omega}{c}. \qquad (8.28)$$

In addition to the response to the bowing (8.27) there will be a superposition of the free oscillations (5.38)-(5.42), which can be used to satisfy initial conditions. We ignore these free oscillations here and concentrate on the solution of the inhomogeneous equation (8.28).

We suppose the ends of the string are fixed

$$u(0) = u(l) = 0. \tag{8.29}$$

This is the special form of case **a)** covered in Section 8.3. The Green's function satisfies the differential equation

$$\frac{d^2 G(x, x')}{dx^2} + k^2 G(x, x') = \delta(x - x') \tag{8.30}$$

and the boundary conditions

$$G(0, x') = G(l, x') = 0. \tag{8.31}$$

For $x \neq x'$ the delta function on the right of (8.30) vanishes and we find

$$G(x, x') = \begin{cases} A \sin kx + B \cos kx, & x < x' \\[2mm] C \sin kx + D \cos kx, & x > x' \end{cases} \tag{8.32}$$

Applying the boundary conditions (8.31) to (8.32) we can reduce the latter to

$$G(x, x') = \begin{cases} A \sin kx, & x < x' \\[2mm] E \sin k(x - l), & x > x' \end{cases} \tag{8.33}$$

According to (8.23), (8.24) the delta function on the right of (8.30) leads to the conditions

$$\left. \frac{d}{dx} G(x, x') \right|_{x=x'-\epsilon}^{x'+\epsilon} = 1, \tag{8.34}$$

i.e.,

$$kE \cos k(x' - l) - kA \cos kx' = 1, \tag{8.35}$$

and

$$G(x' + \epsilon, x') - G(x' - \epsilon, x') = 0, \tag{8.36}$$

i.e.,

$$E \sin k(x' - l) - A \sin kx' = 0. \tag{8.37}$$

The two linear equations (8.35), (8.37) for E and A can be solved and we find, using $\sin kx' \cos k(x' - l)$-$\cos kx' \sin k(x' - l)$=$\sin kl$,

$$A = \frac{1}{k \sin kl} \sin k(x' - l), \tag{8.38}$$

$$E = \frac{1}{k \sin kl} \sin kx'. \tag{8.39}$$

Substituting in (8.33) we thus obtain

$$G(x, x') = \frac{1}{k \sin kl} \begin{cases} \sin kx \sin k(x' - l), & x < x' \\ \\ \sin kx' \sin k(x - l), & x > x' \end{cases}$$

$$= \frac{1}{k \sin kl} \sin kx_< \sin k(x_> - l). \tag{8.40}$$

This turns out to be a symmetric function of x and x' as predicted in (8.18).

Using the formula (8.12) from Section 8.3 case **(a)** we find as the solution of (8.28) with boundary condition (8.29)

$$u(x) = \int_0^l dx' \, G(x, x') f(x')$$

$$= \frac{\sin k(x - l)}{k \sin kl} \int_0^x dx' \, \sin kx' f(x') + \frac{\sin kx}{k \sin kl} \int_x^l dx' \, \sin k(x' - l) f(x'). \tag{8.41}$$

It is instructive to check directly that this is a solution of the differential equation (8.28) and boundary conditions (8.29).

We can use this same Green's function (8.40) to solve a different type of problem. Suppose the bowing force $f(x)$ vanishes but one end of the string is wiggled with frequency ω and amplitude A:

$$f = 0,$$
$$u(0) = 0,$$
$$u(l) = A. \tag{8.42}$$

Employing the general result (8.12), we find

$$u(x) = A \frac{d}{dx'} G(x, x') \Big|_{x'=l}$$

$$= A \frac{1}{k \sin kl} \sin kx \cdot k \cos k(l - l)$$

$$= A \frac{\sin kx}{\sin kl}. \tag{8.43}$$

This is of course a rather trivial application, but the corresponding consideration in the case of partial differential equations is not so simple.

For either of the two cases treated above we note that $u(x) \to \infty$ when $k = \omega/c$ is such that $\sin kl = 0$. This is the equation which determines the normal modes – see (5.40), (5.42). Thus if we shake the system at the frequency of a normal mode, we get a big response.

8.5 EXPANSION OF GREEN'S FUNCTION IN EIGENFUNCTIONS

As a last topic to be considered in this section we study the expansion of the Green's function in the eigenfunctions of the corresponding Sturm-Liouville eigenvalue problem. We found in Section 2.8 that the Sturm-Liouville eigenvalue problem

$$\mathcal{L}u_n(x) \equiv \frac{d}{dx}\left[p(x)\frac{du_n(x)}{dx}\right] - q(x)u_n(x) = -\lambda_n\rho(x)u_n(x) \tag{8.44}$$

with boundary conditions of the general type

$$Au_n(a) + Bu'_n(a) = 0 \tag{8.45}$$
$$Cu_n(b) + Du'_n(b) = 0 \tag{8.46}$$

has a complete set of orthonormal eigenfunctions $u_n(x)$:

$$\int_a^b dx\ \rho(x)u_n(x)u_m(x) = \delta_{nm}. \tag{8.47}$$

With this normalization the completeness relation (2.122) reads

$$\sum_n \rho(x)u_n(x)u_n(x') = \delta(x - x'). \tag{8.48}$$

Suppose now that we wish to solve the inhomogeneous differential equation

$$\mathcal{L}u + \lambda\rho u = \phi(x), \tag{8.49}$$

with boundary conditions

$$Au(a) + Bu'(a) = X,$$
$$Cu(b) + Du'(b) = Y. \tag{8.50}$$

Here $\phi(x)$ is a given function and X and Y are given constants. According to the discussion in Section 8.3 case **(c)**, we need a Green's function $G(x, x')$ satisfying the differential equation

$$\mathcal{L}G(x, x') + \lambda\rho G(x, x') = \delta(x - x') \tag{8.51}$$

and the boundary conditions

$$AG(a, x') + B\left.\frac{d}{dx}G(x, x')\right|_{x=a} = 0,$$

$$CG(b, x') + D\left.\frac{d}{dx}G(x, x')\right|_{x=b} = 0. \tag{8.52}$$

In terms of this Green's function $G(x, x')$, which is known to be symmetric, the solution of (8.49), (8.50) is given by (8.10):

$$u(x) = \int_a^b dx'\ G(x, x')\phi(x')$$

$$- \left[p(x')\left\{G(x, x')\frac{du(x')}{dx'} - u(x')\frac{d}{dx'}G(x, x')\right\}\right]_{x'=a}^{b}; \tag{8.53}$$

and according to (8.16) only the given combinations X and Y, (8.50), will appear in the boundary condition term of (8.53).

It is easy to check that the solution of (8.51) (8.52) for $G(x, x')$ is given by the series

$$G(x, x') = \sum_n \frac{u_n(x)u_n(x')}{\lambda - \lambda_n}, \tag{8.54}$$

where the $u_n(x)$ are the eigenfunctions of (8.44)-(8.46). Using (8.44) and (8.48) we find

$$(\mathcal{L} + \lambda\rho)G(x, x') = \sum_n \frac{(\lambda - \lambda_n)\rho(x)u_n(x)u_n(x')}{\lambda - \lambda_n} = \delta(x - x'); \tag{8.55}$$

and using (8.45), (8.46) we find

$$AG(a, x') + B\left.\frac{d}{dx}G(x, x')\right|_{x=a} = \sum_n \frac{[Au_n(a) + Bu_n'(a)]u_n(x')}{\lambda - \lambda_n} = 0, \tag{8.56}$$

$$CG(b, x') + D\left.\frac{d}{dx}G(x, x')\right|_{x=b} = \sum_n \frac{[Cu_n(b) + Du_n'(b)]u_n(x')}{\lambda - \lambda_n} = 0, \tag{8.57}$$

so that (8.54) is indeed the solution of (8.51), (8.52).

The general formula (8.54) for $G(x, x')$ is a nice result. Unfortunately, it involves an infinite series. If we can find a closed expression for $G(x, x')$, that will generally be more useful. Comparison of a closed expression for $G(x, x')$ with the eigenfunction expansion leads to an expansion theorem.

Let us pursue our example of the bowed string described by the differential equation (8.28). We have already found a closed expression for the Green's function in (8.40). The corresponding eigenvalue problem was discussed in Section 5.3. The eigenvalues are

$$\lambda_n = k_n^2 = \left(\frac{n\pi}{l}\right)^2, \tag{8.58}$$

and the normalized eigenfunctions are

$$u_n(x) = \sqrt{\frac{2}{l}}\sin\frac{n\pi x}{l}. \tag{8.59}$$

Substituting (8.58), (8.59) in the general formula (8.54) and comparing with (8.40) we find

$$G(x, x') = \frac{2}{l}\sum_{n=1}^{\infty}\frac{\sin\frac{n\pi x}{l}\sin\frac{n\pi x'}{l}}{k^2 - \left(\frac{n\pi}{l}\right)^2}$$

$$= \frac{\sin kx_< \sin k(x_> - l)}{k\sin kl}. \tag{8.60}$$

8.6 POISSON'S EQUATION

We want to solve the equation

$$\nabla^2 \psi(\mathbf{r}) = -4\pi \rho(\mathbf{r}) \tag{8.61}$$

in a certain region V with the boundary condition that $\psi(\mathbf{r})$ or its normal derivative $\partial \psi(\mathbf{r})/\partial n$ is given on the surface S of the region. The source density $\rho(\mathbf{r})$ is a given function.

For the operator ∇^2 we have the well-known Green's theorem

$$\int_V d^3x \left[u\nabla^2 v - v\nabla^2 u \right] = \int_S dA \left[u\frac{\partial v}{\partial n} - v\frac{\partial u}{\partial n} \right], \tag{8.62}$$

which follows directly from applying Gauss's theorem to $u\nabla v - v\nabla u$.

We shall introduce a Green's function which is a solution of Poisson's equation with a point source,

$$\nabla^2 G(\mathbf{r}, \mathbf{r}') = \delta^3(\mathbf{r} - \mathbf{r}'), \tag{8.63}$$

in the region V with suitable boundary conditions on the surface S, to be specified below. Applying the Green's theorem (8.62) to $u = \psi(\mathbf{r})$ and $v = G(\mathbf{r}, \mathbf{r}')$ and using (8.61) and (8.63) we find

$$\int_V d^3x \left[\psi(\mathbf{r})\delta^3(\mathbf{r} - \mathbf{r}') - G(\mathbf{r}, \mathbf{r}')(-4\pi)\rho(\mathbf{r}) \right] =$$

$$\int_S dA \left[\psi(\mathbf{r})\frac{\partial G(\mathbf{r}, \mathbf{r}')}{\partial n} - G(\mathbf{r}', \mathbf{r})\frac{\partial \psi(\mathbf{r})}{\partial n} \right]. \tag{8.64}$$

Interchanging $\mathbf{r}$ and $\mathbf{r}'$ this becomes

$$\psi(\mathbf{r}) = -4\pi \int_V d^3x'\, G(\mathbf{r}', \mathbf{r})\rho(\mathbf{r}') + \int_S dA' \left[\psi(\mathbf{r}')\frac{\partial G(\mathbf{r}', \mathbf{r})}{\partial n'} - G(\mathbf{r}', \mathbf{r})\frac{\partial \psi(\mathbf{r}')}{\partial n'} \right]. \tag{8.65}$$

As in the one-dimensional case discussed in Section 8.3, we now choose the boundary conditions on G so as to eliminate unknown quantities from the surface integral on the right of (8.65). This depends on the boundary conditions on $\psi(\mathbf{r})$ and we distinguish two simple cases:

a) $\psi(\mathbf{r}')$ given on S.

For this case choose

$$G(\mathbf{r}', \mathbf{r}) = 0 \quad \text{for } \mathbf{r}' \text{ on } S. \tag{8.66}$$

This eliminates the unknown normal derivative $\dfrac{\partial \psi(\mathbf{r}')}{\partial n'}$ from the right side of (8.13), and we find

$$\psi(\mathbf{r}) = -4\pi \int_V d^3x'\, G(\mathbf{r}', \mathbf{r})\rho(\mathbf{r}') + \int_S dA'\, \psi(\mathbf{r}')\frac{\partial G(\mathbf{r}', \mathbf{r})}{\partial n'}. \tag{8.67}$$

b) $\dfrac{\partial \psi(\mathbf{r}')}{\partial n'}$ given on S.

We cannot choose $\dfrac{\partial G(\mathbf{r}', \mathbf{r})}{n'} = 0$ on S; integrating (8.63) over V and using Gauss's theorem we find

$$\int_S dA' \, \frac{\partial G(\mathbf{r}', \mathbf{r})}{\partial n'} = \int_V d^3x' \, \nabla_{\mathbf{r}'}^2 \, G(\mathbf{r}', \mathbf{r}) = \int d^3x \, \delta^3(\mathbf{r}' - \mathbf{r}) = 1. \qquad (8.68)$$

This is consistent with a boundary condition

$$\frac{\partial G(\mathbf{r}', \mathbf{r})}{\partial n'} = \frac{1}{S}, \qquad (8.69)$$

with S the area of the surface. Substituting (8.69) in (8.65) we find

$$\psi(\mathbf{r}) = \frac{1}{S} \int_S dA' \, \psi(\mathbf{r}') - 4\pi \int_V d^3x' \, G(\mathbf{r}', \mathbf{r}) \rho(\mathbf{r}') - \int_S dA' \, G(\mathbf{r}', \mathbf{r}) \frac{\partial \psi(\mathbf{r}')}{\partial n'}. \qquad (8.70)$$

The first term on the right here is a constant and can be dropped since $\psi(\mathbf{r})$ is undetermined up to an additive constant by the boundary condition.

Applying Green's theorem (8.62) to $u = G(\mathbf{r}, \mathbf{r}')$ and $v = G(\mathbf{r}, \mathbf{r}'')$ and using (8.63) we find

$$G(\mathbf{r}'', \mathbf{r}') - G(\mathbf{r}', \mathbf{r}'') = \int_S dA \left[G(\mathbf{r}, \mathbf{r}') \frac{\partial G(\mathbf{r}, \mathbf{r}'')}{\partial n} - G(\mathbf{r}, \mathbf{r}'') \frac{\partial G(\mathbf{r}, \mathbf{r}')}{\partial n} \right]. \qquad (8.71)$$

For case **(a)**, with the boundary condition (8.66), the right-hand side of (8.71) vanishes and we find that $G(\mathbf{r}', \mathbf{r}'')$ is symmetric under interchange of $\mathbf{r}'$ and $\mathbf{r}''$.

Every physicist knows one solution of (8.63). The solution of Poisson's equation for the electrostatic potential $\phi(\mathbf{r})$ due to a point charge e located at $\mathbf{r}'$,

$$\nabla^2 \phi(\mathbf{r}) = -4\pi e \delta^3(\mathbf{r} - \mathbf{r}'), \qquad (8.72)$$

is the Coulomb potential

$$\phi(\mathbf{r}) = \frac{e}{|\mathbf{r} - \mathbf{r}'|}. \qquad (8.73)$$

Comparing, we see that a solution of (8.63) is

$$G(\mathbf{r}, \mathbf{r}') = -\frac{1}{4\pi} \frac{1}{|\mathbf{r} - \mathbf{r}'|}. \qquad (8.74)$$

This describes the behavior near the singular point $\mathbf{r} \to \mathbf{r}'$ but does not take account of the boundary conditions. Thus in general we will have

$$G(\mathbf{r}, \mathbf{r}') = -\frac{1}{4\pi} \frac{1}{|\mathbf{r} - \mathbf{r}'|} + F(\mathbf{r}, \mathbf{r}'), \qquad (8.75)$$

where

$$\nabla^2 F(\mathbf{r}, \mathbf{r}') = 0, \qquad (8.76)$$

and F is chosen so as to satisfy the boundary condition on S.

We now study some examples.

8.7   POISSON'S EQUATION FOR ALL SPACE

For the case that the surface S recedes to infinity, we can use the Green's function $G(\mathbf{r}, \mathbf{r}')$ given by (8.74). Substituting this in (8.65) we find

$$\psi(\mathbf{r}) = \phi(\mathbf{r}) + \int d^3x' \, \frac{\rho(\mathbf{r}')}{|\mathbf{r} - \mathbf{r}'|}, \tag{8.77}$$

where $\phi(\mathbf{r})$ is the contribution from the surface integral in (8.65). Applying the operator ∇^2 to (8.77) we immediately find that $\phi(\mathbf{r})$ is a solution of Laplace's equation $\nabla^2\phi(\mathbf{r}) = 0$. The integral in (8.77) is the contribution to $\psi(\mathbf{r})$ due to the sources $\rho(\mathbf{r})$. The field $\phi(\mathbf{r})$ is to be thought of as produced by some infinitely remote external sources. An example is the uniform external field of Section 3.10. If there are no such external fields, we take $\phi(\mathbf{r}) = 0$.

8.8   ELECTROSTATICS WITH BOUNDARY CONDITIONS ON SURFACES AT FINITE DISTANCES – THE IMAGE METHOD

When the case (a) covered in Section 8.6 applies; the potential is given by (8.67). We can interpret the differential equation (8.63) and boundary condition (8.66) for $G(\mathbf{r}, \mathbf{r}')$:

$$G(\mathbf{r}, \mathbf{r}') = \text{potential at } \mathbf{r} \text{ due to point charge } e = \left(-\frac{1}{4\pi}\right)$$

at $\mathbf{r}'$ in presence of grounded conductor on the surface S. \qquad (8.78)

For certain simple geometries one can solve for $G(\mathbf{r}, \mathbf{r}')$ by the image method familiar from elementary electricity and magnetism. Consider as an example the interior of a sphere. With a point charge q a distance r' from the center of the sphere we need an image charge $q' = -qa/r'$ a distance $\rho = a^2/r'$ from the center of the sphere as shown in Figure 8.1. With this value of ρ there are two similar triangles in the figure and we have

$$\frac{r'}{a} = \frac{a}{\rho} = \frac{R_1}{R_2} \tag{8.79}$$

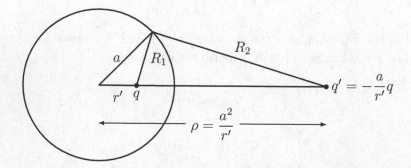

Figure 8.1: Image charge near a conducting sphere

for R_1, R_2 the distances from q to q' to a point on the surface of the sphere. With the given value of the image charge we then find

$$\frac{q}{R_1} + \frac{q'}{R_2} = 0 \tag{8.80}$$

so that the potential due to q and q' vanishes on the surface of the sphere. Setting $q = -1/4\pi$ and evaluating the potential at a general point inside the sphere, as per Figure 8.2, we find the Green's function (8.78):

$$G(\mathbf{r}, \mathbf{r}') = -\frac{1}{4\pi} \left[\frac{1}{R_1} - \frac{a/r'}{R_2} \right]$$

$$= -\frac{1}{4\pi} \left[\frac{1}{\sqrt{r^2 + r'^2 - 2rr' \cos \Theta}} - \frac{a/r'}{\sqrt{r^2 + \left(\frac{a^2}{r'}\right)^2 - 2r\frac{a^2}{r'} \cos \Theta}} \right]$$

$$= -\frac{1}{4\pi} \left[\frac{1}{\sqrt{r^2 + r'^2 - 2rr' \cos \Theta}} - \frac{a}{\sqrt{r^2 r'^2 + a^4 - 2rr'a^2 \cos \Theta}} \right]. \tag{8.81}$$

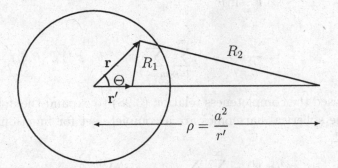

Figure 8.2: Potential within the sphere

To evaluate the surface integral in (8.67) we need

$$\left. \frac{\partial G(\mathbf{r}, \mathbf{r}')}{\partial r'} \right|_{r'=a} = \frac{1}{4\pi} a \left[1 - \frac{r^2}{a^2} \right] \frac{1}{(r^2 + a^2 - 2ar \cos \Theta)^{3/2}}. \tag{8.82}$$

Using these results in (8.67) we find the solution of Poisson's equation for the interior of a sphere:

$$\psi(\mathbf{r}) = -4\pi \int d^3 x' \, G(\mathbf{r}, \mathbf{r}') \rho(\mathbf{r}')$$

$$+ \frac{a^3}{4\pi} \left[1 - \frac{r^2}{a^2} \right] \int_0^\pi \sin \theta' \, d\theta' \int_0^{2\pi} d\varphi' \frac{\psi(a, \theta', \varphi')}{(r^2 + a^2 - 2ra \cos \Theta)^{3/2}}. \tag{8.83}$$

For $\rho = 0$ one has only the second term. As shown in Problem 6, Chapter 3, this second term can also be obtained by summing the series found when Laplace's equation is solved for the interior of a sphere and the arbitrary boundary function $\psi(a, \theta', \varphi')$ is expanded in spherical harmonics.

The example above is a nice one because the image method provides a closed formula for the Green's function $G(\mathbf{r}, \mathbf{r}')$. This image method works equally well for the exterior of a sphere, the interior and exterior of a cylinder when all quantities are z independent, and the half-space with an infinite plane boundary.

8.9 EXPANSION OF THE GREEN'S FUNCTION FOR THE INTERIOR OF A SPHERE IN SERIES

The image method discussed in the previous section is basically a trick. If this trick were not available, we could proceed by the following straightforward method to find the Green's function for the interior of a sphere. We wish to solve

$$\mathbf{\nabla}^2 G(\mathbf{r}, \mathbf{r}') = \delta(\mathbf{r} - \mathbf{r}'). \tag{8.84}$$

In spherical coordinates the δ function can be written

$$\delta(\mathbf{r} - \mathbf{r}') = \frac{1}{r'^2 \sin \theta'} \delta(r - r') \delta(\theta - \theta') \delta(\varphi - \varphi')$$

$$= \frac{1}{r'^2} \delta(r - r') \sum_{l=0}^{\infty} \sum_{m=-l}^{l} Y_l^{m*}(\theta', \varphi') Y_l^m(\theta, \varphi), \tag{8.85}$$

where we have used the completeness relation (3.98) to expand the delta functions of angles. Since the spherical harmonics are a complete set for functions of angles, we can also expand

$$G(\mathbf{r}, \mathbf{r}') = \sum_{l=0}^{\infty} \sum_{m=-l}^{l} G_{lm}(r, r') Y_l^m(\theta, \varphi). \tag{8.86}$$

Substituting (8.85) and (8.86) in (8.84) and using the Laplacian in spherical coordinates (2.32) and the properties of spherical harmonics (3.83), (3.85), we find

$$\sum_{l,m} \left\{ \frac{1}{r} \frac{d^2}{dr^2} (r G_{lm}) - \frac{l(l+1)}{r^2} G_{lm} \right\} Y_l^m(\theta, \varphi)$$

$$= \frac{1}{r'^2} \delta(r - r') \sum_{l,m} Y_l^{m*}(\theta', \varphi') Y_l^m(\theta, \varphi). \tag{8.87}$$

The orthogonality of the $Y_l^m(\theta, \varphi)$ then implies

$$\frac{1}{r} \frac{d^2}{dr^2} (r G_{lm}) - \frac{l(l+1)}{r^2} G_{lm} = \frac{1}{r'^2} \delta(r - r') Y_l^{m*}(\theta', \varphi'). \tag{8.88}$$

We can separate off the θ', φ' dependence of G_{lm},

$$G_{lm} = g_l(r, r') Y_l^{m*}(\theta', \varphi'), \tag{8.89}$$

to obtain

$$\frac{1}{r} \frac{d^2}{dr^2} [r g_l(r, r')] - \frac{l(l+1)}{r^2} g_l(r, r') = \frac{1}{r'^2} \delta(r - r'). \tag{8.90}$$

This is the equation for a one-dimensional Green's function. For $r \neq r'$ the right side of (8.90) vanishes so the solutions of (8.90) are the well-known radial functions for the Laplacian, (3.140):

$$g_l(r, r') = \begin{cases} A r^l + \dfrac{B}{r^{l+1}}, & r < r' \\[3mm] C r^l + \dfrac{D}{r^{l+1}}, & r > r' \end{cases}. \tag{8.91}$$

The boundary conditions that $G(r, r')$ be finite at $r = 0$ and vanish for $r = a$ imply

$$B = 0, \tag{8.92}$$

$$C a^l + \frac{D}{a^{l+1}} = 0, \tag{8.93}$$

so that we have

$$g_l(r, r') = \begin{cases} A r^l, & r < r' \\[3mm] C \left[r^l - \dfrac{a^{2l+1}}{r^{l+1}} \right], & r > r' \end{cases}. \tag{8.94}$$

The matching conditions at $r = r'$ are obtained by integrating (8.90) over a region of infinitesimal length from $r = r' - \epsilon$ to $r = r' + \epsilon$. As in (8.23), (8.24) we find

$$g_l(r' - \epsilon, r') = g_l(r' + \epsilon, r'), \tag{8.95}$$

$$\frac{d}{dr} [r g_l(r, r')] \bigg|_{r=r'-\epsilon}^{r'+\epsilon} = \frac{1}{r'}. \tag{8.96}$$

Using (8.92), (8.93) these become

$$A r'^l = C \left[r'^l - \frac{a^{2l+1}}{r'^{l+1}} \right], \tag{8.97}$$

$$C \left[(l+1) r'^l + l \frac{a^{2l+1}}{r'^{l+1}} \right] - A(l+1) r'^l = \frac{1}{r'}. \tag{8.98}$$

These two equations can be solved for A and C, and we find

$$C = \frac{r'^l}{(2l+1)a^{2l+1}}, \tag{8.99}$$

$$A = \frac{r'^l}{(2l+1)a^{2l+1}}\left[1 - \left(\frac{a}{r'}\right)^{2l+1}\right]. \tag{8.100}$$

Collecting results (8.94), (8.89) and (8.86) we thus obtain

$$g_l(r,r') = \frac{r^l r'^l}{(2l+1)a^{2l+1}} \cdot \begin{cases} \left[1 - \left(\dfrac{a}{r'}\right)^{2l+1}\right], & r < r' \\[3mm] \left[1 - \left(\dfrac{a}{r}\right)^{2l+1}\right], & r > r' \end{cases} \tag{8.101}$$

$$G(\mathbf{r},\mathbf{r}') = \sum_{l=0}^{\infty}\sum_{m=-l}^{l} g_l(r,r') Y_l^{m*}(\theta',\varphi') Y_l^m(\theta,\varphi)$$

$$= \frac{1}{4\pi}\sum_{l=0}^{\infty}(2l+1)g_l(r,r')P_l(\cos\Theta), \tag{8.102}$$

where in the last line we have used the spherical harmonics addition theorem (3.114) with Θ the angle between $\mathbf{r}$ and $\mathbf{r}'$. The two expressions (8.81) and (8.102) are the Green's function for the same problem and so are equal; (8.102) is the expansion of (8.81) in Legendre polynomials.

Naturally one prefers the closed expression (8.81) to the infinite series (8.102). However, the method employed here is more general and can be used in problems where the image method fails. For example, the calculation just above can be easily extended to find the Green's function for the region between two concentric spheres by replacing (8.92) by the condition that $g_l(r,r')$ vanish on the other sphere at, say, $r = b$. The series expansion method can also be employed for the Helmholtz equation.

8.10   THE HELMHOLTZ EQUATION – THE FORCED DRUMHEAD

We start with an example – the forced drumhead. This is similar to the bowed string considered in Section 8.4. If the force per unit area normal to the drumhead is $f(\mathbf{r},t)$ the equation of motion for the drumhead becomes

$$\left[\boldsymbol{\nabla}^2 - \frac{1}{c^2}\frac{\partial^2}{\partial t^2}\right]u(\mathbf{r},t) = -\frac{1}{c^2\sigma}f(\mathbf{r},t) \equiv F(\mathbf{r},t). \tag{8.103}$$

We assume a sinusoidal forcing function

$$F(\mathbf{r},t) = F(\mathbf{r})e^{-i\omega t} \tag{8.104}$$

and consider only the forced response of the drumhead, which will have the same time dependence,

$$u(\mathbf{r}, t) = u(\mathbf{r})e^{-i\omega t}. \tag{8.105}$$

Substituting in (8.103) we find a two-dimensional Helmholtz equation with an inhomogeneous term

$$(\nabla^2 + k^2)u(\mathbf{r}) = F(\mathbf{r}), \quad k = \frac{\omega}{c}. \tag{8.106}$$

The boundary condition is

$$u(\mathbf{r}) = 0, \quad \mathbf{r} \text{ on the edge of drumhead.} \tag{8.107}$$

We shall consider the case of a round drum of radius a as shown in Figure 8.3.

To solve (8.106), (8.107) we introduce a Green's function $G(\mathbf{r}, \mathbf{r}')$ which satisfies

$$(\nabla^2 + k^2)G(\mathbf{r}, \mathbf{r}') = \delta(\mathbf{r} - \mathbf{r}'). \tag{8.108}$$

Using the two-dimensional analogue of the Green's theorem (8.62), we easily find

$$\int d^2x \left[u(\mathbf{r})\nabla^2 G(\mathbf{r}, \mathbf{r}') - G(\mathbf{r}, \mathbf{r}')\nabla^2 u(\mathbf{r})\right]$$

$$= u(\mathbf{r}') - \int d^2x \, G(\mathbf{r}, \mathbf{r}')F(\mathbf{r})$$

$$= \int_{\substack{\text{edge} \\ \text{of drumhead}}} dl \left[u(\mathbf{r})\frac{\partial G(\mathbf{r}, \mathbf{r}')}{\partial n} - G(\mathbf{r}, \mathbf{r}')\frac{\partial u(\mathbf{r})}{\partial n}\right]. \tag{8.109}$$

We choose the boundary condition

$$G(\mathbf{r}, \mathbf{r}') = 0, \quad \mathbf{r} \text{ on edge of drumhead.} \tag{8.110}$$

This eliminates the unknown $\partial du(\mathbf{r})/\partial n$ on the right-hand side of (8.109). Also, with the boundary condition (8.107), the other portion of the right-hand side of (8.109) vanishes and we find

$$u(\mathbf{r}) = \int d^2x' \, G(\mathbf{r}, \mathbf{r}')F(\mathbf{r}'). \tag{8.111}$$

Here we have used the symmetry of $G(\mathbf{r}, \mathbf{r}')$ with respect to interchange of $\mathbf{r}$ and $\mathbf{r}'$, which follows from applying Green's theorem to $G(\mathbf{r}, \mathbf{r}')$ and $G(\mathbf{r}, \mathbf{r}'')$.

In the absence of an image method, we can use a technique analogous to that employed in the previous section.

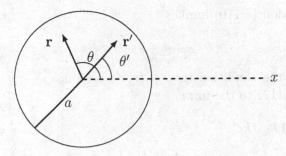

Figure 8.3: Two-dimensional drumhead in polar coordinates

The Green's function can be expanded in a Fourier series in θ:

$$G(\mathbf{r}, \mathbf{r}') = \sum_{m=-\infty}^{\infty} G_m(r, r') e^{im\theta}. \tag{8.112}$$

We need the expression for the two-dimensional delta function on the right of (8.108) in polar coordinates:

$$\delta(\mathbf{r} - \mathbf{r}') = \frac{1}{r}\delta(r - r')\delta(\theta - \theta')$$

$$= \frac{1}{r}\delta(r - r')\frac{1}{2\pi}\sum_{m=-\infty}^{\infty} e^{im(\theta - \theta')}. \tag{8.113}$$

Here we have used the completeness relation for the Fourier series to expand the delta function of the angle. Substituting (8.112) and (8.113) in (8.108), using the form for the Laplacian in polar coordinates (2.20), and the orthogonality of the functions $e^{im\theta}$, we find as equation for the G_m

$$\frac{d^2 G_m}{dr^2} + \frac{1}{r}\frac{dG_m}{dr} + \left[k^2 - \frac{m^2}{r^2}\right] G_m = \frac{1}{r}\delta(r - r')\frac{e^{-im\theta'}}{2\pi}. \tag{8.114}$$

The θ' dependence can be factored off,

$$G_m = \frac{1}{2\pi}e^{-im\theta'} g_m(r, r'), \tag{8.115}$$

and we find

$$\frac{d^2 g_m(r, r')}{dr^2} + \frac{1}{r}\frac{dg_m(r, r')}{dr} + \left[k^2 - \frac{m^2}{r^2}\right] g_m(r, r') = \frac{1}{r}\delta(r - r'). \tag{8.116}$$

For $r \neq r'$ this is of course Bessel's equation, so the solution is of the form

$$g_m(r, r') = \begin{cases} A J_{|m|}(kr) + B N_{|m|}(kr), & r < r' \\ \\ C J_{|m|}(kr) + D N_{|m|}(kr), & r > r' \end{cases} \tag{8.117}$$

The boundary condition (8.110) implies

$$g_m(a, r') = 0. \tag{8.118}$$

In addition we have the condition that $g_m(r, r')$ be finite at $r = 0$. Together these conditions reduce (8.117) to the form

$$g_m(r, r') = \begin{cases} A J_{|m|}(kr), & r < r' \\ \\ E\left[N_{|m|}(ka) J_{|m|}(kr) - J_{|m|}(ka) N_{|m|}(kr)\right], & r > r' \end{cases} \tag{8.119}$$

The delta function on the right side of (8.116) is, as discussed in Section 8.3, equivalent to the joining conditions

$$g_m(r' - \epsilon, r') = g_m(r' + \epsilon, r'), \tag{8.120}$$

$$\left. \frac{dg_m(r, r')}{dr} \right|_{r=r'-\epsilon}^{r'+\epsilon} = \frac{1}{r'}. \tag{8.121}$$

Using (8.119) these become

$$AJ_{|m|}(kr') = E\left[N_{|m|}(ka)J_{|m|}(kr') - J_{|m|}(ka)N_{|m|}(kr') \right], \tag{8.122}$$

$$kE\left[N_{|m|}(ka)J'_{|m|}(kr') - J_{|m|}(ka)N'_{|m|}(kr') \right] - kAJ'_{|m|}(kr') = \frac{1}{r'}. \tag{8.123}$$

These can be solved to yield

$$E = -\frac{J_{|m|}(kr')}{kr'J_{|m|}(ka)} \cdot \frac{1}{J_{|m|}(kr')N'_{|m|}(kr') - N_{|m|}(kr')J'_{|m|}(kr')} \tag{8.124}$$

and A given by (8.122).

To simplify these expressions, we can use the formula for the Wronskian of the solutions of Bessel's equation

$$J_m(x)N'_m(x) - J'_m(x)N_m(x) = \frac{2}{\pi x}. \tag{8.125}$$

This was obtained in Problem 1, Chapter 4. Using (8.125), we find from (8.124), (8.122)

$$E = -\frac{\pi}{2}\frac{J_{|m|}(kr')}{J_{|m|}(ka)}, \tag{8.126}$$

$$A = -\frac{\pi}{2}\frac{N_{|m|}(ka)J_{|m|}(kr') - J_{|m|}(ka)N_{|m|}(kr')}{J_{|m|}(ka)}. \tag{8.127}$$

Combining results (8.112), (8.115), (8.119), (8.126), (8.127), we find for the Green's function

$$G(\mathbf{r}, \mathbf{r}') = \sum_{m=0}^{\infty} \cos m(\theta - \theta') \times$$

$$\begin{cases} \dfrac{J_m(ka)N_m(kr') - N_m(ka)J_m(kr')}{2\epsilon_m J_m(ka)} J_m(kr), & r < r' \\[4mm] \dfrac{J_m(ka)N_m(kr) - N_m(ka)J_m(kr)}{2\epsilon_m J_m(ka)} J_m(kr'), & r > r' \end{cases} \tag{8.128}$$

In this expression the factor

$$\epsilon_m = \begin{cases} 2, & m = 0 \\ 1, & m = 1, 2, 3, \ldots \end{cases} \tag{8.129}$$

was obtained when combining the terms $\pm m$ to get $\cos m(\theta - \theta')$.

8.11 EIGENFUNCTION EXPANSION OF THE GREEN'S FUNCTION FOR THE HELMHOLTZ EQUATION

It is possible to expand the Green's function for the Helmholtz equation in an eigenfunction series similar to that given in Section 8.5 for the one-dimensional case. The relevant eigenfunctions for the Helmholtz equation were discussed in Section 5.2, equation (5.13)ff. These eigenfunctions are the solutions of

$$(\mathbf{\nabla}^2 + k_n^2)u_n(\mathbf{r}) = 0 \tag{8.130}$$

with a boundary condition

$$u_n(\mathbf{r}) = 0, \quad \mathbf{r} \text{ on boundary} \tag{8.131}$$

or

$$\frac{\partial u_n(\mathbf{r})}{\partial n} = 0, \quad \mathbf{r} \text{ on boundary.} \tag{8.132}$$

The eigenfunctions form a complete orthogonal set and can be normalized so that

$$\int d^3x \, u_n(\mathbf{r})u_m(\mathbf{r}) = \delta_{nm}, \tag{8.133}$$

$$\sum_n u_n(\mathbf{r})u_n(\mathbf{r}') = \delta(\mathbf{r} - \mathbf{r}'). \tag{8.134}$$

We can use these eigenfunctions to find an explicit formula for the Green's function $G(\mathbf{r}, \mathbf{r}')$ defined by the partial differential equation

$$(\mathbf{\nabla}^2 + k^2)G(\mathbf{r}, \mathbf{r}') = \delta(\mathbf{r} - \mathbf{r}') \tag{8.135}$$

and boundary condition

$$G(\mathbf{r}, \mathbf{r}') = 0, \quad \mathbf{r} \text{ on boundary} \tag{8.136}$$

or

$$\frac{\partial G(\mathbf{r}, \mathbf{r}')}{\partial n} = 0, \quad \mathbf{r} \text{ on boundary.} \tag{8.137}$$

In fact, using (8.130), (8.131) or (8.132) and (8.134), it is simple to verify that $G(\mathbf{r}, \mathbf{r}')$ is given by the formula

$$G(\mathbf{r}, \mathbf{r}') = \sum_n \frac{u_n(\mathbf{r})u_n(\mathbf{r}')}{k^2 - k_n^2}. \tag{8.138}$$

Note that setting $k^2 = 0$ gives the Green's function for Laplace's equation in terms of the eigenfunctions of the Helmholtz equation.

We can apply the formula (8.138) to the drumhead problem considered in the previous section. The eigenfunctions for the drumhead are given by (5.54),

$$u_{mn1}(r,\theta) = N_{mn}J_m(k_{mn}r)\sin m\theta,$$

$$u_{mn2}(r,\theta) = N_{mn}J_m(k_{mn}r)\cos m\theta, \tag{8.139}$$

and the normalization factor N_{mn} is determined by, using u_{mn1} as an example,

$$\int_0^{2\pi} d\theta \int_0^a r dr\, u_{mn1}^2 = 1,$$

$$N_{mn} \int_0^{2\pi} d\theta\, \sin^2 m\theta \int_0^a r dr\, J_m^2(k_{mn}r) = \frac{2\pi}{2}N_{mn}^2 \int_0^a r dr\, J_m^2(k_{mn}r) = 1$$

so that

$$N_{mn} = \left[\sqrt{\frac{\pi}{2}\epsilon_m}\, aJ_m'(k_{mn}a)\right]^{-1}, \tag{8.140}$$

where (4.82) was used and where ϵ_m is given in (8.129) to be applicable for both u_{mn1} and u_{mn2} as well as $m = 0$. Substituting (8.139) in (8.138) and using $\cos m\theta \cos m\theta' + \sin m\theta \sin m\theta' = \cos m(\theta - \theta')$ we find

$$G(\mathbf{r},\mathbf{r'}) = \sum_{m=0}^{\infty} \sum_{n=1}^{\infty} \frac{N_{mn}^2 J_m(k_{mn}r)J_m(k_{mn}r')\cos m(\theta - \theta')}{k^2 - k_{mn}^2}. \tag{8.141}$$

Comparing this formula with (8.128), which is another expression for the same Green's function, we derive an identity

$$\sum_{n=1}^{\infty} \frac{N_{mn}^2 J_m(k_{mn}r)J_m(k_{mn}r')}{k^2 - k_{mn}^2} =$$

$$\frac{1}{2\epsilon_m J_m(ka)} [J_m(ka)N_m(kr_>) - N_m(ka)J_m(kr_>)]\, J_m(kr_<). \tag{8.142}$$

We have illustrated the Green's function for the Helmholtz equation with a two-dimensional example, the round drumhead. Evidently the same techniques could be employed for three-dimensional problems, e.g., the interior of a cylinder or the interior of a sphere.

PROBLEMS

8.1   a) Solve the inhomogeneous differential equation

$$\mathcal{L}u + \lambda\rho u = \phi(x)$$

with homogeneous boundary conditions

$$Au(a) + Bu'(a) = 0,$$
$$Cu(b) + Du'(b) = 0,$$

by expanding $u(x)$ and $\dfrac{\phi(x)}{\rho(x)}$,

$$u(x) = \sum_n a_n u_n(x),$$

$$\phi(x) = \rho(x) \sum_n b_n u_n(x),$$

in the eigenfunctions of $\mathcal{L}$,

$$\mathcal{L}u_n(x) = -\lambda_n \rho(x) u_n(x),$$

with

$$Au_n(a) + Bu'_n(a) = 0,$$
$$Cu_n(b) + Du'_n(b) = 0.$$

Show that the result can be written in the form

$$u(x) = \int_a^b dx' \; G(x, x')\phi(x'),$$

where $G(x, x')$ is given by

$$G(x, x') = \sum_n \frac{u_n(x)u_n(x')}{\lambda - \lambda_n}.$$

This is an alternate derivation of the formula (8.54) for the Green's function.

b) Show that for the case $\lambda = \lambda_m$ the differential equation

$$\mathcal{L}u + \lambda\rho u = \phi(x)$$

has no solution unless $\phi(x)$ is orthogonal to $u_m(x)$,

$$\int_a^b dx \; u_m(x)\phi(x) = 0,$$

and that in this case,

$$u(x) = cu_m(x) + \int dx' \; G_{\text{gen}}(x, x')\phi(x'),$$

where c is an arbitrary constant and the so-called generalized Green's function $G_{\text{gen}}(x, x')$ is given by

$$G_{\text{gen}}(x, x') = \sum_{n \neq m} \frac{u_n(x)u_n(x')}{\lambda_m - \lambda_n}$$

and satisfies the equation

$$(\mathcal{L} + \lambda_m \rho)G_{\text{gen}}(x, x') = \delta(x - x') - \rho(x)u_m(x)u_m(x')$$

and the condition

$$\int_a^b dx \; \rho(x)u_m(x)G_{\text{gen}}(x, x') = 0.$$

8.2 Discuss the motion of bowed stretched string with both ends clamped when the bowing frequency is equal to one of the resonant frequencies of the string.

  a) Under what conditions will the string not break?

  b) Find the generalized Green's function discussed in Problem 8.1(b) in two ways:

   i) as an eigenfunction expansion.

   ii) by solving the differential equation for $G_{\text{gen}}(x, x')$ explicitly.

   Write the general solution for the motion of the string in terms of $G_{\text{gen}}(x, x')$.

8.3 a) Find the Green's function for Poisson's equation for the exterior of a sphere of radius a and use it to solve Poisson's equation

$$\nabla^2 \psi = -4\pi\rho$$

in the region outside the sphere, i.e., $r > a$, with the boundary condition

$$\psi(r = a, \theta, \varphi) = F(\theta, \varphi),$$

where $F(\theta, \varphi)$ is a given function.

  b) Assume the sphere consists of two metal hemispheres separated by a thin layer of insulator and that the two hemispheres are maintained at potentials $+V$ and $-V$ by a battery inside the sphere as in Figure 8.4. Take $\rho = 0$ outside the sphere. Use the formula of part (a) to find the potential along the axis above the middle of the hemisphere at potential $+V$.

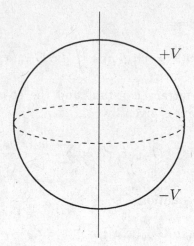

Figure 8.4: Metal hemispheres at two potentials for Problem 8.3

8.4 a) Use the method of images to find the electrostatic potential for the system consisting of a point charge q a distance z' in front of an infinite grounded conducting plane located at $z = 0$.

b) Find the Green's function for Poisson's equation for the half-space $z > 0$ and use it to solve Poisson's equation

$$\nabla^2 \psi = -4\pi\rho$$

in the region $z > 0$ with the boundary condition

$$\psi(x, y, z = 0) = F(x, y),$$

where $F(x, y)$ is a given function.

c) Take

$$F(x, y) = \begin{cases} V = \text{const}, & \rho = \sqrt{x^2 + y^2} < a \\ 0, & \rho > a \end{cases},$$

$$\rho(\mathbf{r}, t) = 0,$$

and use the formula of part (b) to find $\psi(x = 0, y = 0, z)$ along the z axis, $z > 0$. Compare with the result of Problem 4, Chapter 4.

8.5 a) Use the method of images to find the potential for the system consisting of a line charge q/unit length parallel to and a distance r' from the axis of an infinitely long grounded conducting cylinder of radius a as shown in Figure 8.5.

b) Find the Green's function for Poisson's equation for the exterior of an infinitely long cylinder of radius a and use it to solve Poisson's equation

$$\nabla^2 \psi = -4\pi\rho$$

in the region outside the cylinder when the charge density $\rho(r,\theta)$ and the boundary condition

$$\psi(r = a, \theta, z) = F(\theta)$$

are both independent of z.

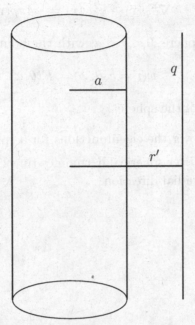

Figure 8.5: Line charge a distance away from conducting cylinder for Problem 8.5

8.6 Find the Green's function for Poisson's equation for the region between two concentric cylinders of radii a and b. Use the Green's function to find the solution of Poisson's equation

$$\nabla^2 \psi = -4\pi\rho(r,\theta)$$

in the region between the two cylinders with boundary conditions

$$\psi(r = a, \theta, z) = F_1(\theta),$$
$$\psi(r = b, \theta, z) = F_2(\theta),$$

on the two cylindrical surfaces. Assume that the cylinders are infinitely long and that all quantities are independent of z.

8.7 Find in two ways the Green's function for the Poisson equation

$$\nabla^2 \psi(\mathbf{r}) = -4\pi\rho(\mathbf{r})$$

for the interior of a cube, $0 \leq x \leq a$, $0 \leq y \leq a$, $0 \leq z \leq a$, with boundary condition $\psi(\mathbf{r})$ given on the sides of the cube:

a) as a triple sum over the eigenfunctions of the Helmholtz equation for the cube,

b) as a double sum over the eigenfunctions for a square multiplied by a one-dimensional Green's function in the third dimension.

8.8 Find in two ways the Green's function for the Helmholtz equation

$$\nabla^2 \psi(\mathbf{r}) + k^2 \psi(\mathbf{r}) = -4\pi\rho(\mathbf{r})$$

for the interior of a sphere, $0 < r < a$, with the boundary condition

$$\psi(r = a, \theta, \varphi) = F(\theta, \varphi)$$

given on the surface of the sphere:

a) as a triple sum over the eigenfunctions for a sphere,

b) as a double sum over spherical harmonics times a one-dimensional Green's function in the radial direction.

CHAPTER 9

Green's Functions: Part Two

9.1 INTRODUCTION

In the previous chapter the time independent consideration of Green's functions for dealing with source terms in differential equations was covered as well as sinusoidal driving and response time dependency. Recall, the partial differential equations

$$\nabla^2 \psi = -4\pi\rho, \tag{9.1}$$

$$\nabla^2 \psi - \frac{1}{\kappa}\frac{\partial\psi}{\partial t} = -4\pi\rho, \tag{9.2}$$

$$\nabla^2 \psi - \frac{1}{c^2}\frac{\partial^2\psi}{\partial t^2} = -4\pi\rho, \tag{9.3}$$

the inhomogeneous terms $4\pi\rho$ on the right are the source terms and are regarded as given. In this chapter, expanding the problem to infinite regions and general time dependency of Green's function solutions are considered.

Green's functions can also be used to convert a homogeneous differential equation such as

$$\nabla^2 \psi(\mathbf{r}) + k^2\psi(\mathbf{r}) = V(\mathbf{r})\psi(\mathbf{r}) \tag{9.4}$$

into an integral equation. This application will be taken up in the following chapter.

9.2 THE HELMHOLTZ EQUATION FOR INFINITE REGIONS, RADIATION, AND THE WAVE EQUATION; SINUSOIDAL TIME DEPENDENCE

Consider the wave equation for a system with a source,

$$\nabla^2 \psi(\mathbf{r},t) - \frac{1}{c^2}\frac{\partial^2\psi(\mathbf{r},t)}{\partial t^2} = -4\pi\rho(\mathbf{r},t), \tag{9.5}$$

and suppose, for the time being, that the source is sinusoidal:

$$\rho(\mathbf{r},t) = \rho(\mathbf{r})e^{-i\omega t}. \tag{9.6}$$

The field generated by this source will have the same time dependence,

$$\psi(\mathbf{r}, t) = \psi(\mathbf{r})e^{-i\omega t}. \tag{9.7}$$

Substituting in (9.5) we find the Helmholtz equation

$$(\boldsymbol{\nabla}^2 + k^2)\psi(\mathbf{r}) = -4\pi\rho(\mathbf{r}), \quad k = \frac{\omega}{c}. \tag{9.8}$$

To solve (9.8) we introduce a Green's function which is a solution of

$$(\boldsymbol{\nabla}^2 + k^2)G(\mathbf{r}, \mathbf{r}') = \delta(\mathbf{r} - \mathbf{r}'). \tag{9.9}$$

From the standard manipulation with Green's theorem,

$$\int_V d^3x \left[\psi(\mathbf{r})\boldsymbol{\nabla}^2 G(\mathbf{r}, \mathbf{r}') - G(\mathbf{r}, \mathbf{r}')\boldsymbol{\nabla}^2\psi(\mathbf{r})\right]$$

$$= \int_S dA \left[\psi(\mathbf{r})\frac{\partial G(\mathbf{r}, \mathbf{r}')}{\partial n} - G(\mathbf{r}, \mathbf{r}')\frac{\partial \psi(\mathbf{r})}{\partial n}\right], \tag{9.10}$$

we obtain

$$\psi(\mathbf{r}) = -4\pi \int_V d^3x'\, G(\mathbf{r}', \mathbf{r})\rho(\mathbf{r}') + \int_S dA' \left[\psi(\mathbf{r}')\frac{\partial G(\mathbf{r}', \mathbf{r})}{\partial n'} - G(\mathbf{r}', \mathbf{r})\frac{\partial \psi(\mathbf{r}')}{\partial n'}\right]. \tag{9.11}$$

We can easily find solutions of (9.9) suitable when there are no boundaries. Because of the homogeneity and isotropy of space, $G(\mathbf{r}, \mathbf{r}')$ for this case will be a function only of the distance $|\mathbf{r} - \mathbf{r}'|$ from the source $\delta(\mathbf{r} - \mathbf{r}')$, and for $\mathbf{r} \neq \mathbf{r}'$ the Green's function is a solution of the Helmholtz equation. Referring to (6.79), (6.80) we find the two Green's functions

$$G_{\text{out}}(\mathbf{r}, \mathbf{r}') = -\frac{1}{4\pi}\frac{e^{ik|\mathbf{r}-\mathbf{r}'|}}{|\mathbf{r} - \mathbf{r}'|} \propto h_0^{(1)}(k\,|\mathbf{r} - \mathbf{r}'|), \tag{9.12}$$

$$G_{\text{in}}(\mathbf{r}, \mathbf{r}') = -\frac{1}{4\pi}\frac{e^{-ik|\mathbf{r}-\mathbf{r}'|}}{|\mathbf{r} - \mathbf{r}'|} \propto h_0^{(2)}(k\,|\mathbf{r} - \mathbf{r}'|). \tag{9.13}$$

For $\mathbf{r} \to \mathbf{r}'$ both of these Green's functions approach

$$G_{\substack{\text{out}\\\text{in}}}(\mathbf{r}, \mathbf{r}') \xrightarrow[\mathbf{r}\to\mathbf{r}']{} -\frac{1}{4\pi}\frac{1}{|\mathbf{r} - \mathbf{r}'|}. \tag{9.14}$$

As discussed in (8.72)ff. this is the correct singularity to account for the delta function on the right of (9.9). With the time dependence $e^{-i\omega t}$ the two solutions (9.12), (9.13) correspond respectively to outgoing and ingoing waves. We could form other Green's functions by taking linear combinations $G = \alpha G_{\text{out}} + (1 - \alpha)G_{\text{in}}$. With $\alpha = 1/2$, the resulting Green's function corresponds to standing waves.

Substituting first one, then the other of the Green's functions (9.12), (9.13) in (9.11), we obtain two forms for the solution $\psi(\mathbf{r}, t)$:

$$\psi(\mathbf{r}, t) = \phi_1(\mathbf{r}) + \int d^3x' \, \frac{e^{ik|\mathbf{r}-\mathbf{r}'|}}{|\mathbf{r}-\mathbf{r}'|} \rho(r') \tag{9.15}$$

$$= \phi_2(\mathbf{r}) + \int d^3x' \, \frac{e^{-ik|\mathbf{r}-\mathbf{r}'|}}{|\mathbf{r}-\mathbf{r}'|} \rho(r'). \tag{9.16}$$

Here $\phi_1(\mathbf{r})$ and $\phi_2(\mathbf{r})$ are the contributions from the surface integrals in (9.11). Applying $\boldsymbol{\nabla}^2 + k^2$ to these formulas we find immediately that $\phi_1(\mathbf{r})$ and $\phi_2(\mathbf{r})$ are both solutions of the Helmholtz equation $(\boldsymbol{\nabla}^2 + k^2)\phi(\mathbf{r}) = 0$. The two possibilities (9.15), (9.16) are equally appropriate choices of the sourceless waves $\phi_1(\mathbf{r})$, $\phi_2(\mathbf{r})$.

Although other possibilities are sometimes of interest, we are often concerned with a source such as a loudspeaker or an antenna emitting waves. In order to obtain purely outgoing waves generated by the source $\rho(\mathbf{r})$, we take a solution written in the form (9.15) with $\phi_1(\mathbf{r}) = 0$

$$\psi(\mathbf{r}, t) = \int d^3x' \, \frac{e^{ik|\mathbf{r}-\mathbf{r}'|}}{|\mathbf{r}-\mathbf{r}'|} \rho(\mathbf{r}'). \tag{9.17}$$

This solution satisfies the so-called <u>Ausstrahlungsbedingung</u> (=out radiation condition). We have used this boundary condition before, in Sections 6.5 and 6.6 to pick the outgoing waves.

For the scattering problem to be discussed in Section 10.2 we shall write the solution in the form (9.15) and $\phi_1(\mathbf{r})$ will be a plane wave representing the incoming wave before scattering. For both this case and (9.17) the solution can be written rather awkwardly in the form (9.16) with a nonvanishing $\phi_2(\mathbf{r})$. A mathematical possibility of little physical interest is the solution (9.16) with $\phi_2(\mathbf{r}) = 0$. This corresponds to purely ingoing waves absorbed by a sink function $\rho(\mathbf{r})$.

Far from the radiating source $\mathbf{r}' \ll \mathbf{r}$ and the radiation field, (9.17) assumes a simpler form:

$$|\mathbf{r}-\mathbf{r}'| \xrightarrow[r\to\infty]{} r - \hat{\mathbf{n}} \cdot \mathbf{r}', \quad \hat{\mathbf{n}} = \frac{\mathbf{r}}{r}, \tag{9.18}$$

$$\frac{e^{ik|\mathbf{r}-\mathbf{r}'|}}{|\mathbf{r}-\mathbf{r}'|} \xrightarrow[r\to\infty]{} \frac{e^{ikr}}{r} e^{-i\mathbf{k}'\cdot\mathbf{r}'}, \quad \mathbf{k}' = k\hat{\mathbf{n}}, \tag{9.19}$$

$$\psi_{\text{out}}(\mathbf{r}) \xrightarrow[r\to\infty]{} \frac{e^{ikr}}{r} f(\mathbf{k}'), \tag{9.20}$$

$$f(\mathbf{k}') = \int d^3x' \, e^{-i\mathbf{k}'\cdot\mathbf{r}'} \rho(\mathbf{r}'). \tag{9.21}$$

This is similar to the result (6.89), (6.90). The similarity can be made closer by using the expansion (6.57) for $e^{-i\mathbf{k}'\cdot\mathbf{r}'}$ to obtain the expansion of $f(\mathbf{k}')$ in spherical harmonics:

$$f(\mathbf{k}') = \sum_{l,m} C_{lm} Y_l^m(\theta_{\mathbf{k}'}, \phi_{\mathbf{k}'}), \tag{9.22}$$

$$C_{lm} = 4\pi(-i)^l \int d^3x' \, j_l(kr') Y_l^{m*}(\theta_{\mathbf{k}'}, \phi_{\mathbf{k}'})\rho(\mathbf{r}'). \tag{9.23}$$

9.3   GENERAL TIME DEPENDENCE

In the previous section we assumed a sinusoidal time dependence. The general case can be treated by Fourier transforming the time dependence:

$$\rho(\mathbf{r}, t) = \int_{-\infty}^{\infty} d\omega \, \widetilde{\rho}(\mathbf{r}, \omega)e^{-i\omega t}, \tag{9.24}$$

$$\widetilde{\rho}(\mathbf{r}, \omega) = \frac{1}{2\pi} \int_{-\infty}^{\infty} dt' \, e^{i\omega t'} \rho(\mathbf{r}, t'), \tag{9.25}$$

$$\psi(\mathbf{r}, t) = \int_{-\infty}^{\infty} d\omega \, \widetilde{\psi}(\mathbf{r}, \omega)e^{-i\omega t}, \tag{9.26}$$

$$\widetilde{\psi}(\mathbf{r}, \omega) = \frac{1}{2\pi} \int_{-\infty}^{\infty} dt' \, e^{i\omega t'} \psi(\mathbf{r}, t'). \tag{9.27}$$

Substituting (9.24), (9.26) in (9.5) we find

$$\int_{-\infty}^{\infty} d\omega \, e^{-i\omega t} \left[\boldsymbol{\nabla}^2 \widetilde{\psi}(\mathbf{r}, \omega) + \frac{\omega^2}{c^2}\widetilde{\psi}(\mathbf{r}, \omega) + 4\pi\widetilde{\rho}(\mathbf{r}, \omega) \right] = 0, \tag{9.28}$$

and inversion of the Fourier transform gives

$$\boldsymbol{\nabla}^2 \widetilde{\psi}(\mathbf{r}, \omega) + \frac{\omega^2}{c^2}\widetilde{\psi}(\mathbf{r}, \omega) = -4\pi\widetilde{\rho}(\mathbf{r}, \omega). \tag{9.29}$$

This is the problem we have just solved in Section 9.2. According to (9.17) and the discussion following, the appropriate solution with outgoing waves is

$$\widetilde{\psi}(\mathbf{r}, \omega) = \int d^3x' \, \frac{\exp\left(i\frac{\omega}{c}|\mathbf{r} - \mathbf{r}'|\right)}{|\mathbf{r} - \mathbf{r}'|} \, \widetilde{\rho}(\mathbf{r}', \omega). \tag{9.30}$$

[Note that for negative ω in (9.26) we need negative ω in (9.30) to obtain outgoing waves.] Substituting (9.30) in (9.26) and using (9.25) for $\widetilde{\rho}(\mathbf{r}', \omega)$ we find

$$\psi(\mathbf{r}, t) = \int_{-\infty}^{\infty} d\omega \, e^{-i\omega t} \int d^3x' \, \frac{\exp\left(i\frac{\omega}{c}|\mathbf{r} - \mathbf{r}'|\right)}{|\mathbf{r} - \mathbf{r}'|} \frac{1}{2\pi} \int_{-\infty}^{\infty} dt' \, e^{i\omega t'} \rho(\mathbf{r}', t')$$

$$= \int d^3x' \int_{-\infty}^{\infty} dt' \frac{1}{|\mathbf{r} - \mathbf{r}'|} \frac{1}{2\pi} \int_{-\infty}^{\infty} d\omega \, \exp\left(i\omega \left[\frac{1}{c}|\mathbf{r} - \mathbf{r}'| - (t - t')\right]\right) \rho(\mathbf{r}', t').$$

(9.31)

We can use the Fourier expansion (2.143) of the delta function,

$$\frac{1}{2\pi} \int_{-\infty}^{\infty} d\omega \, \exp\left(i\omega \left[\frac{1}{c}|\mathbf{r} - \mathbf{r}'| - (t - t')\right]\right) = \delta\left[\frac{1}{c}|\mathbf{r} - \mathbf{r}'| - (t - t')\right],$$

(9.32)

and write (9.31) in the elegant form

$$\psi(\mathbf{r}, t) = -4\pi \int d^3x' \int_{-\infty}^{\infty} dt' \, G(\mathbf{r}, t; \mathbf{r}', t') \rho(\mathbf{r}', t'),$$

(9.33)

with

$$G(\mathbf{r}, t; \mathbf{r}', t') = -\frac{\delta\left[\frac{1}{c}|\mathbf{r} - \mathbf{r}'| - (t - t')\right]}{4\pi |\mathbf{r} - \mathbf{r}'|}.$$

(9.34)

This function $G(\mathbf{r}, t; \mathbf{r}', t')$ is the four-dimensional Green's function for the wave equation. We shall return to it in Section 9.4. Using the delta function in $G(\mathbf{r}, t; \mathbf{r}', t')$ we can carry out the time integration in (9.33) to obtain

$$\psi(\mathbf{r}, t) = \int d^3x' \, \frac{\rho(\mathbf{r}', t - \frac{1}{c}|\mathbf{r} - \mathbf{r}'|)}{|\mathbf{r} - \mathbf{r}'|}.$$

(9.35)

This is the so-called retarded potential solution of the wave equation with sources. The source $\rho(\mathbf{r}', t')$ on the right of (9.35) is evaluated at the retarded time $t' = t - \frac{1}{c}|\mathbf{r} - \mathbf{r}'|$, which is such that a wavelet leaving $\mathbf{r}'$ at t' and traveling with velocity c arrives at $\mathbf{r}$ at time t.

The retarded potential solution is of basic importance in electrodynamics. As discussed briefly in Section 1.8, the equations for the scalar and vector potentials of electrodynamics are

$$\nabla^2 \phi - \frac{1}{c^2}\frac{\partial^2 \phi}{\partial t^2} = -4\pi\rho,$$

(9.36)

$$\nabla^2 \mathbf{A} - \frac{1}{c^2}\frac{\partial^2 \mathbf{A}}{\partial t^2} = -\frac{4\pi}{c}\mathbf{j},$$

(9.37)

$$\nabla \cdot \mathbf{A} + \frac{1}{c}\frac{\partial \phi}{\partial t} = 0.$$

(9.38)

According to the result (9.35) just derived we see that the appropriate solutions of (9.36) and (9.37) with the radiation boundary condition are

$$\phi(\mathbf{r}, t) = \int d^3x' \, \frac{\rho(\mathbf{r}, t - \frac{1}{c}|\mathbf{r} - \mathbf{r}'|)}{|\mathbf{r} - \mathbf{r}'|},$$

(9.39)

$$\mathbf{A}(\mathbf{r}, t) = \frac{1}{c} \int d^3x' \, \frac{\mathbf{j}(\mathbf{r}, t - \frac{1}{c}|\mathbf{r} - \mathbf{r}'|)}{|\mathbf{r} - \mathbf{r}'|}.$$

(9.40)

It is easy to check that the condition (9.38) is satisfied. Writing the solutions in the form (9.33) and noting that G is a function only of coordinate differences we find

$$\nabla \cdot \mathbf{A} + \frac{1}{c}\frac{\partial \phi}{\partial t} =$$

$$-\frac{4\pi}{c}\int d^3x' \int dt' \left[\nabla_\mathbf{r} \cdot G(\mathbf{r},t;\mathbf{r}',t')\mathbf{j}(\mathbf{r}',t') + \frac{\partial}{\partial t}G(\mathbf{r},t;\mathbf{r}',t')\rho(\mathbf{r}',t') \right]$$

$$= +\frac{4\pi}{c}\int d^3x' \int dt' \left[\{\nabla_{\mathbf{r}'}G(\mathbf{r},t;\mathbf{r}',t')\} \cdot \mathbf{j}(\mathbf{r}',t') + \left\{\frac{\partial}{\partial t'}G(\mathbf{r},t;\mathbf{r}',t')\right\} \rho(\mathbf{r}',t') \right]$$

$$= -\frac{4\pi}{c}\int d^3x' \int dt' \, G(\mathbf{r},t;\mathbf{r}',t')\left[\nabla_{\mathbf{r}'} \cdot \mathbf{j}(\mathbf{r}',t') + \frac{\partial\rho(\mathbf{r}',t')}{\partial t'} \right]. \tag{9.41}$$

Here we have integrated by parts and assumed that $\mathbf{j}$ and ρ vanish for $\mathbf{r}$ or $t \to \infty$. Since electric charge is conserved, we have the continuity equation

$$\nabla_{\mathbf{r}'} \cdot \mathbf{j}(\mathbf{r}',t) + \frac{\partial\rho(\mathbf{r}',t)}{\partial t'} = 0, \tag{9.42}$$

and (9.41) reduces to (9.38). The important results (9.39), (9.40) provide the general solution for electrodynamics in terms of given sources.

Finally we note that the ingoing wave solution (9.16) with $\phi_2(\mathbf{r}) = 0$ for (9.29) leads to the so-called advanced potential solutions

$$\psi_{\text{adv.}}(\mathbf{r},t) = \int d^3x' \, \frac{\rho(\mathbf{r}',t + \frac{1}{c}|\mathbf{r} - \mathbf{r}'|)}{|\mathbf{r} - \mathbf{r}'|}. \tag{9.43}$$

This is a solution of (9.5). It does not make sense for standard physical applications, however, since ψ at time t is determined by the behavior of ρ at advanced times $t' = t + \frac{1}{c}|\mathbf{r} - \mathbf{r}'| > t$, in violation of common sense ideas about causality. Further discussion of this point appears at the end of Section 9.5.

9.4   THE WAVE EQUATION

We have already obtained the Green's function (9.34) for the wave equation for an infinite region, in the previous section. We redo the calculation with some increase in generality here. We wish to solve the inhomogeneous wave equation

$$\nabla^2\psi(\mathbf{r},t) - \frac{1}{c^2}\frac{\partial^2\psi(\mathbf{r},t)}{\partial t^2} = -4\pi\rho(\mathbf{r},t) \tag{9.44}$$

with the boundary condition, relative to surface of closed volume as shown in Figure 9.1,

$$\psi(\mathbf{r},t) \quad \text{given for } \mathbf{r} \text{ on a surface } S \text{ of space } V \tag{9.45}$$

<u>or</u>

$$\frac{\partial \psi(\mathbf{r}, t)}{\partial n} \quad \text{given for } \mathbf{r} \text{ on } S \qquad (9.46)$$

plus initial conditions

$$\psi(\mathbf{r}, t) \quad \underline{\text{and}} \quad \frac{\partial \psi(\mathbf{r}, t)}{\partial t} \quad \text{given at an initial time } t = T \text{ throughout } V. \qquad (9.47)$$

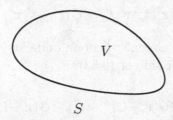

Figure 9.1: Arbitrary closed volume for the wave equation

To solve this problem we introduce a Green's function $G(\mathbf{r}, t; \mathbf{r}', t')$ which is a solution of

$$\boldsymbol{\nabla}^2 G(\mathbf{r}, t; \mathbf{r}', t') - \frac{1}{c^2}\frac{\partial^2}{\partial t^2}G(\mathbf{r}, t; \mathbf{r}', t') = \delta(\mathbf{r} - \mathbf{r}')\delta(t - t') \qquad (9.48)$$

with the boundary condition [corresponding respectively to (9.45) or (9.46)]

$$G(\mathbf{r}, t; \mathbf{r}', t') = 0 \quad \text{for } \mathbf{r} \text{ on } S \qquad (9.49)$$

<u>or</u>

$$\frac{\partial G(\mathbf{r}, t; \mathbf{r}', t')}{\partial n} = 0 \quad \text{for } \mathbf{r} \text{ on } S \qquad (9.50)$$

plus an initial condition

$$G(\mathbf{r}, t; \mathbf{r}', t') = 0 \quad \text{for } t < t'. \qquad (9.51)$$

We shall show below by construction that a Green's function satisfying these conditions actually exists. Other possible initial conditions are considered at the end of this section.

The symmetry properties of $G(\mathbf{r}, t; \mathbf{r}', t')$ are different from the three-dimensional cases considered previously. Using (9.48) and

$$\boldsymbol{\nabla}^2 G(\mathbf{r}, -t; \mathbf{r}'', -t'') - \frac{1}{c^2}\frac{\partial^2}{\partial t^2}G(\mathbf{r}, -t; \mathbf{r}'', -t'') = \delta(\mathbf{r} - \mathbf{r}'')\delta(t - t''), \qquad (9.52)$$

which is obtained from (9.48) by the replacements $\mathbf{r}' \to \mathbf{r}''$, $t \to -t$, $t' \to -t''$, we can evaluate

$$\int_{-\infty}^{\infty} dt \int_V d^3x \left[G(\mathbf{r},t;\mathbf{r}',t')\boldsymbol{\nabla}^2 G(\mathbf{r},-t;\mathbf{r}'',-t'') - G(\mathbf{r},-t;\mathbf{r}'',-t'')\boldsymbol{\nabla}^2 G(\mathbf{r},t;\mathbf{r}',t') \right.$$

$$\left. -\frac{1}{c^2}\frac{\partial}{\partial t}\left\{ G(\mathbf{r},t;\mathbf{r}',t')\frac{\partial}{\partial t}G(\mathbf{r},-t;\mathbf{r}'',-t'') - G(\mathbf{r},-t;\mathbf{r}'',-t'')\frac{\partial}{\partial t}G(\mathbf{r},t;\mathbf{r}',t') \right\} \right]$$

$$= G(\mathbf{r}'',t'';\mathbf{r}',t') - G(\mathbf{r}',-t';\mathbf{r}'',-t''). \tag{9.53}$$

On the other hand, if we use Green's theorem and carry out the integral of the time derivative, we find for the left side of (9.53)

$$\int_{-\infty}^{\infty} dt \int_S dA \left[G(\mathbf{r},t;\mathbf{r}',t')\frac{\partial}{\partial n}G(\mathbf{r},-t;\mathbf{r}'',-t'') - G(\mathbf{r},-t;\mathbf{r}'',-t'')\frac{\partial}{\partial n}G(\mathbf{r},t;\mathbf{r}',t') \right]$$

$$-\frac{1}{c^2}\int_V d^3x \left[G(\mathbf{r},t;\mathbf{r}',t')\frac{\partial}{\partial t}G(\mathbf{r},-t;\mathbf{r}'',-t'') - G(\mathbf{r},-t;\mathbf{r}'',-t'')\frac{\partial}{\partial t}G(\mathbf{r},t;\mathbf{r}',t') \right]_{t=-\infty}^{t=\infty}. \tag{9.54}$$

The surface integral in this expression vanishes because of the boundary conditions (9.49) or (9.50). The volume integral evaluated at $t = \pm\infty$ also vanishes since from the initial conditions (9.51) we see that $G(\mathbf{r},t;\mathbf{r}',t')$ and $\frac{\partial}{\partial t}G(\mathbf{r},t;\mathbf{r}',t')$ vanish for $t = -\infty < t'$, and $G(\mathbf{r},-t;\mathbf{r}'',-t'')$ and $\frac{\partial}{\partial t}G(\mathbf{r},-t;\mathbf{r}'',-t'')$ vanish for $t = +\infty$, i.e., $-t = -\infty < -t''$. Thus the left side of (9.53) vanishes, and we find the symmetry relation

$$\boxed{G(\mathbf{r},t;\mathbf{r}',t') = G(\mathbf{r}',-t';\mathbf{r},-t).} \tag{9.55}$$

To solve the differential equation (9.44) we use the equation for the Green's function written in the form

$$\boldsymbol{\nabla}^2_{\mathbf{r}'}\, G(\mathbf{r},t;\mathbf{r}',t') - \frac{1}{c^2}\frac{\partial^2}{\partial t'^2}G(\mathbf{r},t;\mathbf{r}',t') = \delta(\mathbf{r}-\mathbf{r}')\delta(t-t'). \tag{9.56}$$

This is obtained from (9.48) by replacing $\mathbf{r} \leftrightarrow \mathbf{r}'$, $t \leftrightarrow -t'$ and using (9.55). Using (9.44) with $\mathbf{r} \to \mathbf{r}'$, $t \to t'$ and (9.56), we can evaluate

$$\int_T^{\infty} dt' \int_V d^3x' \left[G(\mathbf{r},t;\mathbf{r}',t')\boldsymbol{\nabla}^2_{\mathbf{r}'}\,\psi(\mathbf{r}',t') - \psi(\mathbf{r}',t')\boldsymbol{\nabla}^2_{\mathbf{r}'}\, G(\mathbf{r},t;\mathbf{r}',t') \right.$$

$$\left. -\frac{1}{c^2}\frac{\partial}{\partial t'}\left\{ G(\mathbf{r},t;\mathbf{r}',t')\frac{\partial\psi(\mathbf{r}',t')}{\partial t'} - \psi(\mathbf{r}',t')\frac{\partial}{\partial t'}G(\mathbf{r},t;\mathbf{r}',t') \right\} \right]$$

$$= -4\pi \int_T^{\infty} dt' \int_V d^3x'\, G(\mathbf{r},t;\mathbf{r}',t')\rho(\mathbf{r}',t') - \psi(\mathbf{r},t). \tag{9.57}$$

Using Green's theorem and integrating the time derivative on the left-hand side of (9.57), we find

$$\psi(r,t) = -4\pi \int_T^t dt' \int d^3x' \; G(\mathbf{r},t;\mathbf{r}',t')\rho(\mathbf{r}',t')$$

$$- \int_T^t dt' \int_S dA' \left[G(\mathbf{r},t;\mathbf{r}',t')\frac{\partial \psi(\mathbf{r}',t)}{\partial n'} - \psi(\mathbf{r}',t')\frac{\partial}{\partial n'}G(\mathbf{r},t;\mathbf{r}',t') \right]$$

$$- \frac{1}{c^2} \int_V d^3x' \left[G(\mathbf{r},t;\mathbf{r}',T)\frac{\partial \psi(\mathbf{r}',t')}{\partial t'}\bigg|_{t'=T} - \psi(\mathbf{r}',T)\frac{\partial}{\partial t'}G(\mathbf{r},t;\mathbf{r}',t')\bigg|_{t'=T} \right]. \quad (9.58)$$

Here we have used the initial condition (9.51) on $G(\mathbf{r},t;\mathbf{r}',t')$ to discard terms involving $t' > t$. In the second integral on the right of (9.58), one or the other of the terms will vanish because of the boundary condition (9.49) or (9.50) on G, and the other term will be determined by the boundary condition (9.45) or (9.46) on ψ. The third integral on the right of (9.58) is determined by the initial conditions (9.47) on ψ.

9.5 THE WAVE EQUATION FOR ALL SPACE, NO BOUNDARIES AT FINITE DISTANCES

As an example we shall content ourselves with the case of all space, i.e., $S \to \infty$. With no finite boundaries and with a delta function source which is a function only of coordinate differences, it is clear that G will be a function only of coordinate differences so that (9.48) becomes

$$\left[\boldsymbol{\nabla}^2 - \frac{1}{c^2}\frac{\partial^2}{\partial t^2} \right] G(\mathbf{r}-\mathbf{r}',t-t') = \delta(\mathbf{r}-\mathbf{r}')\delta(t-t'). \quad (9.59)$$

With no loss of generality we can now set $\mathbf{r}' = 0$, $t' = 0$:

$$\left[\boldsymbol{\nabla}^2 - \frac{1}{c^2}\frac{\partial^2}{\partial t^2} \right] G(\mathbf{r},t) = \delta(\mathbf{r})\delta(t). \quad (9.60)$$

The initial condition (9.51) becomes

$$G(\mathbf{r},t) = 0, \quad t < 0. \quad (9.61)$$

To solve these equations we Fourier transform in the space coordinates:

$$G(\mathbf{r},t) = \frac{1}{(2\pi)^3} \int d^3x \; e^{i\mathbf{k}\cdot\mathbf{r}}\widetilde{G}(\mathbf{k},t), \quad (9.62)$$

$$\widetilde{G}(\mathbf{k},t) = \int d^3x \; e^{-i\mathbf{k}\cdot\mathbf{r}} G(\mathbf{r},t), \quad (9.63)$$

$$\delta(\mathbf{r}) = \frac{1}{(2\pi)^3} \int d^3k \; e^{i\mathbf{k}\cdot\mathbf{r}}. \quad (9.64)$$

Substituting these expressions in (9.60) we find

$$-k^2 \widetilde{G}(\mathbf{k}, t) - \frac{1}{c^2} \frac{\partial^2}{\partial t^2} \widetilde{G}(\mathbf{k}, t) = \delta(t). \tag{9.65}$$

For $t > 0$ this has the solution

$$\widetilde{G}(\mathbf{k}, t) = Ae^{ikct} + Be^{-ikct}, \quad t > 0. \tag{9.66}$$

For $t < 0$ the condition (9.61) implies

$$\widetilde{G}(\mathbf{k}, t) = 0, \quad t < 0. \tag{9.67}$$

As in the one-dimensional case discussed in Section 8.3, the delta function on the right of (9.65) is equivalent to joining conditions at $t = 0$:

$$\widetilde{G}(\mathbf{k}, 0 - \epsilon) = \widetilde{G}(\mathbf{k}, 0 + \epsilon), \tag{9.68}$$

$$\left. \frac{\partial \widetilde{G}(\mathbf{k}, t)}{\partial t} \right|_{t=0-\epsilon}^{t=0+\epsilon} = -c^2. \tag{9.69}$$

Using (9.66), (9.67), we see that these imply $A = -B = -c/2ik$, so that we find

$$\widetilde{G}(\mathbf{k}, t) = -\frac{c}{2ik} \left[e^{ikct} - e^{-ikct} \right], \quad t > 0. \tag{9.70}$$

We can substitute this in (9.62) and carry out the integration over the angles of $\mathbf{k}$ to obtain

$$G(\mathbf{r}, t) = \frac{1}{(2\pi)^3} \int d^3k \, e^{ikr\cos\theta} \left[-\frac{c}{2ik} \right] \left[e^{ikct} - e^{-ikct} \right]$$

$$= \frac{1}{(2\pi)^3} \frac{2\pi}{ir} \int_0^\infty k \, dk \, \left[e^{ikr} - e^{-ikr} \right] \left[-\frac{c}{2ik} \right] \left[e^{ikct} - e^{-ikct} \right]$$

$$= \frac{c}{8\pi^2 r} \int_0^\infty dk \, \left[e^{ik(r+ct)} - e^{ik(r-ct)} - e^{-ik(r-ct)} + e^{-ik(r+ct)} \right]$$

$$= \frac{c}{8\pi^2 r} \int_{-\infty}^\infty dk \, \left[e^{ik(r+ct)} - e^{ik(r-ct)} \right]$$

$$= \frac{c}{4\pi r} \left[\delta(r + ct) - \delta(r - ct) \right], \quad t > 0, \tag{9.71}$$

where we have used the Fourier expansion of the δ function. For $t > 0$ and $r > 0$, $r + ct > 0$ and the first δ function in (9.59) vanishes. We thus find

$$G(\mathbf{r}, t) = \begin{cases} 0, & t < 0 \\ \\ -\dfrac{1}{4\pi r} \delta \left(\dfrac{r}{c} - t \right), & t > 0 \end{cases}, \tag{9.72}$$

or equivalently

$$G(\mathbf{r}, t; \mathbf{r}', t') - \begin{cases} 0, & t < t' \\ \\ -\dfrac{1}{4\pi} \dfrac{\delta \left[\frac{1}{c} |\mathbf{r} - \mathbf{r}'| - (t - t')\right]}{|\mathbf{r} - \mathbf{r}'|}, & t > t' \end{cases} \tag{9.73}$$

This is the same as the expression (9.34) obtained previously. Note that the symmetry relation (9.55) is satisfied.

The Green's function (9.73) for the wave equation has a direct physical interpretation. It is the wavelet produced by the delta function disturbance at time $t = t'$ at the point $\mathbf{r} = \mathbf{r}'$. At earlier times $t < t'$ there is no wavelet. After the initial disturbance the spherical wavelet spreads out and at time $t > t'$ is located at $|\mathbf{r} - \mathbf{r}'| = c(t - t')$. As the wavelet spreads, its amplitude decreases as $1/|\mathbf{r} - \mathbf{r}'|$. The analogy immediately springs to mind of the wavelet produced by dropping a pebble in a quiet pool of water.

The formula (9.58) tells us how to superpose all the elementary wavelets, some originating from the source function $\rho(\mathbf{r}', t)$, some from the boundary and initial conditions. We can simplify (9.58) considerably for the case $S \to \infty$ when the Green's function (9.73) containing the delta function $\delta \left[\frac{1}{c} |\mathbf{r} - \mathbf{r}'| - (t - t')\right]$ applies. The second integral on the right of (9.58) vanishes since with $\mathbf{r}, t$ finite and t' in the range $T < t' < t$, the delta function in $G(\mathbf{r}, t; \mathbf{r}', t')$ vanishes identically for $\mathbf{r}' \to \infty$. The third integral on the right of (9.58) reduces to a surface integral over a sphere of radius $|\mathbf{r}' - \mathbf{r}| = c(t - T)$, and the first integral on the right of (9.58) reduces to a volume integral over the interior of this sphere. Thus we find, after some integrations by parts in the surface integral term,

$$\psi(\mathbf{r}, t) = \int_{|\mathbf{r}' - \mathbf{r}| \le c(t-T)} d^3 x' \, \frac{\rho(\mathbf{r}', t - \frac{1}{c}|\mathbf{r} - \mathbf{r}'|)}{|\mathbf{r} - \mathbf{r}'|}$$

$$+ \frac{1}{4\pi} \int d\Omega' \left[(t - T) \frac{\partial \psi(\mathbf{r}', t')}{\partial t'} \bigg|_{t'=T, \, |\mathbf{r}'-\mathbf{r}|=c(t-T)} \right.$$

$$\left. + \frac{\partial}{\partial |\mathbf{r} - \mathbf{r}'|} \{ |\mathbf{r}' - \mathbf{r}| \, \psi(\mathbf{r}', T) \} \bigg|_{|\mathbf{r}-\mathbf{r}'|=c(t-T)} \right]. \tag{9.74}$$

The behavior of the system before time T does not enter explicitly in the formula (9.74); all such information is contained in the second integral on the right involving the initial conditions. The first integral on the right is over the interior of a sphere of radius $c(t - T)$ about $\mathbf{r}$. Waves originating outside this sphere would have to start before time T to reach $\mathbf{r}$ at time t; their effect is felt only implicitly through the surface integral term involving the initial conditions. We can go to the limit $T \to -\infty$. The surface integral in (9.74) is then over an infinitely large sphere at remote times in the past. If we assume $\psi(\mathbf{r}', t') \to 0$ for $\mathbf{r}' \to \infty$ and $t' \to -\infty$, the surface integral vanishes and we recover the retarded potential solution of the wave equation,

$$\psi(\mathbf{r}, t) = \int d^3x' \, \frac{\rho(\mathbf{r}', t - \frac{1}{c}|\mathbf{r} - \mathbf{r}'|)}{|\mathbf{r} - \mathbf{r}'|}, \tag{9.75}$$

which we have already obtained in the previous section – see (9.35). More generally, since the surface integral describes waves generated by infinitely remote sources, it is a solution of the wave equation without sources and we have

$$\psi(\mathbf{r}, t) = \psi_{\text{in}}(\mathbf{r}, t) + \int d^3x' \, \frac{\rho(\mathbf{r}', t - \frac{1}{c}|\mathbf{r} - \mathbf{r}'|)}{|\mathbf{r} - \mathbf{r}'|}, \tag{9.76}$$

where $\psi_{\text{in}}(\mathbf{r}, t)$ is a solution of

$$\left[\nabla^2 - \frac{1}{c^2} \frac{\partial^2}{\partial t^2} \right] \psi_{\text{in}}(\mathbf{r}, t) = 0. \tag{9.77}$$

We have used in this section the so-called retarded Green's function which satisfies the initial condition

$$G_{\text{ret}}(\mathbf{r}, t; \mathbf{r}', t') = 0, \quad t < t'. \tag{9.78}$$

Another possibility is the advanced Green's function satisfying the "final condition"

$$G_{\text{adv}}(\mathbf{r}, t; \mathbf{r}', t') = 0, \quad t > t'. \tag{9.79}$$

For the case of all space, a calculation analogous to (9.59)ff. yields

$$G_{\text{adv}}(\mathbf{r}, t; \mathbf{r}', t') - \begin{cases} -\dfrac{1}{4\pi} \dfrac{\delta\left[\frac{1}{c}|\mathbf{r} - \mathbf{r}'| + (t - t')\right]}{|\mathbf{r} - \mathbf{r}'|}, & t < t' \\[4mm] 0, & t > t' \end{cases} \tag{9.80}$$

With the advanced Green's function one obtains a formula like (9.58) except that the integral involving the initial conditions on ψ is replaced by an integral involving "final conditions" at some final time $T > t$. For the case of all space and $T \to \infty$ we obtain an analogy to (9.76)

$$\psi(\mathbf{r}, t) = \psi_{\text{out}}(\mathbf{r}, t) + \int d^3x' \, \frac{\rho(\mathbf{r}', t + \frac{1}{c}|\mathbf{r} - \mathbf{r}'|)}{|\mathbf{r} - \mathbf{r}'|}, \tag{9.81}$$

where $\psi_{\text{out}}(\mathbf{r}, t)$ is a solution of the wave equation without sources,

$$\left[\nabla^2 - \frac{1}{c^2} \frac{\partial^2}{\partial t^2} \right] \psi_{\text{out}}(\mathbf{r}, t) = 0. \tag{9.82}$$

The two expressions (9.76) and (9.81) are both valid; in fact they are equal when the two sourceless fields $\psi_{\text{in}}(\mathbf{r}, t)$ and $\psi_{\text{out}}(\mathbf{r}, t)$ are correctly chosen. In (9.76) it is possible to choose

$$\psi_{\text{in}}(\mathbf{r}, t) = 0. \tag{9.83}$$

The retarded potential contribution in (9.76) then gives the radiation due to the source $\rho(\mathbf{r}, t)$. With $\psi_{\text{in}}(\mathbf{r}, t) = 0$, $\psi_{\text{out}}(\mathbf{r}, t)$ is given by

$$\psi_{\text{out}}(\mathbf{r}, t) = \int d^3x' \, \frac{\rho(\mathbf{r}', t - \frac{1}{c}|\mathbf{r} - \mathbf{r}'|) - \rho(\mathbf{r}', t + \frac{1}{c}|\mathbf{r} - \mathbf{r}'|)}{|\mathbf{r} - \mathbf{r}'|} \tag{9.84}$$

and will in general not vanish.

It is not possible, however, to set $\psi_{\text{out}}(\mathbf{r}, t) = 0$. This leads to the advanced potential solution

$$\psi_{\text{adv}}(\mathbf{r}, t) = \int d^3x' \, \frac{\rho(\mathbf{r}', t + \frac{1}{c}|\mathbf{r} - \mathbf{r}'|)}{|\mathbf{r} - \mathbf{r}'|}, \tag{9.85}$$

which is not acceptable since it violates the principle of causality, which states that causes should precede their effects. This condition is used here as a boundary condition, like the Ausstrahlungsbedingung, to pick out the correct solution.

9.6   FIELD DUE TO A POINT SOURCE

As an example of the retarded potential (9.75) we consider the field produced by a point source:

$$\rho(\mathbf{r}, t) = e\delta[\mathbf{r} - \mathbf{x}(t)]. \tag{9.86}$$

Here $\mathbf{x}(t)$ is the prescribed orbit of the point source. To carry out the integrals, it is easiest to write the retarded potential solution in terms of the Green's function:

$$\psi(\mathbf{r}, t) = -4\pi \int d^3x' \int dt' \, G_{\text{ret}}(\mathbf{r}, t; \mathbf{r}', t')\rho(\mathbf{r}', t')$$

$$= e \int d^3x' \int dt' \, \frac{\delta[\frac{1}{c}|\mathbf{r} - \mathbf{r}'| - (t - t')]}{|\mathbf{r} - \mathbf{r}'|} \delta[\mathbf{r}' - \mathbf{x}(t')]. \tag{9.87}$$

The delta function in ρ can be used to do the integral over $\mathbf{r}'$:

$$\psi(\mathbf{r}, t) = e \int dt' \frac{\delta[\frac{1}{c}|\mathbf{r} - \mathbf{x}(t')| - (t - t')]}{|\mathbf{r} - \mathbf{x}(t')|}. \tag{9.88}$$

We can use the delta function in (9.88) to do the remaining integral, but we must be careful to take care of all the time dependence in the argument of the delta function. Introduce a new variable

$$u = t' + \frac{1}{c}|\mathbf{r} - \mathbf{x}(t')|. \tag{9.89}$$

Then when we change variables we find

$$dt' = du \left(\frac{1}{\frac{du}{dt'}} \right),$$ (9.90)

$$\frac{du}{dt'} = 1 - \frac{1}{c}\hat{\mathbf{n}}(t') \cdot \mathbf{v}(t'),$$ (9.91)

where $\mathbf{v}(t') = \frac{d\mathbf{x}(t')}{dt'}$ is the velocity of the particle at time t', and $\hat{\mathbf{n}}(t')$ is a unit vector pointing from the position $\mathbf{x}(t')$ of the particle at time t' to the observation point $\mathbf{r}$ as shown in Figure 9.2.

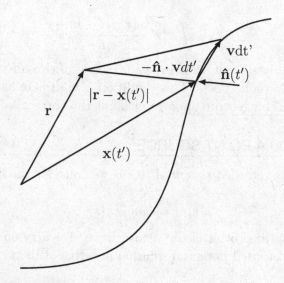

Figure 9.2: Particle path relative to observer at $\mathbf{r}$

The formula (9.88) becomes

$$\psi(\mathbf{r}, t) = e \int du \, \frac{1}{1 - \frac{1}{c}\hat{\mathbf{n}}(t') \cdot \mathbf{v}(t')} \, \delta(u - t) \, \frac{1}{|\mathbf{r} - \mathbf{x}(t')|}$$

$$= e \, \frac{1}{1 - \frac{1}{c}\hat{\mathbf{n}}(t') \cdot \mathbf{v}(t')} \, \frac{1}{|\mathbf{r} - \mathbf{x}(t')|} \bigg|_{u=t}$$

$$= e \, \frac{1}{1 - \frac{1}{c}\hat{\mathbf{n}}(t') \cdot \mathbf{v}(t')} \, \frac{1}{|\mathbf{r} - \mathbf{x}(t')|} \bigg|_{t'=t-\frac{1}{c}|\mathbf{r} - \mathbf{x}(t')|}$$

$$\boxed{\psi(\mathbf{r}, t) \equiv \frac{e}{\kappa R}\bigg|_{\text{ret.}},}$$ (9.92)

with

$$\kappa = 1 - \frac{1}{c}\hat{\mathbf{n}}(t') \cdot \mathbf{v}(t') \tag{9.93}$$

and

$$R = |\mathbf{r} - \mathbf{x}(t')| \tag{9.94}$$

evaluated at the retarded time t' which is a solution of

$$t' = t - \frac{1}{c}|\mathbf{r} - \mathbf{x}(t')|. \tag{9.95}$$

The result (9.92) is a famous one known as the Lienard Wiechert potential. In the application to electrodynamics we have both a scalar and vector potential

$$\left[\boldsymbol{\nabla}^2 - \frac{1}{c^2}\frac{\partial^2}{\partial t^2}\right]\phi(\mathbf{r},t) = -4\pi\rho(\mathbf{r},t), \tag{9.96}$$

$$\left[\boldsymbol{\nabla}^2 - \frac{1}{c^2}\frac{\partial^2}{\partial t^2}\right]\mathbf{A}(\mathbf{r},t) = -\frac{4\pi}{c}\mathbf{j}(\mathbf{r},t), \tag{9.97}$$

and the solutions for a point source,

$$\rho(\mathbf{r},t) = e\delta\left[\mathbf{r} - \mathbf{x}(t)\right], \tag{9.98}$$
$$\mathbf{j}(\mathbf{r},t) = e\mathbf{v}\delta\left[\mathbf{r} - \mathbf{x}(t)\right], \tag{9.99}$$

are

$$\boxed{\phi(\mathbf{r},t) = \left.\frac{e}{\kappa R}\right|_{\text{ret.}},} \tag{9.100}$$

$$\boxed{\mathbf{A}(\mathbf{r},t) = \left.\frac{e}{c}\frac{\mathbf{v}}{\kappa R}\right|_{\text{ret.}}.} \tag{9.101}$$

9.6.1 Point Source Moving with Constant Velocity, $v < c$

The simplest application of (9.92) is to the case of a particle moving with a constant velocity in a straight line:

$$\begin{aligned} x(t') &= vt', \\ y(t') &= 0, \\ z(t') &= 0. \end{aligned} \tag{9.102}$$

The equation for the retarded time becomes

$$t' = t - \frac{1}{c}|\mathbf{r} - \mathbf{x}(t')|$$

$$= t - \frac{1}{c}\sqrt{(x - vt')^2 + \rho^2}, \quad \rho^2 = y^2 + z^2. \tag{9.103}$$

Rearranging and squaring this leads to a quadratic equation in t' with two roots:

$$t' = \frac{1}{1 - \frac{v^2}{c^2}} \left[t - \frac{vx}{c^2} \pm \frac{1}{c} \sqrt{(x - vt)^2 + \left(1 - \frac{v^2}{c^2} \right) \rho^2} \right]. \tag{9.104}$$

Writing (9.104) in the form

$$t' - t = \frac{1}{1 - \frac{v^2}{c^2}} \left[-\frac{v}{c^2}(x - vt) \pm \frac{1}{c} \sqrt{(x - vt)^2 + \left(1 - \frac{v^2}{c^2} \right) \rho^2} \right], \tag{9.105}$$

we see that for the case $v/c < 1$ one of these roots has $t' < t$ in agreement with (9.103) while the other root with $t' > t$ is an extraneous root introduced by squaring (9.103). For the case $v/c < 1$ we thus keep only the solution with the minus sign in (9.104). The quantity which appears in the retarded potential (9.92) is

$$\kappa R \Big|_{\text{ret.}} = \left[1 - \frac{\hat{\mathbf{n}}(t') \cdot \mathbf{v}(t')}{c} \right] |\mathbf{r} - \mathbf{x}(t')|$$

$$= \left[1 - \frac{v}{c} \frac{x - vt'}{\sqrt{(x - vt')^2 + \rho^2}} \right] \sqrt{(x - vt')^2 + \rho^2}$$

$$= \sqrt{(x - vt')^2 + \rho^2} - \frac{v}{c}(x - vt')$$

$$= c(t - t') - \frac{v}{c}(x - vt')$$

$$= ct - \frac{v}{c}x - c \left[t - \frac{v}{c^2}x - \frac{1}{c} \sqrt{(x - vt)^2 + \left(1 - \frac{v^2}{c^2} \right) \rho^2} \right]$$

$$= \sqrt{(x - vt)^2 + \left(1 - \frac{v^2}{c^2} \right) \rho^2}, \tag{9.106}$$

so that we finally obtain

$$\psi(\mathbf{r}, t) = \frac{e}{\sqrt{(x - vt)^2 + \left(1 - \frac{v^2}{c^2} \right)(y^2 + z^2)}}. \tag{9.107}$$

9.6.2 Point Source Moving with Constant Velocity, $v > c$

It is interesting to consider the case $v > c$. For the electromagnetic application this can be achieved, without violating special relativity, by a particle traveling through a dielectric at a speed greater than the velocity of light in the dielectric (Čerenkov

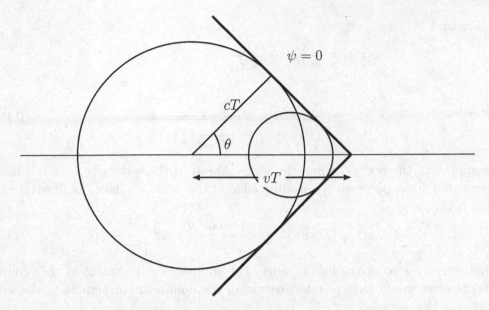

Figure 9.3: Wake of wave as sum of two wavelets

effect). For the acoustic application there is of course nothing unusual about the case $v > c$. When the particle emitting the waves travels at a speed v greater than the velocity c of the wave, a wake forms behind the particle. Since the waves travel a distance cT while the particle travels a distance vT, the angle made by the normal to the leading edge of the wake with the direction of motion of the particle is

$$\cos \theta = \frac{cT}{vT} = \frac{c}{v} < 1. \tag{9.108}$$

The field $\psi = 0$ ahead of the leading edge of the wake since there is not time for the disturbance traveling at velocity c to get there. The field at time t at a point $\mathbf{r}$ behind the leading edge of the wake is the sum of the contributions of two wavelets emitted at different retarded times t' – represented by the intersections of the two circles in Figure 9.3. This can also be seen from the algebra. With $v/c > 1$ it is easy to see from (9.105) that behind the wake both solutions of the quadratic equation satisfy the condition $t' < t$ and hence are solutions of (9.103). The same algebra as in (9.106) leads to the two values

$$\kappa R \Big|_{\text{ret.}} = \pm \sqrt{(vt - x)^2 - \left(\frac{v^2}{c^2} - 1\right)\rho^2} \tag{9.109}$$

corresponding to these two solutions. We must take account of both these contributions in the calculation (9.88) - (9.92). We note from (9.109) that for one of these contributions κ is negative since R is always positive. For this contribution a minus sign enters when we change variables according to (9.90), (9.91):

$$\int_{-\infty}^{\infty} dt' = \int_{\infty}^{-\infty} \frac{du}{\kappa} = -\int_{-\infty}^{\infty} \frac{du}{\kappa}. \tag{9.110}$$

We thus find

$$\psi(\mathbf{r},t) = \frac{e}{(\kappa R)_1} - \frac{e}{(\kappa R)_2}$$

$$= \frac{2e}{\sqrt{(vt-x)^2 - \left(\frac{v^2}{c^2}-1\right)(y^2+z^2)}}. \qquad (9.111)$$

This applies to the region behind the wake. Ahead of the wake $\psi(\mathbf{r},t) = 0$. Finally we note that for a point on the leading edge of the wake we have, as illustrated in Figure 9.4

$$(vt-x)^2 = (\rho\tan\theta)^2 = \rho^2\frac{1-\cos^2\theta}{\cos^2\theta} = \rho^2\left[\frac{v^2}{c^2}-1\right], \qquad (9.112)$$

so that $\psi(\mathbf{r},t) \to \infty$ at the leading edge. For an accurate treatment of the acoustic application we would have to take account of the nonlinear corrections to the wave equation in this region.

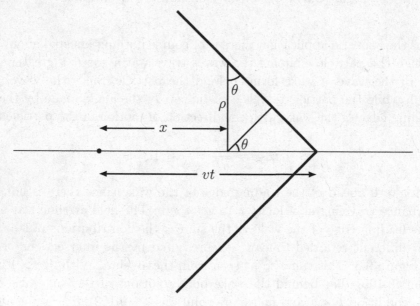

Figure 9.4: Point at leading edge of wake

9.7   THE DIFFUSION EQUATION

Our considerations here parallel those in the preceding section. We wish to solve the inhomogeneous diffusion equation

$$\nabla^2\psi(\mathbf{r},t) - \frac{1}{\kappa}\frac{\partial\psi(\mathbf{r},t)}{\partial t} = -4\pi\rho(\mathbf{r},t) \qquad (9.113)$$

with boundary condition for a closed volume as shown in Figure 9.5

$$\psi(\mathbf{r},t) \text{ given for } \mathbf{r} \text{ on surface } S \text{ of } V \qquad (9.114)$$

<u>or</u>

$$\frac{\partial \psi(\mathbf{r}, t)}{\partial n} \text{ given for } \mathbf{r} \text{ on } S \qquad (9.115)$$

plus an initial condition

$$\psi(\mathbf{r}, t) \text{ given at an initial time } t = T. \qquad (9.116)$$

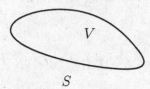

Figure 9.5: Arbitrary closed volume for diffusion equation

To solve this problem we introduce a Green's function $G(\mathbf{r}, t; \mathbf{r}', t')$ which is a solution of

$$\boldsymbol{\nabla}^2 G(\mathbf{r}, t; \mathbf{r}', t') - \frac{1}{\kappa}\frac{\partial}{\partial t}G(\mathbf{r}, t; \mathbf{r}', t') = \delta(\mathbf{r} - \mathbf{r}')\delta(t - t') \qquad (9.117)$$

with the boundary condition [corresponding respectively to (9.114) or (9.115)]

$$G(\mathbf{r}, t; \mathbf{r}', t') = 0 \text{ for } \mathbf{r} \text{ on } S \qquad (9.118)$$

<u>or</u>

$$\frac{\partial G(\mathbf{r}, t; \mathbf{r}', t')}{\partial n} = 0 \text{ for } \mathbf{r} \text{ on } S \qquad (9.119)$$

plus an initial condition

$$G(\mathbf{r}, t; \mathbf{r}', t') = 0 \text{ for } t < t'. \qquad (9.120)$$

To establish the symmetry properties of G use (9.117) and

$$\boldsymbol{\nabla}^2 G(\mathbf{r}, -t; \mathbf{r}'', -t'') + \frac{1}{\kappa}\frac{\partial}{\partial t}G(\mathbf{r}, -t; \mathbf{r}'', -t'') = \delta(\mathbf{r} - \mathbf{r}'')\delta(t - t'') \qquad (9.121)$$

which is obtained from (9.117) by the replacements $t \to -t$, $\mathbf{r}' \to \mathbf{r}''$, $t' \to -t''$, to evaluate

$$\int_{-\infty}^{\infty} dt \int_V d^3x \left[G(\mathbf{r}, t; \mathbf{r}', t')\boldsymbol{\nabla}^2 G(\mathbf{r}, -t; \mathbf{r}'', -t'') \right.$$

$$- G(\mathbf{r}, -t; \mathbf{r}'', -t'')\boldsymbol{\nabla}^2 G(\mathbf{r}, t; \mathbf{r}', t')$$

$$\left. + \frac{1}{\kappa}\frac{\partial}{\partial t}\left\{ G(\mathbf{r}, t; \mathbf{r}', t')G(\mathbf{r}, -t; \mathbf{r}'', -t'') \right\} \right]$$

$$= G(\mathbf{r}'', t''; \mathbf{r}', t') - G(\mathbf{r}', -t; \mathbf{r}'', -t''). \qquad (9.122)$$

On the other hand, using Green's theorem and carrying out the integral of the time derivative, we find that the left side of (9.122) can be written

$$
\int_{-\infty}^{\infty} dt \int_S dA \left[G(\mathbf{r}, t; \mathbf{r}', t') \frac{\partial}{\partial n} G(\mathbf{r}, -t; \mathbf{r}'', -t'') \right.
$$

$$
\left. - G(\mathbf{r}, -t; \mathbf{r}'', -t'') \frac{\partial}{\partial n} G(\mathbf{r}, t; \mathbf{r}', t') \right]
$$

$$
+ \frac{1}{\kappa} \int_V d^3x \; G(\mathbf{r}, t; \mathbf{r}', t') G(\mathbf{r}, -t; \mathbf{r}'', -t'') \Big|_{t=-\infty}^{t=\infty}. \tag{9.123}
$$

The first integral in (9.123) vanishes as a consequence of the boundary conditions (9.118) or (9.119), and the second integral vanishes as a consequence of the initial condition (9.120), so we find from (9.122) the symmetry relation

$$
\boxed{G(\mathbf{r}, t; \mathbf{r}', t') = G(\mathbf{r}', -t'; \mathbf{r}, -t).} \tag{9.124}
$$

To solve the differential equation (9.113) we use the equation for the Green's function written in the form

$$
\boldsymbol{\nabla}_{\mathbf{r}'}^2 \, G(\mathbf{r}, t; \mathbf{r}', t') + \frac{1}{\kappa} \frac{\partial}{\partial t'} G(\mathbf{r}, t; \mathbf{r}', t') = \delta(\mathbf{r} - \mathbf{r}') \delta(t - t'). \tag{9.125}
$$

This is obtained from (9.117) by replacing $\mathbf{r} \leftrightarrow \mathbf{r}'$, $t \leftrightarrow -t'$, and using the symmetry relation (9.124). Using (9.113) with $\mathbf{r} \to \mathbf{r}'$, $t \to t'$ and (9.125), we can evaluate

$$
\int_T^{\infty} dt' \int_V d^3x' \left[G(\mathbf{r}, t; \mathbf{r}', t') \boldsymbol{\nabla}_{\mathbf{r}'}^2 \psi(\mathbf{r}', t') - \psi(\mathbf{r}', t') \boldsymbol{\nabla}_{\mathbf{r}'}^2 G(\mathbf{r}, t; \mathbf{r}', t') \right.
$$

$$
\left. - \frac{1}{\kappa} \frac{\partial}{\partial t'} \left\{ G(\mathbf{r}, t; \mathbf{r}', t') \psi(\mathbf{r}', t) \right\} \right]
$$

$$
= -4\pi \int_T^{\infty} dt' \int_V d^3x' \; G(\mathbf{r}, t; \mathbf{r}', t') \rho(\mathbf{r}', t') - \psi(\mathbf{r}, t). \tag{9.126}
$$

Using Green's theorem and carrying out the integral of the time derivative on the left side of (9.126), we finally obtain

$$
\psi(\mathbf{r}, t) = -4\pi \int_T^t dt' \int d^3x' \; G(\mathbf{r}, t; \mathbf{r}', t') \rho(\mathbf{r}', t)
$$

$$
- \int_T^t dt' \int_S dA' \left[G(\mathbf{r}, t; \mathbf{r}', t') \frac{\partial}{\partial n'} \psi(\mathbf{r}', t) - \psi(\mathbf{r}', t) \frac{\partial}{\partial n'} G(\mathbf{r}, t; \mathbf{r}', t') \right]
$$

$$
- \frac{1}{\kappa} \int_V d^3x' \; G(\mathbf{r}, t; \mathbf{r}', T) \psi(\mathbf{r}', T). \tag{9.127}
$$

Here we have used the initial condition (9.120) on $G(\mathbf{r}, t; \mathbf{r}', t')$ to discard contributions with $t' > t$. The second integral in (9.127) is determined by the boundary condition (9.114) or (9.115) on $\psi(\mathbf{r}, t)$, one or the other of the terms in this second integral vanishing because of the boundary condition (9.118) or (9.119) on $G(\mathbf{r}, t; \mathbf{r}', t')$. The third integral in (9.127) is determined by the initial condition (9.116) on $\psi(\mathbf{r}, t)$.

9.8 THE DIFFUSION EQUATION FOR ALL SPACE, NO BOUNDARIES AT FINITE DISTANCES

For the case of all space, $S \to \infty$, we can find the Green's function for the diffusion equation by the same method used in Section 9.5 for the wave equation. With no finite boundaries the Green's function will be a function only of coordinate differences, $\mathbf{r} - \mathbf{r}'$, $t - t'$ so that (9.117) becomes

$$\boldsymbol{\nabla}^2 G(\mathbf{r} - \mathbf{r}', t - t') - \frac{1}{\kappa}\frac{\partial}{\partial t}G(\mathbf{r} - \mathbf{r}', t - t') = \delta(\mathbf{r} - \mathbf{r}')\delta(t - t'). \tag{9.128}$$

With no loss of generality one can set $\mathbf{r}' = 0$, $t' = 0$ to obtain

$$\boldsymbol{\nabla}^2 G(\mathbf{r}, t) - \frac{1}{\kappa}\frac{\partial}{\partial t}G(\mathbf{r}, t) = \delta(\mathbf{r})\delta(t). \tag{9.129}$$

The initial condition (9.120) becomes

$$G(\mathbf{r}, t) = 0, \quad t < 0. \tag{9.130}$$

We can solve (9.129), (9.130) by Fourier transforming in the space coordinates:

$$G(\mathbf{r}, t) = \frac{1}{(2\pi)^3}\int d^3k \; e^{i\mathbf{k}\cdot\mathbf{r}}\widetilde{G}(\mathbf{k}, t), \tag{9.131}$$

$$\widetilde{G}(\mathbf{k}, t) = \int d^3r \; e^{-i\mathbf{k}\cdot\mathbf{r}}G(\mathbf{r}, t), \tag{9.132}$$

$$\delta(\mathbf{r}) = \frac{1}{(2\pi)^3}\int d^3k \; e^{i\mathbf{k}\cdot\mathbf{r}}. \tag{9.133}$$

Substituting these expressions in (9.129), we find

$$-k^2\widetilde{G}(\mathbf{k}, t) - \frac{1}{\kappa}\frac{\partial}{\partial t}\widetilde{G}(\mathbf{k}, t) = \delta(t). \tag{9.134}$$

For $t > 0$ the solution of this equation is

$$\widetilde{G}(\mathbf{k}, t) = Ae^{-\kappa k^2 t}, \tag{9.135}$$

and for $t < 0$ we have the initial condition (9.130). Integrating (9.134) with respect to t over a time interval of vanishingly small length about $t = 0$ we find that the delta function on the right of the equation is equivalent to the discontinuity condition

$$\widetilde{G}(\mathbf{k}, 0 + \epsilon) - \widetilde{G}(\mathbf{k}, 0 - \epsilon) = -\kappa. \tag{9.136}$$

Using (9.130) and (9.135), we see that this implies $A = -\kappa$ so that we have

$$\widetilde{G}(\mathbf{k}, t) = \begin{cases} 0, & t < 0 \\ -\kappa e^{-\kappa k^2 t}, & t > 0 \end{cases}. \tag{9.137}$$

Substituting this in (9.131) and doing the Gaussian integral by completing the square, we find

$$G(\mathbf{r}, t) = \frac{-\kappa}{(2\pi)^3} \int d^3 k \; e^{i\mathbf{k}\cdot\mathbf{r} - \kappa k^2 t}$$

$$= \frac{-\kappa}{(2\pi)^3} \exp\left(-\frac{r^2}{4\kappa t}\right) \int d^3 k \; \exp\left(-\kappa t \left(\mathbf{k} - \frac{i\mathbf{r}}{2\kappa t}\right)^2\right)$$

$$= -\frac{\kappa}{(4\pi\kappa t)^{3/2}} \exp\left(-\frac{r^2}{4\kappa t}\right), \quad t > 0. \tag{9.138}$$

In summary, we have

$$G(\mathbf{r}, t; \mathbf{r}', t') = \begin{cases} 0, & t < t' \\ -\dfrac{\kappa}{[4\pi\kappa(t - t')]^{3/2}} \exp\left(-\dfrac{|\mathbf{r} - \mathbf{r}'|^2}{4\kappa(t - t')}\right), & t > t' \end{cases}. \tag{9.139}$$

This Green's function has a characteristic bell shape in $|\mathbf{r} - \mathbf{r}'|$ with a width $\sim \{\kappa(t - t')\}^{1/2}$ growing with time and a height $\sim \{\kappa(t - t')\}^{-3/2}$ decreasing with time in such a way that the volume under the curve

$$\int d^3 x \; G(\mathbf{r}, t; \mathbf{r}', t') = -\kappa \tag{9.140}$$

is constant. At $t = t'$ the disturbance starts as a delta function:

$$G(\mathbf{r}, t; \mathbf{r}', t') = -\kappa\delta(\mathbf{r} - \mathbf{r}'). \tag{9.141}$$

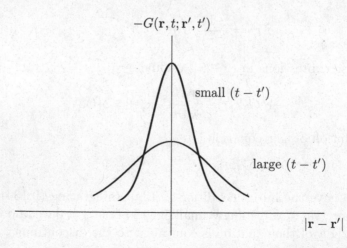

Figure 9.6: Delta function at two different times

This Green's function describes the way a δ function impulse at $t = t'$ diffuses through the medium at later times as shown in Figure 9.6.

We can now substitute the Green's function (9.139) for all space into our solution (9.127). The surface integral in (9.127) does not contribute for $\mathbf{r}' \to \infty$ because of the rapidly decreasing exponential in the Green's function, and we find

$$\psi(\mathbf{r}, t) = \frac{1}{\sqrt{4\pi\kappa}} \int_T^t dt' \int d^3x' \, \frac{\exp\left(-\frac{|\mathbf{r}-\mathbf{r}'|^2}{4\kappa(t-t')}\right)}{(t-t')^{3/2}} \, \rho(\mathbf{r}', t')$$

$$+ \frac{1}{[4\pi\kappa(t-T)]^{3/2}} \int d^3x' \, \exp\left(-\frac{|\mathbf{r}-\mathbf{r}'|^2}{4\kappa(t-T)}\right) \, \psi(\mathbf{r}', T). \qquad (9.142)$$

PROBLEMS

9.1 A monochromatic wave

$$\psi(\mathbf{r}, t) = u(\mathbf{r})e^{-i\omega t}$$

is emitted by a point source in front of a reflecting plane as shown in Figure 9.7, on which the boundary condition is

$$\psi(\mathbf{r}, t) = 0, \quad \mathbf{r} \text{ on a plane } z = 0.$$

The wave function satisfies the equation

$$\nabla^2 \psi(\mathbf{r}, t) - \frac{1}{c^2}\frac{\partial^2 \psi(\mathbf{r}, t)}{\partial t^2} = 0.$$

a) Use the image method to find $u(\mathbf{r})$ in the region $z > 0$.

b) Use this result to solve the more general problem

$$\nabla^2 \phi(\mathbf{r}) + k^2 \phi(\mathbf{r}) = -4\pi\rho(\mathbf{r})$$

in the region $z > 0$ with the boundary condition

$$\phi(x, y, z = 0) = F(x, y),$$

where $F(x, y)$ is a given function. Assume outgoing waves.

c) Apply your formula to the special case

$$\rho(\mathbf{r}) = 0,$$

$$F(x, y) = \begin{cases} V = \text{constant}, & \rho = \sqrt{x^2 + y^2} < a \\ 0, & \rho > a \end{cases}.$$

Find an explicit integral expression along the line $x = y = 0$. Compare with the result of Problem 8.4 (c).

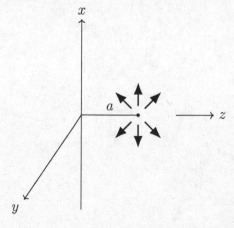

Figure 9.7: Monochromatic wave emitted from point source for Problem 9.1

9.2 a) Find the four-dimensional Green's function for the half-space $z > 0$ for the inhomogeneous wave equation

$$\nabla^2 \psi(\mathbf{r}, t) - \frac{1}{c^2} \frac{\partial^2 \psi(\mathbf{r}, t)}{\partial t^2} = -4\pi \rho(\mathbf{r}, t)$$

with the boundary condition

$$\psi(x, y, z = 0, t) = F(x, y, t) \text{ given,}$$

and use the Green's function to express $\psi(\mathbf{r}, t)$ in terms of $F(x, y, t)$ and $\rho(\mathbf{r}, t)$.

b) Apply your formula to the special case

$$\rho(\mathbf{r}, t) = 0,$$

$$F(x, y, t) = \begin{cases} Ve^{-i\omega t}, & \rho = \sqrt{x^2 + y^2} < a \\ 0, & \rho > a \end{cases}.$$

Find an explicit expression along the line $x = y = 0$. Compare with the result of Problem 8.9(c).

9.3 a) Find the four-dimensional Green's function for the half-space $z > 0$ for the inhomogeneous diffusion equation

$$\nabla^2 \psi(\mathbf{r}, t) - \frac{1}{\kappa} \frac{\partial \psi(vr, t)}{\partial t} = -4\pi \rho(\mathbf{r}, t)$$

with the boundary condition

$$\psi(x, y, z = 0, t) = F(x, y, t) \text{ given,}$$

and use the Green's function to express $\psi(\mathbf{r}, t)$ in terms of $F(x, y, t)$ and $\rho(\mathbf{r}, t)$.

b) Apply your formula to the special case

$$\rho(\mathbf{r}, t) = 0,$$

$$F(x, y, t) = \begin{cases} V = \text{constant}, & \rho = \sqrt{x^2 + y^2} < a \\ 0, & \rho > a \end{cases}.$$

Evaluate the integrals along the line $x = y = 0$ and compare with the previous solutions of this problem.

9.4 Find the four-dimensional retarded Green's function $G(\mathbf{r}, t; \mathbf{r}', t')$ for the inhomogeneous Schrödinger equation

$$-\frac{\hbar^2}{2m} \nabla^2 \psi(\mathbf{r}, t) - i\hbar \frac{\partial \psi(\mathbf{r}, t)}{\partial t} = \frac{\hbar^2}{2m} 4\pi \rho(\mathbf{r}, t)$$

and use it to solve the initial value problem

$$\psi(\mathbf{r}, 0) = f(\mathbf{r}),$$

where $f(\mathbf{r})$ is a given function, i.e., find $\psi(\mathbf{r}, t)$ for $t > 0$ in terms of $\psi(\mathbf{r}, 0) = f(\mathbf{r})$ and $\rho(\mathbf{r}, t)$. Assume all space is available; there are no boundaries.

CHAPTER 10

Integral Equations

10.1   INTRODUCTION

An integral equation arises if one applies Green's function techniques to a partial differential equation such as

$$(\boldsymbol{\nabla}^2 + k^2)\psi(\mathbf{r}) = -4\pi U(\mathbf{r})\psi(\mathbf{r}), \tag{10.1}$$

in which the source density $\rho(\mathbf{r}) = U(\mathbf{r})\psi(\mathbf{r})$ involves the unknown quantity $\psi(\mathbf{r})$. An application of this sort in the quantum theory of scattering is discussed in Section 10.2 in order to motivate the study of integral equations.

The remainder of the chapter is devoted to a brief discussion of types of integral equations and the methods available for their solution. We restrict ourselves for the most part to the study of one-dimensional integral equations. Usually, by expansions in spherical harmonics or some other complete set, the multidimensional case can be reduced to the one-dimensional case.

10.2   QUANTUM THEORY OF SCATTERING

The time independent Schrödinger equation

$$-\frac{\hbar^2}{2m}\boldsymbol{\nabla}^2\psi(\mathbf{r}) + V(\mathbf{r})\psi(\mathbf{r}) = E\psi(\mathbf{r}) \tag{10.2}$$

for the interaction of a particle with a potential well $V(\mathbf{r})$ can be written in the form (10.1) with

$$E = \frac{\hbar^2 k^2}{2m}, \tag{10.3}$$

$$U(\mathbf{r}) = -\frac{m}{2\pi\hbar^2}V(\mathbf{r}). \tag{10.4}$$

We shall assume in what follows that $V(\mathbf{r})$ is a short range potential, i.e., that $V(\mathbf{r})$ is effectively negligible for $\mathbf{r}$ greater than some finite range. This excludes the coulomb potential which falls off so slowly that it requires special treatment.

For the scattering problem we want the solution of (10.2) or (10.1) which satisfies the boundary condition for large r

$$\psi(\mathbf{r}) \xrightarrow[r\to\infty]{} e^{i\mathbf{k_0}\cdot\mathbf{r}} + f(\theta,\varphi)\frac{e^{ikr}}{r}. \tag{10.5}$$

Here the first term on the right is the incoming plane wave, whose wave number $\mathbf{k_0}$ has a magnitude $|\mathbf{k_0}| = k$ related to the energy E by (10.3). The second term is the outgoing spherical scattered wave produced by the interaction of the plane wave with the potential well $V(\mathbf{r})$. To check that the scattered wave is outgoing, we can put back in the time dependence to obtain

$$\psi(\mathbf{r})e^{-i\omega t} \to e^{i(\mathbf{k_0}\cdot\mathbf{r}-\omega t)} + f(\theta,\varphi)\frac{e^{i(kr-\omega t)}}{r}, \tag{10.6}$$

where $\omega = E/\hbar$.

We can convert (10.1) to an integral equation, using the outgoing wave Green's function (9.12),

$$G(\mathbf{r},\mathbf{r}') = -\frac{1}{4\pi}\frac{e^{ik|\mathbf{r}-\mathbf{r}'|}}{|\mathbf{r}-\mathbf{r}'|}, \tag{10.7}$$

which satisfies

$$(\boldsymbol{\nabla}^2 + k^2)G(\mathbf{r},\mathbf{r}') = \delta(\mathbf{r}-\mathbf{r}'). \tag{10.8}$$

Thus the partial differential equation (10.1) plus the boundary condition (10.5) can be combined in the integral equation

$$\psi(\mathbf{r}) = e^{i\mathbf{k_0}\cdot\mathbf{r}} + \int d^3x' \frac{e^{ik|\mathbf{r}-\mathbf{r}'|}}{|\mathbf{r}-\mathbf{r}'|}U(\mathbf{r}')\psi(\mathbf{r}'). \tag{10.9}$$

We can check (10.9). Applying the operator $\boldsymbol{\nabla}^2 + k^2$ to it and using (10.8) and $(\boldsymbol{\nabla}^2 + k^2)e^{i\mathbf{k_0}\cdot\mathbf{r}} = 0$, we immediately obtain (10.1). On the other hand for large r we have

$$|\mathbf{r}-\mathbf{r}'| \to r - \hat{\mathbf{n}}\cdot\mathbf{r}', \quad \hat{\mathbf{n}} = \frac{\mathbf{r}}{r},$$

$$\frac{e^{ik|\mathbf{r}-\mathbf{r}'|}}{|\mathbf{r}-\mathbf{r}'|} \to \frac{e^{ikr}}{r}e^{-i\mathbf{k}\cdot\mathbf{r}'}, \quad \mathbf{k} = k\hat{\mathbf{n}}, \tag{10.10}$$

so that the asymptotic form of (10.9) becomes

$$\psi(\mathbf{r}) \xrightarrow[r\to\infty]{} e^{i\mathbf{k_0}\cdot\mathbf{r}} + \frac{e^{ikr}}{r}f(\mathbf{k}), \tag{10.11}$$

with

$$f(\mathbf{k}) = \int d^3x' \, e^{-i\mathbf{k}\cdot\mathbf{r}'}U(\mathbf{r}')\psi(\mathbf{r}'). \tag{10.12}$$

This verifies that we have the correct type of asymptotic behavior and also gives us an explicit formula (10.12) for the scattering amplitude, which can be evaluated once we have solved the integral equation (10.9) for $\psi(\mathbf{r})$. We note that in this integral equation formulation the boundary condition (outgoing spherical scattered wave) is built into the equation (10.9) and does not need to be separately stated as in the differential equation formulation.

As formulated above, equation (10.9), we have a three-dimensional integral equation in coordinate space. For many applications it is desirable to transform to momentum space, but we shall forego that step in this brief treatment. We shall, however, reduce the three-dimensional integral equation to an infinite set of one-dimensional integral equations by using the identities (6.56) and (6.66) derived earlier:

$$e^{i\mathbf{k_0}\cdot\mathbf{r}} = e^{ikr\cos\theta} = \sum_{l=0}^{\infty}(2l+1)i^l j_l(kr)P_l(\cos\theta), \tag{10.13}$$

$$\frac{e^{ik|\mathbf{r}-\mathbf{r}'|}}{|\mathbf{r}-\mathbf{r}'|} = 4\pi ik\sum_{l=0}^{\infty}\sum_{m=-l}^{l} j_l(kr_<)h_l^{(1)}(kr_>)Y_l^m(\theta,\varphi)Y_l^{m*}(\theta',\varphi'). \tag{10.14}$$

We shall assume that $U(\mathbf{r}) = U(r)$ is spherically symmetric; $\psi(\mathbf{r})$ will then be independent of φ and can be expanded in the form

$$\psi(\mathbf{r}) = \sum_{l=0}^{\infty}(2l+1)i^l\,\psi_l(r)P_l(\cos\theta)$$

$$= \sum_{l=0}^{\infty}(2l+1)i^l\,\psi_l(r)\sqrt{\frac{4\pi}{2l+1}}\,Y_l^0(\theta). \tag{10.15}$$

In all these expansions the z axis has been taken along the direction of the incoming wave vector $\mathbf{k_0}$. Using (10.14) and (10.15) and the orthogonality properties of spherical harmonics we can calculate the integral on the right side of (10.9):

$$\int d^3x'\,\frac{e^{ik|\mathbf{r}-\mathbf{r}'|}}{|\mathbf{r}-\mathbf{r}'|}\,U(r')\psi(\mathbf{r}')$$

$$= \int_0^{\infty} r'^2 dr'\int d\Omega'\,4\pi ik\left[\sum_{lm} j_l(kr_<)h_l^{(1)}(kr_>)Y_l^m(\theta,\varphi)Y_l^{m*}(\theta'\varphi')U(r')\times\right.$$

$$\left.\sum_{l'}(2l'+1)i^{l'}\psi_{l'}(r')\sqrt{\frac{4\pi}{2l'+1}}\,Y_{l'}^0(\theta')\right]$$

$$= \sum_l(2l+1)i^l\,P_l(\cos\theta)4\pi ik\int_0^{\infty} r'^2 dr'\,j_l(kr_<)h_l^{(1)}(kr_>)U(r')\psi_l(r'). \tag{10.16}$$

Substituting the expansions (10.13), (10.15) and (10.16) in the integral equation (10.9) and using the orthogonality of the $P_l(\cos\theta)$, we then find in place of (10.9) the infinite set of one-dimensional integral equations

$$\psi_l(r) = j_l(kr) + 4\pi ik\int_0^{\infty} r'^2 dr'\,j_l(kr_<)h_l^{(1)}(kr_>)U(r')\psi_l(r'). \tag{10.17}$$

Using in addition the spherical harmonics addition theorem (3.114) we find for the scattering amplitude (10.12)

$$f(k,\theta) = \int d^3x' \sum_{lm} 4\pi(-i)^l j_l(kr')Y_l^m(\theta,\varphi)Y_l^{m*}(\theta',\varphi')U(r') \times$$

$$\sum_{l'}(2l'+1)i^{l'}\psi_{l'}(r')\sqrt{\frac{4\pi}{2l'+1}}\,Y_{l'}^0(\theta')$$

$$f(k,\theta) = \sum_l (2l+1)f_l(k)P_l(\cos\theta), \tag{10.18}$$

with

$$f_l(k) = 4\pi \int_0^\infty r'^2 dr'\, j_l(kr')U(r')\psi_l(r'). \tag{10.19}$$

The formulation (very incomplete) of the scattering problem presented here perhaps makes it clear that it is worthwhile for a physicist to learn something about how to solve integral equations. Although the scattering problem with simple central potentials can be handled most easily with a differential equation approach, many problems in many body and particle physics lead naturally to integral equations in momentum space which cannot be transformed into simple differential equations in coordinate space.

10.3 TYPES OF INTEGRAL EQUATIONS

10.3.1 First Kind

An integral equation of the type

$$\int_a^b dy\, K(x,y)f(y) = \phi(x) \tag{10.20}$$

with the kernel $K(x,y)$ and the function $\phi(x)$ given is called an integral equation of the first kind. The problem is to solve for the unknown function $f(y)$.

10.3.2 Second Kind

An integral equation such as

$$f(x) = g(x) + \lambda \int_a^b dy\, K(x,y)f(y) \tag{10.21}$$

with $g(x)$, λ and $K(x,y)$ given and $f(x)$ the unknown function is said to be of the second kind. The integral equations of scattering theory, e.g., (10.17), are of this type.

10.3.3 Volterra

In a Volterra equation one of the limits on the integral is the independent variable. For example the equation

$$f(x) = g(x) + \int_a^x dy\, K(x,y)f(y) \tag{10.22}$$

is a Volterra equation of the second kind.

10.3.4   Eigenvalue Problem

An integral equation of the type

$$f_n(x) = \lambda_n \int_a^b dy\ K(x,y) f_n(y) \tag{10.23}$$

is an eigenvalue problem. It has nontrivial (i.e., not identically zero) solutions $f_n(x)$, the eigenfunctions, only for certain values of λ_n, the eigenvalues. Note that the eigenvalue problem (10.23) is obtained from the integral equation of the second kind (10.21) by dropping the inhomogeneous term $g(x)$.

10.4   INTEGRAL EQUATIONS WITH SEPARABLE KERNELS

If the kernel $K(x,y)$ of the integral equation is separable, i.e., of the special form

$$K(x,y) = u(x)v(y), \tag{10.24}$$

types from Sections 10.3.2 and 10.3.4 can be easily solved in closed form. Thus if we substitute (10.24) in (10.21) we find

$$f(x) = g(x) + \lambda \int_a^b dy\ K(x,y) f(y)$$

$$= g(x) + \lambda \int_a^b dy\ u(x)v(y) f(y)$$

$$= g(x) + \lambda A u(x), \tag{10.25}$$

where the constant A is given by the expression

$$A = \int_a^b dy\ v(y) f(y). \tag{10.26}$$

The functional form of $f(x)$ is given by the last line of (10.25). Substituting (10.25) in (10.26) we obtain a linear equation for A:

$$A = \int_a^b dy\ v(y)[g(y) + \lambda A u(y)], \tag{10.27}$$

$$A\left[1 - \lambda \int_a^b dy\ u(y)v(y)\right] = \int_a^b dy\ v(y)g(y), \tag{10.28}$$

$$A = \frac{\int_a^b dy\ v(y)g(y)}{1 - \lambda \int_a^b dy\ u(y)v(y)}. \tag{10.29}$$

The solution of the integral equation is given by (10.25) with A given by (10.29).

The eigenvalue equation (10.23) can also be solved with the kernel (10.24). We find

$$f(x) = \lambda \int_a^b dy \; K(x,y) f(y)$$

$$= \lambda u(x) \int_a^b dy \; v(y) f(y), \tag{10.30}$$

or

$$f(x) = \text{const. } u(x). \tag{10.31}$$

This gives the functional form of $f(x)$; the constant is undetermined by the homogeneous linear equation. Substituting (10.31) in (10.30) we find a formula for λ:

$$\frac{1}{\lambda} = \int_a^b dy \; v(y) u(y). \tag{10.32}$$

Note that with the kernel (10.24) there is only one solution of the eigenvalue problem, i.e., one eigenfunction given by (10.31) and one corresponding eigenvalue given by (10.32).

Note also that if λ is equal to the eigenvalue, as given by (10.32), A as given by (10.29) is infinite and the inhomogeneous equation (10.29) is infinite and the inhomogeneous equation (10.25) has no solution, except for the special case when $g(y)$ and $v(y)$ are such that $\int_a^b dy \; v(y) g(y) = 0$, in which case the solution is $f(x) = g(x) + cu(x)$, with c an arbitrary constant.

Integral equations involving a separable kernel which is a sum of separable pieces,

$$K(x,y) = \sum_{i=1}^N u_i(x) v_i(y), \tag{10.33}$$

can also be solved explicitly. Substituting (10.33) in the integral equation (10.21) of the second kind we find

$$f(x) = g(x) + \lambda \int_a^b dy \; K(x,y) f(y)$$

$$= g(x) + \lambda \sum_{i=1}^N u_i(x) \int_a^b dy \; v_i(y) f(y)$$

$$= g(x) + \lambda \sum_{i=1}^N A_i u_i(x), \tag{10.34}$$

where the constants A_i are given by

$$A_i = \int_a^b dy \; v_i(y) f(y). \tag{10.35}$$

The functional form of $f(x)$ is given by the last line of (10.34). Substituting this expression on the right of (10.35), we find a set of linear equations to determine the A_i:

$$A_i = \int_a^b dy \, v_i(y)g(y) + \lambda \sum_{j=1}^N \left[\int_a^b dy \, v_i(y)u_j(y) \right] A_j. \tag{10.36}$$

Introducing an $N \times N$ matrix $\mathbb{M}$ whose elements are

$$M_{ij} = \int_a^b dy \, v_i(y)u_j(y) \tag{10.37}$$

and a column vector $\mathbf{G}$ whose elements are

$$G_i = \int_a^b dy \, v_i(y)g(y), \tag{10.38}$$

we can write (10.36) as

$$\sum_j \left[\delta_{ij} - \lambda M_{ij} \right] A_j = G_i \tag{10.39}$$

or, in matrix notation,

$$(\mathbb{I} - \lambda \mathbb{M})\mathbf{A} = \mathbf{G}, \tag{10.40}$$

where $\mathbb{I}$ is the identity matrix. For the inhomogeneous integral equation $[g(x) \neq 0]$, the A_i of (10.34) are determined by inverting the matrix $(\mathbb{I} - \lambda \mathbb{M})$:

$$\mathbf{A} = (\mathbb{I} - \lambda \mathbb{M})^{-1}\mathbf{G}. \tag{10.41}$$

For the eigenvalue problem (10.34) still applies with $g(x) \equiv 0$. Setting $\mathbf{G} = 0$ in (10.40) we find an $N \times N$ matrix eigenvalue problem,

$$(\mathbb{I} - \lambda \mathbb{M})\mathbf{A} = 0, \tag{10.42}$$

or

$$\mathbb{M}\mathbf{A} = \frac{1}{\lambda}\mathbf{A}, \tag{10.43}$$

to determine the eigenvalues λ_n and the relative values of the $A_i^{(n)}$ in the various eigenfunctions:

$$f_n = \text{const.} \sum_{i=1}^N A_i^{(n)} u_i(x). \tag{10.44}$$

As is well-known from matrix theory, (10.42) has a nontrivial solution (not all A_i equal zero) if and only if

$$\det(\mathbb{I} - \lambda \mathbb{M}) = 0. \tag{10.45}$$

This equation determines at least one and at most N eigenvalues. Substituting these eigenvalues λ_n back in (10.42), one can solve for the ratios of the $A_i^{(n)}$, $i = 1, 2, \ldots$.

We note that just as for the simplest case (10.25) the inhomogeneous equation (10.34) will in general have no solution for λ equal to an eigenvalue of the corresponding homogeneous equation. When $\det(\mathbb{I} - \lambda \mathbb{M}) = 0$ the matrix $(\mathbb{I} - \lambda \mathbb{M})$ has no inverse and the expression (10.41) for **A** has infinite components unless **G** has special properties. We return to a discussion of this point in Section 10.10.

Separable kernels are important just because it is so easy to obtain an explicit solution. Often a nonseparable kernel can be adequately approximated by a separable kernel. Of course, if we choose for the $u_i(x)$ an appropriate complete set of functions, we can expand $K(x, y)$ in an infinite series

$$K(x, y) = \sum_{i=1}^{\infty} u_i(x) v_i(y). \tag{10.46}$$

In this case the matrix equations (10.40), (10.41), and (10.45) involve $\infty \times \infty$ matrices. In practice this may only be useful if we can approximate these by finite $N \times N$ matrices.

10.5   CONVOLUTION INTEGRAL EQUATIONS

An integral equation of the type

$$f(x) = g(x) + \int_{-\infty}^{\infty} dy \ K(x - y) f(y), \tag{10.47}$$

with a kernel which is a function only of the difference of its two variables and with infinite limits, can be solved by Fourier transformation:

$$f(x) = \int_{-\infty}^{\infty} dk \ \widetilde{f}(k) e^{ikx}, \tag{10.48}$$

$$\widetilde{f}(k) = \frac{1}{2\pi} \int_{-\infty}^{\infty} dx \ f(x) e^{-ikx}, \tag{10.49}$$

$$g(x) = \int_{-\infty}^{\infty} dk \ \widetilde{g}(k) e^{ikx}, \tag{10.50}$$

$$K(x - y) = \int_{-\infty}^{\infty} dk \ \widetilde{K}(k) e^{ik(x-y)}. \tag{10.51}$$

Fourier transforming (10.47) and using the expression

$$\delta(k - k') = \frac{1}{2\pi} \int_{-\infty}^{\infty} dx \ e^{-i(k-k')x} \tag{10.52}$$

for the delta function, we find

$$\widetilde{f}(k) = \widetilde{g}(k) + \frac{1}{2\pi} \int_{-\infty}^{\infty} dx \ e^{-ikx} \int_{-\infty}^{\infty} dy \int_{-\infty}^{\infty} dk' \ \widetilde{K}(k') e^{ik'(x-y)} \int_{-\infty}^{\infty} dk'' \ \widetilde{f}(k'') e^{ik''y}$$

$$= \widetilde{g}(k) + 2\pi \int_{-\infty}^{\infty} dk' \int_{-\infty}^{\infty} dk'' \frac{1}{2\pi} \int_{-\infty}^{\infty} dx \ e^{-i(k-k')x} \frac{1}{2\pi} \int_{-\infty}^{\infty} dy \ e^{-i(k'-k'')y} \ \widetilde{K}(k') \widetilde{f}(k'')$$

or

$$\widetilde{f}(k) = \widetilde{g}(k) + 2\pi \int_{-\infty}^{\infty} dk' \int_{-\infty}^{\infty} dk'' \, \delta(k - k')\delta(k' - k'')\widetilde{K}(k')\widetilde{f}(k'')$$

$$= \widetilde{g}(k) + 2\pi\widetilde{K}(k)\widetilde{f}(k). \tag{10.53}$$

The Fourier transformation has reduced the convolution integral in (10.47) to a product, and we can solve (10.53) algebraically:

$$\boxed{\widetilde{f}(k) = \frac{\widetilde{g}(k)}{1 - 2\pi\widetilde{K}(k)}.} \tag{10.54}$$

The solution of the original integral equation is then obtained by substituting (10.54) in (10.48).

10.6   ITERATION – LIOUVILLE-NEUMANN SERIES

Integral equations with separable kernels and integral equations of the convolution form have simple solutions which can be found in closed form. Other types of integral equations cannot be solved so easily. We consider in this and the succeeding two sections some general methods.

We start with the iteration solution of the equation

$$f(x) = g(x) + \lambda \int_a^b dy \, K(x, y)f(y). \tag{10.55}$$

If $f(y)$ is approximated by $g(y)$ in the integral, we obtain

$$f^{(1)}(x) \approx g(x) + \lambda \int_a^b dy \, K(x, y)g(y). \tag{10.56}$$

A better approximation is obtained by using the approximation $f^{(1)}(y)$ for $f(y)$ in the integral in (10.55). This leads to

$$f^{(2)}(x) \approx g(x) + \lambda \int_a^b dy \, K(x, y)g(y) + \lambda^2 \int_a^b dy \int_a^b dz \, K(x, y)K(y, z)g(z). \tag{10.57}$$

Continuing in this way we generate an infinite series:

$$f(x) = g(x) + \lambda \int_a^b dy \, K(x, y)g(y)$$

$$+ \lambda^2 \int_a^b dy \int_a^b dy_1 \, K(x, y_1)K(y_1, y)g(y)$$

$$+ \lambda^3 \int_a^b dy \int_a^b dy_1 \int_a^b dy_2 \, K(x, y_2)K(y_2, y_1)K(y_1, y)g(y)$$

$$+ \cdots . \tag{10.58}$$

One can easily check by substituting this series on both sides of (10.55) and comparing the terms that this infinite series is formally an exact solution of the integral equation (10.55).

The formal manipulations above say nothing about the convergence of the series (10.58). In fact the series will converge only for sufficiently small values of λ. We return to a discussion of this point in Section 10.9.

Perturbation theory can often be formulated in terms of the iteration solution of an integral equation. For example for the scattering problem discussed in Section 10.2 the integral equation to be solved was (10.9),

$$\psi(\mathbf{r}) = e^{i\mathbf{k_0}\cdot\mathbf{r}} + \int d^3x' \frac{e^{ik|\mathbf{r}-\mathbf{r}'|}}{|\mathbf{r}-\mathbf{r}'|} U(r')\psi(\mathbf{r}'), \tag{10.59}$$

and the scattering amplitude was given in terms of the solution $\psi(\mathbf{r})$ of this integral equation by (10.12),

$$f(\mathbf{k}) = \int d^3x\, e^{-i\mathbf{k}\cdot\mathbf{r}} U(r)\psi(\mathbf{r}). \tag{10.60}$$

The iteration solution of (10.59) is

$$\psi(\mathbf{r}) = e^{i\mathbf{k_0}\cdot\mathbf{r}} + \int d^3x' \frac{e^{ik|\mathbf{r}-\mathbf{r}'|}}{|\mathbf{r}-\mathbf{r}'|} U(r') e^{i\mathbf{k_0}\cdot\mathbf{r}'}$$

$$+ \int d^3x' \int d^3x'' \frac{e^{ik|\mathbf{r}-\mathbf{r}'|}}{|\mathbf{r}-\mathbf{r}'|} U(r') \frac{e^{ik|\mathbf{r}'-\mathbf{r}''|}}{|\mathbf{r}'-\mathbf{r}''|} U(r'') e^{i\mathbf{k_0}\cdot\mathbf{r}''}$$

$$+\dots. \tag{10.61}$$

Substituting this in (10.60) we find the perturbation series for $f(\mathbf{k})$ in powers of the strength of the potential $U(r)$:

$$f(\mathbf{k}) = \int d^3x\, e^{-i\mathbf{k}\cdot\mathbf{r}} U(r) e^{i\mathbf{k_0}\cdot\mathbf{r}}$$

$$+ \int d^3x \int d^3x' e^{-i\mathbf{k}\cdot\mathbf{r}} U(r) \frac{e^{ik|\mathbf{r}-\mathbf{r}'|}}{|\mathbf{r}-\mathbf{r}'|} U(r') e^{i\mathbf{k_0}\cdot\mathbf{r}'}$$

$$+ \int d^3x \int d^3x' \int d^3x'' \, e^{-i\mathbf{k}\cdot\mathbf{r}} U(r) \frac{e^{ik|\mathbf{r}-\mathbf{r}'|}}{|\mathbf{r}-\mathbf{r}'|} U(r') \frac{e^{ik|\mathbf{r}'-\mathbf{r}''|}}{|\mathbf{r}'-\mathbf{r}''|} U(r'') e^{i\mathbf{k_0}\cdot\mathbf{r}''}$$

$$+\dots. \tag{10.62}$$

For a strong potential $U(r)$ this series may converge slowly or not at all.

10.7 NUMERICAL SOLUTION

It is fairly easy to solve integral equations numerically on a computer. We first need to recall how to calculate an integral numerically. With a suitable set of points x_i and weights W_i, as shown in Figure 10.1, an integral can be approximated by

$$\int_a^b dx\, f(x) \approx \sum_{i=1}^n W_i f(x_i). \tag{10.63}$$

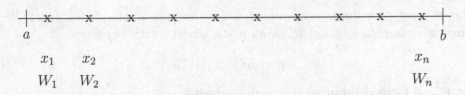

Figure 10.1: Set of points with associated weights

There are various possibilities for the x_i and W_i:

1) Equally spaced points and equal weights $W_i = (a - b)/n$. In the limit $n \to \infty$ (10.63) then reduces to the Riemann definition of the integral. With a finite number of points it is possible to achieve higher accuracy with other sets of points and weights.

2) Simpson's rule. The points x_i are equally spaced. The weights W_i are chosen in such a way that the procedure is equivalent to fitting parabolas to the function $f(x)$ at successive sets of three neighboring points and then doing the integrals of the parabolas exactly.

3) Gaussian quadrature. The points are no longer equally spaced. The points and weights are chosen in an optimum way so that n point Gaussian quadrature is exact for any polynomial of degree $2n - 1$ or less.

Details of these and other choices of points and weights (for integrals with infinite limits, etc.) can be found in texts on numerical methods.[1] All we need here is the basic formula (10.63) which has the same form for any choice of points and weights.

To solve an integral equation of the form

$$f(x) = g(x) + \lambda \int_a^b dy\, K(x,y)f(y) \tag{10.64}$$

we thus introduce an appropriate set of points x_i and weights W_i and approximate

---

[1] F. B. Hildebrand, Introduction to Numerical Analysis (McGraw Hill, New York, 1956) Chapter 8, p. 312;

M. Abramowitz and I. A. Stegun, Handbook of Mathematical Functions, National Bureau of Standards Applied Mathematics Series. 55. (U.S. Government Printing Office, Washington, D.C., 1965) p. 887.

the integral according to (10.63):

$$\int_a^b dy \, K(x,y)f(y) \approx \sum_{j=1}^n K(x,x_j)W_j f(x_j). \tag{10.65}$$

With this approximation for the integral the equation (10.64) is evaluated at the points x_i:

$$f(x_i) = g(x_i) + \lambda \sum_{j=1}^n K(x_i,x_j)W_j f(x_j). \tag{10.66}$$

This is a set of n linear algebraic equations for the n quantities $f(x_i)$. We can introduce matrix notation. Let $\mathbb{M}$ be an $n \times n$ matrix with elements

$$M_{ij} = K(x_i,x_j)W_j, \tag{10.67}$$

and let $\mathbf{F}$ and $\mathbf{G}$ be column vectors with elements

$$F_i = f(x_i), \tag{10.68}$$
$$G_i = g(x_i), \tag{10.69}$$

The equation (10.66) can then be written as

$$\mathbf{F} = \mathbf{G} + \lambda \mathbb{M}\mathbf{F} \tag{10.70}$$

or

$$(\mathbb{I} - \lambda \mathbb{M})\mathbf{F} = \mathbf{G}. \tag{10.71}$$

The solution of this equation is given by

$$\mathbf{F} = (\mathbb{I} - \lambda \mathbb{M})^{-1}\mathbf{G}, \tag{10.72}$$

where $(\mathbb{I} - \lambda \mathbb{M})^{-1}$ is the inverse of the matrix $(\mathbb{I} - \lambda \mathbb{M})$.

It is straightforward to invert matrices on a computer. The accuracy of the results can be checked by changing the number of integration points.

The eigenvalue problem can be handled in a similar way. The integral equation

$$f(x) = \lambda \int_a^b dy \, K(x,y)f(y) \tag{10.73}$$

is approximated by the set of n linear algebraic equations

$$f(x_i) = \lambda \sum_j K(x_i,x_j)W_j f(x_j), \tag{10.74}$$

or in the matrix notation introduced above

$$(\mathbb{I} - \lambda \mathbb{M})\mathbf{F} = 0. \tag{10.75}$$

The eigenvalues are the values of λ for which this has a nonzero solution $\mathbf{F}$. According to matrix theory these are the values of λ for which

$$\boxed{\det(\mathbb{I} - \lambda \mathbb{M}) = 0.} \tag{10.76}$$

It is straightforward to evaluate the determinant $\det(\mathbb{I} - \lambda\mathbb{M})$ on a computer for any given value of λ. One then uses Newton's method to search for values of λ for which $\det(\mathbb{I} - \lambda\mathbb{M}) = 0$.

We note that for λ an eigenvalue the matrix $(\mathbb{I} - \lambda\mathbb{M})$ has no inverse so that the inhomogeneous equation has no solution (10.72) unless **G** has special properties.

We defer to Section 10.9 a discussion of the conditions under which it is legitimate to approximate an integral equation by a finite dimensional matrix equation. As mentioned above we can check the stability of the procedure by changing the number of integration points.

It is noteworthy that the matrix algebra in this section is identical to that encountered in the study of separable kernels in Section 10.4 – see equation (10.39)ff. In fact, another method for obtaining numerical solutions of integral equations is to approximate the kernel by a separable kernel such as (10.33). A systematic way to do this is to introduce an appropriate complete orthonormal set $\phi_i(x)$,

$$\int_a^b dx \; \phi_i(x)\phi_j(x) = \delta_{ij}, \tag{10.77}$$

and expand $f(x)$, $g(x)$, and $K(x, y)$ in the $\phi_i(x)$:

$$f(x) = \sum_{i=1}^{\infty} f_i\phi_i(x), \tag{10.78}$$

$$g(x) = \sum_{i=1}^{\infty} g_i\phi_i(x), \tag{10.79}$$

$$K(x, y) = \sum_{i,j=1}^{\infty} K_{ij}\phi_i(x)\phi_j(y). \tag{10.80}$$

Substituting these expansions in the integral equation (10.64) and using the orthonormality property (10.77), we find that (10.64) is equivalent to the set of algebraic equations

$$f_i = g_i + \lambda \sum_{j=1}^{\infty} K_{ij}f_j. \tag{10.81}$$

This is an equation of the same form as (10.70) except that the indices run from 1 to ∞. However if the complete set $\phi_i(x)$ is chosen <u>appropriately</u>, one can obtain a good approximation keeping only a finite number of terms in (10.78)-(10.80). In this case (10.81) becomes a finite matrix equation which can be solved on a computer in the same way as indicated above for (10.70). Again the accuracy of results can be checked by increasing the number of terms kept in the series (10.78)-(10.80), i.e., increasing the size of the matrix K_{ij}.

10.8   FREDHOLM'S FORMULAS

These formulas provide an explicit, though computationally awkward, solution for an integral equation of the type

$$f(x) = g(x) + \lambda \int_a^b dy\, K(x,y) f(y). \tag{10.82}$$

The method is a theoretical version of the numerical method discussed in the previous section. The interval $a \le x \le b$ is divided into n subintervals of equal length δ as in Figure 10.2, and the integral equation (10.82) is approximated by a set of algebraic equations:

$$f(x_i) = g(x_i) + \lambda \delta \sum_{j=1}^n K(x_i, x_j) f(x_j) \tag{10.83}$$

or

$$\sum_{j=1}^n \left[\delta_{ij} - \lambda \delta K(x_i, x_j) \right] f(x_j) = g(x_i). \tag{10.84}$$

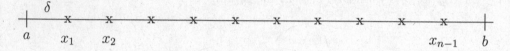

Figure 10.2: Set of points equally spaced

We now solve this set of algebraic equations by Cramer's rule and then in the result take the limit $\delta \to 0$, $n \to \infty$ so as to obtain a solution of the integral equation (10.82).

According to Cramer's rule the solution of (10.84) is given by

$$f(x_p) = \frac{\sum_{q=1}^n \mathrm{Cof}[\delta_{qp} - \lambda \delta K(x_q, x_p)] g(x_q)}{\det |\delta_{ij} - \lambda \delta K(x_i, x_j)|}. \tag{10.85}$$

In the denominator of this expression we have the determinant of the matrix which appears on the left of (10.84). In the numerator we have the determinant of this matrix with the pth column replaced by $g(x_q)$, $q = 1, \ldots, n$. Cof is the cofactor of the indicated matrix element. To proceed, we expand (10.85) and take the limit $\delta \to 0$. It is fairly straightforward to expand the denominator in powers of λ:

$$\det |\delta_{ij} - \lambda \delta K(x_i, x_j)| = \begin{vmatrix} \{1 - \lambda \delta K(x_1, x_1)\} & -\lambda \delta K(x_1, x_2) & -\lambda \delta K(x_1, x_3) & \cdots \\ -\lambda \delta K(x_2, x_1) & \{1 - \lambda \delta K(x_2, x_2)\} & \cdots & \cdots \\ \vdots & \vdots & \vdots & \vdots \\ -\lambda \delta K(x_n, x_1) & \cdots & \cdots & \cdots \end{vmatrix}$$

or

$$\det|\delta_{ij} - \lambda\delta K(x_i, x_j)| = 1 - \lambda\delta\sum_{i=1}^{n} K(x_i, x_i) + \frac{\lambda^2}{2!}\delta^2 \sum_{i,j=1}^{n} \begin{vmatrix} K(x_i, x_i) & K(x_i, x_j) \\ K(x_j, x_i) & K(x_j, x_j) \end{vmatrix}$$

$$- \frac{\lambda^3}{3!}\delta^3 \sum_{i,j,k=1}^{n} \begin{vmatrix} K(x_i, x_i) & K(x_i, x_j) & K(x_i, x_k) \\ K(x_j, x_i) & K(x_j, x_j) & K(x_j, x_k) \\ K(x_k, x_i) & K(x_k, x_j) & K(x_k, x_k) \end{vmatrix} + \dots \quad (10.86)$$

In the limit $\delta \to 0$ this becomes

$$\det|\delta_{ij} - \lambda\delta K(x_i, x_j)| \xrightarrow[\delta \to 0]{} D(\lambda) = 1 - \lambda\int_a^b dx_1\, K(x_1, x_1)$$

$$+ \frac{\lambda^2}{2!}\int_a^b dx_1 \int_a^b dx_2 \begin{vmatrix} K(x_1, x_1) & K(x_1, x_2) \\ K(x_2, x_1) & K(x_2, x_2) \end{vmatrix}$$

$$- \frac{\lambda^3}{3!}\int_a^b dx_1 \int_a^b dx_2 \int_a^b dx_3 \begin{vmatrix} K(x_1, x_1) & K(x_1, x_2) & K(x_1, x_3) \\ K(x_2, x_1) & K(x_2, x_2) & K(x_2, x_3) \\ K(x_3, x_1) & K(x_3, x_2) & K(x_3, x_3) \end{vmatrix} + \dots \quad (10.87)$$

As for the numerator of (10.85), we find

$$\sum_{q=1}^{n} \text{Cof}[\delta_{qp} - \lambda\delta K(x_q, x_p)]g(x_q)$$

$$= \text{Cof}[1 - \lambda\delta K(x_p, x_p)]g(x_p)$$

$$+ \sum_{q\neq p} \text{Cof}[-\lambda\delta K(x_q, x_p)]g(x_q). \quad (10.88)$$

In the limit $\delta \to 0$, $n \to \infty$ it does not matter if we leave out the pth row and column from the matrix (10.86) so that we find

$$\boxed{\text{Cof}[1 - \lambda\delta K(x_p, x_p)] \xrightarrow[\delta \to 0]{} D(\lambda).} \quad (10.89)$$

To calculate $\text{Cof}[-\lambda\delta K(x_q, x_p)]$, which appears in the second term of (10.88), first move the pth row up until it is the first row of the determinant, then move the qth column over until it is the first column of the determinant. Counting the minus signs which arise every time two adjacent rows or columns are interchanged, we find

$\text{Cof}[-\lambda\delta K(x_q,x_p)]=$

$$-\begin{vmatrix} \begin{matrix} -\lambda\delta K(x_p,x_q) & -\lambda\delta K(x_p,x_1) & \cdots & -\lambda\delta K(x_p,x_{q-1}) & -\lambda\delta K(x_p,x_{q+1}) & \cdots \\ -\lambda\delta K(x_1,x_q) & & & & & \\ \vdots & \ddots & & & & \\ -\lambda\delta K(x_{p-1},x_q) & & \text{Matrix} & \delta_{ij}-\lambda\delta K(x_i,x_j) & \text{with } p^{th} & \\ -\lambda\delta K(x_{p+1},x_q) & & \text{and } q^{th} \text{ rows} & \text{and } p^{th} \text{ and } q^{th} & \text{columns} & \\ \vdots & & & \text{removed.} & & \end{matrix} \end{vmatrix}$$

$$\tag{10.90}$$

Expanding this determinant in powers of λ, we find

$$\text{Cof}[-\lambda\delta K(x_q, x_p)] = \lambda\delta K(x_p, x_q) - \lambda^2\delta^2 \sum_i \begin{vmatrix} K(x_p,x_q) & K(x_p,x_i) \\ K(x_i,x_q) & K(x_i,x_i) \end{vmatrix}$$

$$+ \frac{\lambda^3}{2!}\delta^3 \sum_{i,j} \begin{vmatrix} K(x_p,x_q) & K(x_p,x_i) & K(x_p,x_j) \\ K(x_i,x_q) & K(x_i,x_i) & K(x_i,x_j) \\ K(x_j,x_q) & K(x_j,x_i) & K(x_j,x_j) \end{vmatrix} + \dots . \tag{10.91}$$

In the limit $\delta \to 0$ this becomes

$$\frac{1}{\lambda\delta}\text{Cof}[-\lambda\delta K(x_q,x_p)] \xrightarrow[\delta\to 0]{} D(x_p,x_q;\lambda), \tag{10.92}$$

where $D(x,y;\lambda)$ is given by

$$D(x,y;\lambda) = K(x,y) - \lambda \int_a^b dx_1 \begin{vmatrix} K(x,y) & K(x,x_1) \\ K(x_1,y) & K(x_1,x_1) \end{vmatrix}$$

$$+ \frac{\lambda^2}{2!} \int_a^b dx_1 \int_a^b dx_2 \begin{vmatrix} K(x,y) & K(x,x_1) & K(x,x_2) \\ K(x_1,y) & K(x_1,x_1) & K(x_1,x_2) \\ K(x_2,y) & K(x_2,x_1) & K(x_2,x_2) \end{vmatrix} + \dots .$$

Substituting (10.92) in (10.88) and the result in (10.85), we find for the solution of the inhomogeneous integral equation (10.82) the result

$$\boxed{f(x) = g(x) + \frac{\lambda}{D(\lambda)} \int_a^b dy \; D(x,y;\lambda)g(y),} \tag{10.93}$$

where $D(\lambda)$ and $D(x,y;\lambda)$ are given by the infinite series (10.87) and (10.92).

If we set $g(x) = 0$ in equation (10.82), we obtain the homogeneous integral equation, i.e., the eigenvalue problem. The condition that (10.84) has a nontrivial solution when $g(x_i) = 0$ is that

$$\det |\delta_{ij} - \lambda\delta K(x_i,x_j)| = 0. \tag{10.94}$$

In the limit $\delta \to 0$ this becomes, according to (10.87),

$$D(\lambda) = 0. \tag{10.95}$$

The eigenvalues λ_n are the solutions of this equation.

In order to find a formula for the eigenfunctions we first go back to the inhomogeneous case $[g(x) \neq 0]$ and derive an identity for arbitrary values of λ by substituting the solution (10.93) in the integral equation (10.82). We find

$$\frac{\lambda}{D(\lambda)} \int_a^b dy \, D(x, y; \lambda) g(y) = \lambda \int_a^b dy \, K(x, y) g(y)$$

$$+ \frac{\lambda^2}{D(\lambda)} \int_a^b dz \, K(x, z) \int_a^b dy \, D(z, y; \lambda) g(y). \tag{10.96}$$

Since $g(y)$ is an arbitrary function this implies

$$D(x, y; \lambda) = D(\lambda) K(x, y) + \lambda \int_a^b dz \, K(x, z) D(z, y; \lambda). \tag{10.97}$$

This is a general identity. Specializing by setting λ equal to an eigenvalue λ_n, for which $D(\lambda_n) = 0$, we find

$$D(x, y; \lambda_n) = \lambda_n \int_a^b dz \, K(x, z) D(z, y; \lambda_n). \tag{10.98}$$

Thus $D(x, y; \lambda_n)$ is an eigenfunction belonging to the eigenvalue λ_n. This is true for any value of the parameter y. For a nondegenerate eigenvalue (only one eigenfunction) this implies that with $\lambda = \lambda_n$, $D(x, y; \lambda)$ has the form

$$D(x, y; \lambda_n) = F_n(x) G_n(y). \tag{10.99}$$

The factor $G_n(y)$ is just a multiplicative constant; the eigenfunction is $F_n(x)$.*

From our results (10.93), (10.95), (10.98) we can read off the so-called Fredholm alternative. If λ is not an eigenvalue so that $D(\lambda) \neq 0$, (10.93) provides a unique solution for the inhomogeneous equation (10.82) and the homogeneous equation with $g = 0$ has no solution. On the other hand, for λ equal to an eigenvalue, i.e., a root of $D(\lambda) = 0$, the eigenfunction $D(x, y; \lambda_n)$ provides a solution of the homogeneous equation, but the inhomogeneous equation has no solution in general since $D(\lambda_n) = 0$ in (10.93). The inhomogeneous equation will have a solution for $\lambda = \lambda_n$ only if $\int_a^b dy \, D(x, y; \lambda_n) g(y) = 0$, i.e., only if $g(x)$ is orthogonal to the eigenfunction $D(y, x; \lambda_n)$ for the transposed kernel $\overline{K}(x, y) = K(y, x)$.[2]

---

[2]The statements in these two paragraphs are valid for nondegenerate eigenvalues. For degenerate eigenvalues, similar but more complex statements can be made.

10.9 CONDITIONS FOR VALIDITY OF FREDHOLM'S FORMULAS

The results obtained in the preceding three sections have been obtained by purely formal manipulations. No attention has been paid to questions such as the convergence of series or conditions for the validity of the limiting processes or approximation procedures used. The reader who wants a thorough treatment of these mathematical questions will have to consult the mathematical literature on the subject.[3] Here we can only indicate a few of the main points which are sometimes essential, even for the person interested strictly in applications.

The first three references listed in the footnote develop rigorously the theory of integral equations of the type

$$f(x) = g(x) + \lambda \int_a^b dy \ K(x,y)f(y), \tag{10.100}$$

when a and b are finite and $K(x,y)$ is a continuous or piecewise continuous function. In fact, the theory can be developed with less restrictive assumptions, and this is done in references 4 through 8. The limits on the integral can be infinite and the kernel is allowed to have certain types of weak singularities. The essential condition for the validity of the theory is that the kernel $K(x,y)$ be square integrable,

$$\int_a^b dx \int_a^b dy \ [K(x,y)]^2 \equiv ||K||^2 < \infty, \tag{10.101}$$

and we must also assume $g(x)$ to be square integrable,

$$\int_a^b dx \ [g(x)]^2 = ||g||^2 < \infty. \tag{10.102}$$

If the condition (10.101) is satisfied, the Fredholm series (10.87) for $D(\lambda)$ and (10.92) for $D(x,y;\lambda)$ converge for all values of λ so that $D(\lambda)$ and $D(x,y;\lambda)$ are analytic functions throughout the entire complex λ plane. The so-called resolvent

$$H(x,y;\lambda) = \frac{D(x,y;\lambda)}{D(\lambda)}, \tag{10.103}$$

which appears in the solution (10.93) of the integral equation,

$$f(x) = g(x) + \lambda \int_a^b dy \ H(x,y;\lambda)g(y), \tag{10.104}$$

[3] 1. E. T. Whittaker and G. N. Watson, op. cit., Chap. XI.
2. R. Courant and D. Hilbert, op. cit., Vol. I, Chap. III.
3. W. V. Lovitt, Linear Integral Equations (McGraw-Hill, New York, 1924).
4. F. G. Tricomi, Integral Equations (Interscience, New York, 1957).
5. S. G. Mikhlin, Integral Equations (Pergamon, Oxford, 1957).
6. F. Smithies, Integral Equations (Cambridge University Press, Cambridge, 1958).
7. F. Riesz and B. Sz.-Nagy, Functional Analysis (Ungar, New York, 1955).
8. I. Stakgold, Boundary Value Problems of Mathematical Physics (Macmillan, New York, 1967), Vol. I, Chap. III.
9. L. V. Kantorovich and V. I. Krylov, Approximate Methods of Higher Analysis (John Wiley & Sons, New York, 1964), Chap. II.

is thus the ratio of these two analytic functions. Moreover the expression (10.104) really is the unique square integrable solution of the integral equation (10.100) for λ not a zero of $D(\lambda)$, when conditions (10.101) and (10.102) are satisfied.

At the eigenvalues $\lambda = \lambda_n$, where $D(\lambda)$ has zeros, the resolvent $H(x, y; \lambda)$ (10.103) has poles. For $|\lambda| < |\lambda_0|$, where λ_0 is the eigenvalue with the smallest absolute value $|\lambda_0|$, $H(x, y; \lambda)$ is an analytic function which can be expanded in a convergent power series in λ. This power series in λ yields the iteration solution discussed in Section 10.6. Thus we see that the iteration solution power series converges for $|\lambda| < |\lambda_0|$ and diverges for $|\lambda| > |\lambda_0|$.

Finally, when the condition (10.101) is satisfied, the approximation schemes discussed in Section 10.7, in connection with numerical solution of the integral equation, are justified and we can expect higher order approximations (with larger matrices) to converge to the correct result.

On the other hand, if (10.101) is not satisfied, there is no guarantee that anything will work. The theoretical formulas may break down and the numerical procedures of Section 10.7 may diverge as the number of integration points is increased. Thus before writing a computer program to solve an integral equation one has to check the condition (10.101).

The power of the square integrability condition (10.101) derives from the fact that it ensures that the kernel $K(x, y)$ can be approximated arbitrarily closely in the mean by a separable kernel. To see this, let us expand $K(x, y)$ in some arbitrary complete orthonormal set $\phi_i(x)$ on the interval $a \leq x \leq b$:

$$\int_a^b dx \, \phi_i(x)\phi_j(x) = \delta_{ij}, \tag{10.105}$$

$$\sum_i \phi_i(x)\phi_i(x') = \delta(x - x'). \tag{10.106}$$

We find

$$K(x, y) = \sum_{i=1}^{\infty} a_i(y)\phi_i(x), \tag{10.107}$$

where the $a_i(y)$ are given by

$$a_i(y) = \int_a^b dx \, \phi_i(x)K(x, y). \tag{10.108}$$

Using the completeness relation (10.106) we then find

$$\sum_{i=1}^{\infty} \int_a^b dy \, [a_i(y)]^2 = \int_a^b dx \int_a^b dy \int_a^b dz \left[\sum_i \phi_i(x)\phi_i(z) \right] K(x, y)K(z, y)$$

$$= \int_a^b dx \int_a^b dy \, [K(x, y)]^2. \tag{10.109}$$

If the condition (10.101) is satisfied, the series of positive constants on the left of

(10.109) converges:

$$\sum_i \int_a^b dy \, [a_i(y)]^2 = \|K\|^2 < \infty. \tag{10.110}$$

On the other hand, if we consider approximating $K(x, y)$ by the first N terms on the right of (10.107), we find

$$\int_a^b dx \int_a^b dy \left[K(x, y) - \sum_{i=1}^{N} a_i(y)\phi_i(x) \right]^2 = \int_a^b dx \int_a^b dy \left[\sum_{i=N+1}^{\infty} a_i(y)\phi_i(x) \right]^2$$

$$= \sum_{i=N+1}^{\infty} \int_a^b dy \, [a_i(y)]^2. \tag{10.111}$$

Given the convergence of the series (10.110), it follows that the right-hand side of (10.111) can be made arbitrarily small by taking N sufficiently large, so that $K(x, y)$ can be approximated arbitrarily closely in the mean by the separable kernel

$$K(x, y) \approx \sum_{i=1}^{N} a_i(y)\phi_i(x). \tag{10.112}$$

In Section 10.4 we showed how to reduce the solution of an integral equation with a kernel such as this to a finite dimensional matrix problem.

Thus if the kernel is square integrable in the sense of (10.101), the integral equation is essentially reduced to a finite dimensional matrix problem for which there are no convergence problems.

10.10   HILBERT-SCHMIDT THEORY

In Hilbert-Schmidt theory we concentrate on the eigenvalue problem

$$f_n(x) = \lambda_n \int_a^b dy \, K(x, y) f_n(y) \tag{10.113}$$

and assume at the outset that $K(x, y)$ is a real symmetric kernel,

$$K(x, y) = K(y, x). \tag{10.114}$$

Consider first the numerical approximation to (10.113) by a finite dimensional matrix equation, as discussed in the first part of Section 10.7. With integration points and weights x_i and W_i, respectively, and the notation $f_i = f(x_i)$, $K_{ij} = K(x_i, x_j)$, (10.113) becomes

$$f_i = \lambda \sum_{j=1}^{N} K_{ij} W_j f_j. \tag{10.115}$$

The matrix is symmetric, i.e., $K_{ij} = K_{ji}$. To write (10.115) in a symmetric form multiply by $\sqrt{W_i}$ and introduce

$$M_{ij} = \sqrt{W_i}\, K_{ij} \sqrt{W_j} = M_{ji} \qquad (10.116)$$

$$F_i = \sqrt{W_i}\, f_i. \qquad (10.117)$$

In terms of these quantities (10.115) assumes the form

$$\sum_{j=1}^{N} M_{ij} F_j = \frac{1}{\lambda} F_i. \qquad (10.118)$$

We know from matrix theory that the eigenfunctions of a real symmetric matrix form a complete orthonormal set in the N-dimensional vector space. Suppose the eigenvalues are

$$\lambda_n, \quad n = 1, \ldots, N \qquad (10.119)$$

and the corresponding normalized eigenvectors are

$$F_i^n, \quad \text{with} \sum_{i=1}^{N} F_i^n F_i^m = \delta_{nm}. \qquad (10.120)$$

An arbitrary vector ϕ_i in the N-dimensional space can be expanded in terms of the set F_i^n:

$$\phi_i = \sum_n c_n F_i^n, \qquad (10.121)$$

$$c_n = \sum_i F_i^n \phi_i. \qquad (10.122)$$

Applying the matrix (with elements M_{ij}) to the arbitrary vector (10.121) and using the eigenvalue equation, we find

$$\sum_j M_{ij} \phi_j = \sum_n c_n \sum_j M_{ij} F_j^n = \sum_n c_n \frac{1}{\lambda_n} F_i^n.$$

This same result is obtained by applying the operator

$$\sum_n \frac{1}{\lambda_n} F_i^n F_j^n \qquad (10.123)$$

to ϕ_j and using (10.72). Thus since ϕ_i is arbitrary we obtain the expansion

$$M_{ij} = \sum_n \frac{1}{\lambda_n} F_i^n F_j^n \qquad (10.124)$$

for the symmetric matrix $\mathbb{M}$ in terms of its eigenvectors. Using (10.116) and (10.117) to return to K_{ij} and f_i, (10.124) becomes

$$\boxed{K_{ij} = \sum_n \frac{1}{\lambda_n} f_i^n f_j^n.} \qquad (10.125)$$

Let us return now to the integral equation (10.113). We must be a bit careful. For a separable kernel $K(x,y)) = u(x)u(y)$ we showed in Section 10.4 that there is only one eigenfunction $u(x)$ with a corresponding eigenvalue $\lambda = \left[\int_a^b dy\, u^2(y)\right]^{-1}$. Evidently, the eigenfunctions do not form a complete set in this case. According to the mathematical works cited in the footnote in Section 10.9, page 258, the analogue of the formula (10.125) above, the so-called bilinear formula,

$$K(x,y) = \sum_n \frac{1}{\lambda_n} f_n(x) f_n(y), \tag{10.126}$$

is not in general valid for a symmetric kernel. There is only a weaker result known as Mercer's theorem:

Theorem 10.10.1. *If the symmetric quadratically integrable kernel $K(x,y)$ is continuous and has only positive eigenvalues (or at most a finite number of negative eigenvalues), then the series*

$$\sum_{n=1}^{\infty} \frac{1}{\lambda_n} f_n(x) f_n(y) \tag{10.127}$$

converges absolutely and uniformly and the bilinear formula

$$K(x,y) = \sum_{n=1}^{\infty} \frac{1}{\lambda_n} f_n(x) f_n(y) \tag{10.128}$$

holds.

If the bilinear expansion is valid, then it is obvious that any function of the form

$$\psi(x) = \int_a^b dy\, K(x,y)\phi(y) \tag{10.129}$$

can be expanded in terms of the eigenfunctions $f_n(x)$:

$$\psi(x) = \sum_n c_n f_n(x), \tag{10.130}$$

$$c_n = \frac{1}{\lambda_n} \int_a^b dy\, f_n(x)\phi(y). \tag{10.131}$$

Fortunately the result is valid even in the more general case when the bilinear formula breaks down. According to the Hilbert-Schmidt theorem:

Theorem 10.10.2. *If $\psi(x)$ can be written in the form*

$$\psi(x) = \int_a^b dy\, K(x,y)\phi(y), \tag{10.132}$$

where $K(x,y)$ is a symmetric quadratically integrable kernel and $\phi(y)$ is a quadratically integrable function, then $\psi(x)$ can be expanded in the orthonormal set of eigenfunctions $f_n(x)$ of $K(x,y)$, i.e.,

$$\psi(x) = \sum_{n=1}^{\infty} c_n f_n(x), \tag{10.133}$$

with

$$c_n = \int_a^b dx\ f_n(x)\psi(x).$$ (10.134)

Moreover, if

$$\int_a^b dy\ [K(x,y)]^2 = A^2(x) \le const.$$ (10.135)

is bounded, the series (10.133) *converges absolutely and uniformly.*[4]

We can use these eigenfunction expansions to solve the inhomogeneous equation

$$f(x) = g(x) + \lambda \int_a^b dy\ K(x,y)f(y)$$ (10.136)

in terms of the eigenfunctions $f_n(x)$ which satisfy

$$f_n(x) = \lambda_n \int_a^b dy\ K(x,y)f_n(y),$$ (10.137)

$$\int_a^b dy\ f_n(y)f_m(y) = \delta_{nm}.$$ (10.138)

Since $f(x) - g(x)$ is, according to (10.136), a function of the type (10.132), we can use the Hilbert-Schmidt theorem to expand it in the form

$$f(x) - g(x) = \sum_n c_n f_n(x),$$ (10.139)

where c_n is given by

$$c_n = \int_a^b dx\ f_n(x)[f(x) - g(x)]$$ (10.140)

$$= \int_a^b dx\ f_n(x)\lambda \int_a^b dy\ K(x,y)f(y)$$

$$= \lambda \int_a^b dy\ \left\{ \int_a^b dx\ f_n(x)K(x,y) \right\} f(y).$$ (10.141)

Using (10.137) and the assumed symmetry property (10.114), this becomes

$$c_n = \frac{\lambda}{\lambda_n} \int_a^b dy\ f_n(y)f(y)$$

$$= \frac{\lambda}{\lambda_n} \int_a^b dy\ f_n(y)[f(y) - g(y) + g(y)]$$

$$= \frac{\lambda}{\lambda_n} c_n + \frac{\lambda}{\lambda_n} \int_a^b dy\ f_n(y)g(y),$$ (10.142)

---

[4]Proofs of Mercer's theorem and the Hilbert-Schmidt theorem are given, for example, in Tricomi, op. cit., pp. 124, 110.

or

$$(\lambda_n - \lambda)c_n = \lambda \int_a^b dy \; f_n(y)g(y). \qquad (10.143)$$

If $\lambda \neq \lambda_n$, we find

$$c_n = \frac{\lambda}{\lambda_n - \lambda} \int_a^b dy \; f_n(y)g(y), \qquad (10.144)$$

and the solution of the integral equation (10.136) is

$$\boxed{f(x) = g(x) + \sum_n f_n(x) \frac{\lambda}{\lambda_n - \lambda} \int_a^b dy \; f_n(y)g(y).} \qquad (10.145)$$

If $\lambda =$ one of the eigenvalues λ_m, we see from (10.143) that there is no solution of the inhomogeneous equation unless $g(y)$ is orthogonal to each of the eigenfunctions $f_m(y)$ belonging to that eigenvalue:

$$\int_a^b dy \; f_m(y)g(y) = 0. \qquad (10.146)$$

If (10.146) is satisfied, the c_m are undetermined by (10.143) for this eigenvalue. Thus for $\lambda = \lambda_m$ and (10.146) satisfied, the solution of the inhomogeneous integral equation is

$$\boxed{f(x) = g(x) + \sum_{\lambda_n \neq \lambda_m} f_n(x) \frac{\lambda_m}{\lambda_n - \lambda_m} \int_a^b dy \; f_n(y)g(y) + \sum_{\lambda_n = \lambda_m} c_n f_n(x),} \qquad (10.147)$$

where the c_n with $\lambda_n = \lambda_m$ are arbitrary.

As a final result, comparison of (10.93) and (10.145) yields an eigenfunction expansion for the Fredholm resolvent of a symmetric kernel:

$$\boxed{H(x, y; \lambda) = \frac{D(x, y; \lambda)}{D(\lambda)} = \sum_{n=1}^{\infty} \frac{f_n(x)f_n(y)}{\lambda_n - \lambda}.} \qquad (10.148)$$

PROBLEMS

10.1 Solve the integral equation

$$f(x) = 1 + \lambda \int_0^1 dy \; xy^2 f(y)$$

in three ways:

a) using the method for separable kernels,

b) using the Fredholm series,

c) by integration. Expand the result obtained in (a) or (b) in powers of λ and compare with the iteration solution. For what values of λ does the iteration series converge?

Find the eigenvalues and eigenfunctions for the integral equation

$$f(x) = \lambda \int_0^1 dy \; xy^2 f(y)$$

in two ways:

a) using the method for separable kernels,

b) from the Fredholm series.

10.2 Solve the integral equation

$$f(x) = 1 + \lambda \int_0^1 dy \; [xy + x^3 y^2] f(y)$$

in two ways:

a) using the method for separable kernels,

b) using the Fredholm series.

Find the eigenvalues and eigenfunctions for the integral equation

$$f(x) = \lambda \int_0^1 dy \; [xy + x^3 y^2] f(y)$$

by the methods of (a) and (b).

10.3 Find the eigenfunctions and eigenvalues for the integral equation

$$f(x) = \lambda \int_0^{2\pi} dy \; \sin(x + y) f(y).$$

10.4 Find the solutions of the integral equations

a) $f(x) = e^{-|x|} + \lambda \int_{-\infty}^{\infty} dy \; e^{-|x-y|} f(y), \quad \lambda < \frac{1}{2},$

b) $f(x) = e^{-x^2} + \lambda \int_{-\infty}^{\infty} dy \ e^{-(x-y)^2} f(y), \quad \lambda < \dfrac{1}{\sqrt{\pi}}.$

10.5 Find the solution of the integral equation

$$f(x) = g(x) + \lambda \int_0^{\infty} dy \ \cos(xy) f(y).$$

10.6 Show that the Volterra equation

$$f(x) = g(x) + \lambda \int_0^x dy \ u(x)v(y)f(y),$$

where $g(x)$, $u(x)$ and $v(x)$ are given and $f(x)$ is unknown, can be converted to a differential equation. Apply this method to solve the equation

$$f(x) = x + \lambda \int_0^x dy \ xy^2 f(y).$$

Find the eigenfunctions and eigenvalues, if any, for the associated equation

$$f(x) = \lambda \int_0^x dy \ xy^2 f(y).$$

10.7 Find a solution of the integral equation of the first kind

$$\int_a^b dy \ K(x,y)f(y) = \phi(x),$$

where $\phi(x)$ is a given function and $K(x,y)$ is symmetric, by expanding $\phi(x)$ and $f(y)$ in the eigenfunctions of $K(x,y)$.

10.8 Show that the Sturm-Liouville eigenvalue problem on the interval $a < x < b$ defined by the differential equation

$$\mathcal{L}u_n \equiv \frac{d}{dx}\left[p(x)\frac{du_n(x)}{dx}\right] - q(x)u_n(x) = -\lambda_n \rho(x)u_n(x)$$

and the boundary conditions

$$Au_n(a) + B\frac{du_n(x)}{dx}\bigg|_{x=a} = 0,$$

$$Cu_n(b) + D\frac{du_n(x)}{dx}\bigg|_{x=b} = 0,$$

is equivalent to the integral equation

$$u_n(x) = -\lambda_n \int_a^b dx' \ G(x,x')\rho(x')u_n(x'),$$

where the Green's function $G(x, x')$ is determined by the differential equation

$$\mathcal{L}G(x, x') = \delta(x - x')$$

and the boundary conditions

$$AG(a, x') + B\frac{dG(x, x')}{dx}\bigg|_{x=a} = 0,$$

$$CG(b, x') + D\frac{dG(x, x')}{dx}\bigg|_{x=b} = 0.$$

Note that the boundary conditions are incorporated in the integral equation formulation of the eigenvalue problem and need not be separately stated. Mathematical treatments of the Sturm-Liouville eigenvalue problem often proceed by converting it to an integral equation since it is easier to provide the necessary theorems for the integral equations.

10.9   a) Convert the eigenvalue problem for the stretched string,

$$\frac{d^2u_n(x)}{dx^2} = -k_n^2 u_n(x),$$

$$u_n(0) = u_n(l) = 0,$$

to an integral equation. What are the solutions of the integral equation?

b) Convert the eigenvalue problem for the radial functions for the drumhead,

$$\frac{d^2u_n(r)}{dr^2} + \frac{1}{r}\frac{du_n(r)}{dr} - \frac{m^2}{r^2}u_n(r) = -k_n^2 u_n(r),$$

$$u_n(0) \text{ finite},$$
$$u_n(a) = 0,$$

to an integral equation. What are the solutions of the integral equation?

c) Convert the eigenvalue problem for the radial functions for the sound oscillations in a spherical cavity,

$$\frac{1}{r}\frac{d^2}{dr^2}[ru_n(r)] - \frac{l(l+1)}{r^2}u_n(r) = -k_n^2 u_n(r),$$

$$u_n(0) \text{ finite},$$
$$\frac{du_n(r)}{dr}\bigg|_{r=a} = 0,$$

to an integral equation. What are the solutions of the integral equation?

Note that for the case $l = 0$ you need the generalized Green's function of Problem 8.1(b).

The following two problems will require a great deal of time and effort as well as the availability of a computer.

10.10 A relativistic Schrödinger equation for the bound states of two particles of masses m_1 and m_2 has the form in momentum space

$$\psi(\mathbf{p}) = \int \frac{d^3 p'}{(2\pi)^3} \, V(\mathbf{p}, \mathbf{p}') \frac{1}{W - E_1' - E_2'} \psi(\mathbf{p}').$$

Here we work in the center of mass system so that the momenta of the two particles are $\mathbf{p}$ and $-\mathbf{p}$ or $\mathbf{p}'$ and $-\mathbf{p}'$ and their kinetic energies are

$$E_1 = \sqrt{p^2 + m_1^2},$$
$$E_2 = \sqrt{p^2 + m_2^2},$$
$$E_1' = \sqrt{p'^2 + m_1^2},$$
$$E_2' = \sqrt{p'^2 + m_2^2}.$$

The bound state energy eigenvalue of the system is W. The potential $V(\mathbf{p}, \mathbf{p}')$, due to exchange of a particle of mass μ, is

$$V(\mathbf{p}, \mathbf{p}') = -\frac{g^2}{4} \frac{m^2}{\sqrt{E_1 E_2 E_1' E_2'}} \frac{1}{|\mathbf{p} - \mathbf{p}'|^2 + \mu^2},$$

where g is a dimensionless coupling constant. Because of the relativistic kinematic factors the Schrödinger equation above cannot be converted to a differential equation.

a) Make an angular momentum decomposition of the potential,

$$V(\mathbf{p}, \mathbf{p}') = \sum_l (2l + 1) V_l(p, p') P_l(\cos \Theta),$$

where Θ is the angle between $\mathbf{p}$ and $\mathbf{p}'$, $\cos \Theta = \frac{\mathbf{p} \cdot \mathbf{p}'}{pp'}$. Show that

$$V_l(p, p') = -\frac{g^2}{4} \frac{m^2}{\sqrt{E_1 E_2 E_1' E_2'}} \frac{1}{2pp'} Q_l \left[\frac{p^2 + p'^2 + \mu^2}{2pp'} \right],$$

where $Q_l(z)$ is the second solution of Legendre's equation, discussed in Section 3.4:

$$Q_0(z) = \frac{1}{2} \ln \frac{z+1}{z-1},$$

$$Q_1(z) = \frac{z}{2} \ln \frac{z+1}{z-1} - 1,$$

$$Q_2(z) = \frac{1}{4}(3z^2 - 1) \ln \frac{z+1}{z-1} - \frac{3}{2} z.$$

Show that the Schrödinger equation for a bound state of angular momentum l,

$$\psi(\mathbf{p}) = \psi_l(p) Y_l^m(\theta, \varphi),$$

reduces to the one-dimensional integral equation

$$\psi_l(p) = \frac{1}{2\pi^2} \int_0^\infty p'^2 dp' \, V_l(p, p') \frac{1}{W - E_1' - E_2'} \psi_l(p').$$

Define

$$\psi_l'(p) = \sqrt{\frac{pE_1E_2}{E_1 + E_2}} \, \psi_l(p),$$

$$V_l'(p, p') = -\frac{1}{2\pi} \sqrt{\frac{pE_1E_2}{E_1 + E_2}} \sqrt{\frac{p'E_1'E_2'}{E_1' + E_2'}} \, V_l(p, p'),$$

and write the Schrödinger equation in the more symmetric form

$$\psi_l'(p) = -\frac{1}{\pi} \int_{m_1+m_2}^\infty d(E_1' + E_2') \, V_l'(p, p') \frac{1}{W - E_1' - E_2'} \psi_l'(p').$$

b) Solve the integral equation of part (a) numerically on a computer by converting it to a matrix equation. Consider the special case $m_1 = m_2 = m$, $\mu = fm$, $f = 1/4$. Change variables to x defined by

$$E = m \left[1 + \frac{\alpha x}{1 - x} \right], \quad 0 \le x \le 1.$$

The parameter α can be chosen to facilitate the numerical work. Use Gaussian integration points for the x variable with an order of 10-20 points for the intervals $0 \le x \le 1$. It is easiest to generate curves of eigenvalue W vs. coupling constant g by picking W and solving the integral equation for g regarded as eigenvalue. Notice that for the bound state problem $W < 2m$ so that the energy denominator in the integral equation never vanishes. Generate curves of W vs. g for several bound states for each angular momentum value $l = 0, 1, 2$.

10.11 The scattering problem ($W > 2m$) for the system of Problem 9.10 is described by the integral equation

$$K_l'(p, q) = V_l'(p, q) - \frac{1}{\pi} \mathcal{P} \int_{m_1+m_2}^\infty d(E_1' + E_2') V_l'(p, p') \frac{1}{W - E_1' - E_2'} K_l'(p', q).$$

Here q, which enters the integral equation as a parameter, is the magnitude of the relative momentum of the particles before scattering and the total energy W is related to q:

$$W = \sqrt{q^2 + m_1^2} + \sqrt{q^2 + m_2^2}.$$

After the integral equation has been solved for $K_l'(p, q)$ as a function of p, the phase shift $\delta_l(q)$ can be obtained from

$$\tan \delta_l(q) = K_l'(q, q).$$

a) The capital $\mathcal{P}$ in the integral equation above indicates that the integral over the singularity when $E_1' + E_2' = W$ is to be taken as the principal value integral. This singularity prohibits directly solving the integral equation by the matrix inversion method. Show that this problem can be avoided by introducing a subsidiary function $U_l(p, q)$ which satisfies the integral equation

$$U_l(p, q) = V_l'(p, q) - \frac{1}{\pi} \int_{m_1+m_2}^{\infty} d(E_1' + E_2') \frac{V(p, p') - V(p, q)}{W - E_1' - E_2'} U(p', q),$$

and that $K_l'(p, q)$ is related to $U_l(p, q)$ by

$$K_l(p, q) = \frac{U_l(p, q)}{1 + \frac{1}{\pi} \int_{m_1+m_2}^{\infty} d(E_1' + E_2') \frac{1}{W - E_1' - E_2'} U_l(p', q)}.$$

b) Solve the integral equation of part (a) numerically on a computer by converting it to a matrix equation. Consider the same special case as in Problem 9.10(b) and make the same change of variable. Find the phase shifts δ_l for $l = 0, 1, 2$ as functions of the energy W for several interesting values of the coupling constant g.

III

Complex Variable Techniques

CHAPTER 11

Complex Variables; Basic Theory

11.1 INTRODUCTION

In the previous sections of this book we have, by and large, used only real variable theory, although we have assumed the reader knows the meaning of the symbols and relations

$$i = \sqrt{-1}, \quad i^2 = -1 \tag{11.1}$$

and

$$e^{ix} = \cos x + i \sin x. \tag{11.2}$$

In this and the succeeding chapters we give a brief introduction to the theory of analytic functions of a complex variable and its applications in physics. An analytic function of a complex variable is one which has a unique derivative. As we shall see in more detail below, this simple requirement picks out of the totality of functions of a complex variable a very restricted subset of functions which have fantastic, almost incredible, properties.

Fortunately, most of the functions of interest in physics belong to this restricted set and so have these beautiful properties which can then be exploited in practical calculations. These applications range from using the residue theorem as a trick for doing integrals to the S matrix theory of elementary particle interactions, where the function theoretic assumptions form a basic ingredient of the physical theory.

As in previous sections of this book we shall minimize the rigor in our treatment of the subject. The student interested in rigorous proofs should consult the mathematics department or one of the texts listed below. In quoting exact theorems we shall often use the little books by Knopp as a convenient, concise source. The physics student should realize that he can use a good mathematics book as a gold mine of exactly stated theorems, even if he is not interested in the rigorous proofs of the theorems. The book by Titchmarsh listed below is, for example, often used in this way by physicists.

We append here a partial bibliography of books on complex variable theory. Other works can be found consulting the bibliographies of the books listed.

References on Complex Variables

- L. V. Ahlfors, Complex Analysis (McGraw-Hill, New York, 1953).

- G. F. Carrier, M. Krook and C. E. Pearson, Functions of a Complex Variable (McGraw-Hill, New York, 1966).

- R. V. Churchill, Complex Variables and Applications (McGraw-Hill, New York, 1960).

- E. T. Copson, An Introduction to the Theory of Functions of a Complex Variable (Oxford, London, 1935).

- J. Cunningham, Complex Variable Methods in Science and Technology (Van Nostrand, London, 1965).

- P. Dennery and A. Krzywicki, Mathematics for Physicists (Harper and Row, New York, 1967).

- E. Hille, Analytic Function Theory, Vols. I and II (Ginn and Co., Boston, 1959).

- K. Knopp, Elements of the Theory of Functions, Theory of Functions, Parts I and II, English translation by F. Bagemihl (Dover, New York, 1952).

- A. Kyrala, Applied Functions of a Complex Variable (Wiley, New York, 1972).

- E. C. Titchmarsh, The Theory of Functions (Oxford, London, 1939).

- H. Wayland, Complex Variables Applied in Science and Engineering (Van Nostrand Reinhold, New York, 1970).

- E. T. Whittaker and G. N. Watson, A Course of Modern Analysis (Cambridge, London, 1952).

11.2 ANALYTIC FUNCTIONS; THE CAUCHY-RIEMANN EQUATIONS

We consider functions $f(z)$ of a complex variable

$$z = x + iy$$
$$= r\cos\theta + ir\sin\theta$$
$$= re^{i\theta}. \tag{11.3}$$

The values of z are in one-to-one correspondence with the points in a plane (the z plane) as shown in Figure 11.1, and it is often convenient to visualize our considerations geometrically. It is sometimes useful to decompose f into its real and imaginary parts,

$$f(z) = u(x,y) + iv(x,y), \tag{11.4}$$

whence we see that in general we are dealing with two real functions u and v of two real variables x and y.

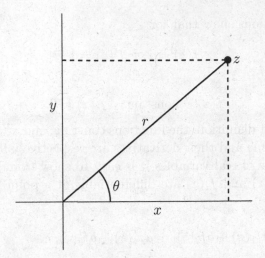

Figure 11.1: Point, z, on the complex plane

We now greatly restrict the class of functions to those which are analytic.

Definition 11.2.1. A function is said to be analytic (or regular) in a region R if it is defined and differentiable at each point of the region.

Definition 11.2.2. The function is differentiable at the point z if the limit

$$\lim_{\zeta \to z} \frac{f(\zeta) - f(z)}{\zeta - z} \tag{11.5}$$

exists and is independent of the manner in which $\zeta \to z$.

This derivative is designated by $f'(z)$ or $\frac{df(z)}{dz}$. Technically speaking, given an arbitrary $\epsilon > 0$, there must exist a δ such that

$$\left| \frac{f(\zeta) - f(z)}{\zeta - z} - f'(z) \right| < \epsilon \tag{11.6}$$

for all ζ in the region $|\zeta - z| < \delta$.

As a simple example of the definition we can consider

$$f(z) = z^2. \tag{11.7}$$

Then we find

$$\frac{f(\zeta) - f(z)}{\zeta - z} = \frac{\zeta^2 - z^2}{\zeta - z} = \zeta + z. \tag{11.8}$$

Evidently we have

$$|\zeta + z - 2z| = |\zeta - z| < \epsilon, \tag{11.9}$$

provided $|\zeta - z| < \delta = \epsilon$, so that the derivative exists for this case and is given by

$$f'(z) = 2z. \tag{11.10}$$

Even simpler calculations show that for

$$f(z) = z, \quad f'(z) = 1 \tag{11.11}$$

and for

$$f(z) = \text{constant}, \quad f'(z) = 0. \tag{11.12}$$

Thus, according to our definition, the functions const., z, and z^2 are analytic throughout the complex z plane and their derivatives are respectively 0, 1, and $2z$.

Just as in the case of real variables it is easy to show from the definition that if the two functions $f_1(z)$ and $f_2(z)$ are differentiable at a point z and if a and b are two constants then

$$\frac{d}{dz}\left[af_1(z) + bf_2(z)\right] = af_1'(z) + bf_2'(z), \tag{11.13}$$

$$\frac{d}{dz}\left[f_1(z)f_2(z)\right] = f_1(z)f_2'(z) + f_1'(z)f_2(z), \tag{11.14}$$

$$\frac{d}{dz}\left[\frac{f_1(z)}{f_2(z)}\right] = \frac{f_2(z)f_1'(z) - f_1(z)f_2'(z)}{f_2^2(z)}, \quad f_2(z) \neq 0, \tag{11.15}$$

For example, to check (11.14) we apply the condition (11.6):

$$
\begin{aligned}
f_1(\zeta)f_2(\zeta) - f_1(z)f_2(z) &= f_1(\zeta)\left[f_2(\zeta) - f_2(z)\right] + f_2(z)\left[f_1(\zeta) - f_1(z)\right] \\
&= f_1(z)\left[f_2(\zeta) - f_2(z)\right] + f_2(z)\left[f_1(\zeta) - f_1(z)\right] \\
&\quad + \left[f_1(\zeta) - f_1(z)\right]\left[f_2(\zeta) - f_2(z)\right],
\end{aligned} \tag{11.16}
$$

so that

$$\left| \frac{f_1(\zeta)f_2(\zeta) - f_1(z)f_2(z)}{\zeta - z} - f_1(z)f_2'(z) - f_1'(z)f_2(z) \right|$$

$$= \left| f_1(z)\left\{ \frac{f_2(\zeta) - f_2(z)}{\zeta - z} - f_2'(z) \right\} + f_2(z)\left\{ \frac{f_1(\zeta) - f_1(z)}{\zeta - z} - f_1'(z) \right\} \right.$$

$$\left. + (\zeta - z)\frac{f_1(\zeta) - f_1(z)}{\zeta - z}\frac{f_2(\zeta) - f_2(z)}{\zeta - z} \right|. \tag{11.17}$$

Given that $f_1(z)$ and $f_2(z)$ have derivatives, we see that the three terms inside the absolute value on the right can be made arbitrarily small by keeping ζ sufficiently close to z. This verifies the formula (11.14).

Using our previous results (11.10)-(11.12) and applying the formula (11.14) repeatedly, we can establish the general formula

$$\boxed{\frac{d}{dz}z^n = nz^{n-1}} \tag{11.18}$$

for n any positive integer. All results so far are the same as in the case of real variables. This similarity can be extended to the whole class of rational functions, i.e., ratios of two polynomials. First, applying (11.13) and (11.18) we see that any polynomial

$$P(z) = a_0 + a_1 z + \ldots a_N z^N$$

$$= \sum_{n=0}^{N} a_n z^n \tag{11.19}$$

is analytic throughout the entire z plane, and its derivative is given by the usual formula

$$P'(z) = \sum_{n=1}^{N} n a_n z^{n-1}. \tag{11.20}$$

Applying (11.15) we can extend our results to the ratio of two polynomials

$$R(z) = \frac{P(z)}{Q(z)}, \tag{11.21}$$

with $P(z)$ given above and $Q(z)$ is a different polynomial,

$$Q(z) = \sum_{m=0}^{M} b_m z^m. \tag{11.22}$$

We see that $R(z)$ is analytic throughout the z plane <u>except at the points where $Q(z) = 0$</u>, i.e., at the zeros of the denominator, and the derivative of $R(z)$ is given by the usual formulas of real variable theory.

Near one of the zeros z_0 of the denominator of (11.21), $R(z)$ will behave like

$$R(z) \sim F(z) = \frac{c}{(z - z_0)^p}, \tag{11.23}$$

where p is an integer which determines the order of the zero of $Q(z)$. $R(z)$ is not defined at $z = z_0$. Consider for example the simple case $z_0 = 0$, $p = 1$, $F(z) = 1/z$, and approach the point $z = 0$ from the four directions $z = \pm r$, $z = \pm ir$, $r \to 0$. We find

$$\frac{1}{z} \to \begin{cases} \dfrac{1}{r} \to +\infty \\[2ex] \dfrac{1}{-r} \to -\infty \\[2ex] \dfrac{1}{ir} \to -i\infty \\[2ex] \dfrac{1}{-ir} \to +i\infty \end{cases} . \tag{11.24}$$

Thus at $z = z_0$, $R(z)$ is not defined and so according to the definition given above is not analytic. We say there is a <u>singularity</u> at $z = z_0$. The particular kind of singularity given by (11.23) is called a <u>pole of order p</u>. Of course, the function $F(z)$ of (11.23) is analytic in any region excluding the pole. Applying the formula (11.15) to calculate its derivative in such a region, we find

$$\frac{d}{dz}\frac{c}{(z-z_0)^p} = \frac{-pc}{(z-z_0)^{p+1}}. \tag{11.25}$$

For the rational functions we have used directly the definition (11.5), (11.6) of analyticity. It is also possible to translate this definition into conditions on the real and imaginary parts $u(x,y)$ and $v(x,y)$ of the function $f(z)$,

$$f(z) = u(x,y) + iv(x,y). \tag{11.26}$$

The derivative is defined by

$$\frac{df(z)}{dz} = \lim_{h\to 0}\frac{f(z+h)-f(z)}{h}, \tag{11.27}$$

and for an analytic function is independent of the manner in which $h \to 0$. Writing h in terms of its real and imaginary parts,

$$h = h_1 + ih_2, \tag{11.28}$$

and using (11.26), we can rewrite (11.27) as

$$\frac{df(z)}{dz} = \lim_{\substack{h_1\to 0 \\ h_2\to 0}}\frac{1}{h_1+ih_2}\left[u(x+h_1,y+h_2)-u(x,y)+i\{v(x+h_1,y+h_2)-v(x,y)\}\right]. \tag{11.29}$$

By setting $h_2 = 0$ and letting h approach 0 along the real axis, we obtain

$$\frac{df(z)}{dz} = \frac{\partial u(x,y)}{\partial x} + i\frac{\partial v(x,y)}{\partial x}. \tag{11.30}$$

By setting $h_1 = 0$ and letting h approach 0 along the imaginary axis, we obtain a different expression:

$$\frac{df(z)}{dz} = \frac{\partial v(x,y)}{\partial y} - i\frac{\partial u(x,y)}{\partial y}. \tag{11.31}$$

For an analytic function these two ways of approaching the limit must lead to the same answer. Equating (11.30) and (11.31) we then find the Cauchy-Riemann equations:

$$\boxed{\frac{\partial u(x,y)}{\partial x} = \frac{\partial v(x,y)}{\partial y},} \tag{11.32}$$

$$\boxed{\frac{\partial u(x,y)}{\partial y} = -\frac{\partial v(x,y)}{\partial x}.} \tag{11.33}$$

The above calculation demonstrates the necessity of the Cauchy-Riemann equations for an analytic function. To demonstrate the sufficiency, we use the formulas from real variable theory for total differentials:

$$du = \frac{\partial u}{\partial x}dx + \frac{\partial u}{\partial y}dy, \tag{11.34}$$

$$dv = \frac{\partial v}{\partial x}dx + \frac{\partial v}{\partial y}dy. \tag{11.35}$$

Then we find

$$df(z) = du + idv$$

$$df(z) = \frac{\partial u}{\partial x}dx + \frac{\partial u}{\partial y}dy + i\left\{\frac{\partial v}{\partial x}dx + \frac{\partial v}{\partial y}dy\right\}. \tag{11.36}$$

Using the Cauchy-Riemann equations (11.32) and (11.33), this can be rewritten in the form

$$df(z) = \frac{\partial u}{\partial x}(dx + idy) + i\frac{\partial v}{\partial x}(dx + idy), \tag{11.37}$$

so that

$$\frac{df(z)}{dz} = \frac{\partial u}{\partial x} + i\frac{\partial v}{\partial x}, \tag{11.38}$$

independently of the way in which $dz = dx + idy \to 0$. A careful treatment along the lines above shows that one must assume the continuity of the partial derivatives of u and v as well as their existence and the validity of the Cauchy-Riemann equations in order to ensure that $f = u + iv$ is an analytic function.[1]

As a simple example consider the complex conjugate of z,

$$f(z) = z^* = x - iy. \tag{11.39}$$

With $u = x$, $v = -y$ the Cauchy-Riemann equation (11.32) is not satisfied, so that z^* is not an analytic function of z. Neither is the absolute square

$$|z|^2 = zz^* = x^2 + y^2. \tag{11.40}$$

In this case $u = x^2 + y^2$, $v = 0$, and neither equation (11.32) nor (11.33) is satisfied. In a similar way we find that the phase of z,

$$\phi(z) = \tan^{-1}\frac{y}{x}, \tag{11.41}$$

is not analytic.

---

[1]See, for example, Titchmarsh, op. cit. p. 66, Knopp; op. cit., Part I, p. 30.

Returning to the case of analytic functions, for which the Cauchy-Riemann equations (11.32), (11.33) are satisfied, we find by differentiating these equations once more and combining results

$$\frac{\partial^2 u}{\partial x^2} + \frac{\partial^2 u}{\partial y^2} = 0, \tag{11.42}$$

$$\frac{\partial^2 v}{\partial x^2} + \frac{\partial^2 v}{\partial y^2} = 0. \tag{11.43}$$

Thus neither the real nor the imaginary part of an analytic function can be chosen arbitrarily. Each must satisfy Laplace's equation and together they must satisfy the Cauchy-Riemann equations. The restricted character and special properties of this class of functions are beginning to emerge; there is much more to follow.

11.3 POWER SERIES

Many analytic functions are defined by means of power series; examples are the exponential, the Bessel functions, and the hypergeometric functions. We begin our discussion with some general considerations on infinite series and their convergence. Our treatment will be incomplete and we shall quote some theorems without proof. The reader interested in proofs will find an elementary but complete treatment in the Elements of the Theory of Functions by Knopp listed in the bibliography in Section 11.1 or in the introductory sections of the other books listed there.

A series

$$\sum_{n=0}^{\infty} c_n \tag{11.44}$$

is said to converge to a limit f if, given an arbitrary $\epsilon > 0$, there exists an integer N_0 such that

$$\left| \sum_{n=0}^{N} c_n - f \right| < \epsilon \tag{11.45}$$

for all $N \geq N_0$. If this condition is satisfied, we write

$$f = \sum_{n=0}^{\infty} c_n. \tag{11.46}$$

An equivalent convergence criterion due to Cauchy eliminates the limit f from the definition:

Definition 11.3.1. A necessary and sufficient condition for the series (11.44) to converge to a limit is that, given an arbitrary $\epsilon > 0$, there exists an N_0 such that

$$\left| \sum_{n=N}^{N+p} c_n \right| = |c_N + c_{N+1} + \ldots + c_{N+p}| < \epsilon \tag{11.47}$$

for all $N > N_0$ and all integers $p \geq 0$.

Definition 11.3.2. The series (11.44) is said to be absolutely convergent if the series of absolute values

$$\sum_{n=0}^{\infty} |c_n| \tag{11.48}$$

converges. Since

$$|c_N + c_{N+1} + \ldots + c_{N+p}| \leq |c_N| + |c_{N+1}| + \ldots + |c_{N+p}|, \tag{11.49}$$

it follows that if a series is absolutely convergent, it is also convergent. The converse is not true.

Using the inequality (11.49) we can easily derive the so-called comparison test:

Theorem 11.3.1. *If $\sum S_n$ is a convergent series of positive real terms and if $|c_n| < S_n$, the series $\sum_n c_n$ is absolutely convergent.*

From the comparison test we can derive the ratio test:

Theorem 11.3.2. *Assuming the limit exists[2], let*

$$\lim_{n \to \infty} \left| \frac{c_{n+1}}{c_n} \right| = \lambda. \tag{11.50}$$

If $\lambda < 1$, $\sum_n c_n$ converges absolutely; if $\lambda > 1$, $\sum_n c_n$ diverges.

Proof. If $\lambda > 1$, $|c_n|$ grows with increasing n; this contradicts (11.47) for $p = 0$ which says $c_n \xrightarrow[n \to \infty]{} 0$. If $\lambda < 1$, then if we take N sufficiently large, we can find a $\lambda' > \lambda$ but still satisfying $\lambda' < 1$, such that

$$|c_N| + |c_{N+1}| + \ldots + |c_{N+p}| < |c_N| \left[1 + \lambda' + \lambda'^2 + \ldots + \lambda'^p \right]$$

$$= |c_N| \frac{1 - \lambda'^{p+1}}{1 - \lambda'}$$

$$\leq |c_N| \frac{1}{1 - \lambda'}. \tag{11.51}$$

Since $|c_N| \xrightarrow[N \to \infty]{} 0$, this can be made arbitrarily small for $N >$ some N_0 and all p. $\qquad \square$

---

[2]For the case when there are two or more limit points of $|c_{n+1}/c_n|$ see for example Bromwich, <u>An Introduction to the Theory of Infinite Series</u> (Macmillan, London, 1926), p. 36.

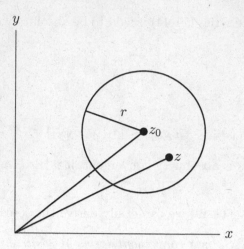

Figure 11.2: Circle of convergence about point z_0

Let us apply these results to a power series by setting $c_n = a_n(z - z_0)^n$:

$$\sum c_n = \sum a_n(z - z_0)^n. \qquad (11.52)$$

According to the ratio test, this series will converge for z such that

$$\lim_{n \to \infty} \left| \frac{a_{n+1}(z - z_0)^{n+1}}{a_n(z - z_0)^n} \right| = |z - z_0| \lim_{n \to \infty} \left| \frac{a_{n+1}}{a_n} \right| < 1, \qquad (11.53)$$

i.e., when

$$|z - z_0| < r, \quad r = \frac{1}{\displaystyle\lim_{n \to \infty} \left| \frac{a_{n+1}}{a_n} \right|}.^3 \qquad (11.54)$$

Thus for r a finite nonzero number the power series (11.52) converges for all z inside a circle of radius r about the point z_0 as shown in Figure 11.2.

This is the circle of convergence of the power series and r is the radius of convergence of the power series. For $|z - z_0| < r$

$$f(z) = \sum_{n=0}^{\infty} a_n(z - z_0)^n \qquad (11.55)$$

is a well-defined function of z. For $|z - z_0| > r$ the series diverges and is probably useless. The series may or may not converge for points $|z - z_0| = r$ on the circle of convergence; it may converge for some points on the circle but not for others. For $r = 0$ the series converges nowhere. For $r = \infty$ the series converges in the entire z plane.

---

[3]This discussion assumes $\lim_{n \to \infty} \left| \frac{a_{n+1}}{a_n} \right|$ exists. If there are two or more points, r^{-1} lies between the maximum and minimum limit points. A better formula in this case is $r^{-1} = \overline{\lim} \sqrt[n]{|a_n|}$, where $\overline{\lim}$ is the maximum limit point. See Knopp, op. cit., Elements, p. 84.

Inside its circle of convergence a power series has simple, in fact ideal, properties. Consider a power series,

$$f(z) = \sum_{n=0}^{\infty} a_n z^n, \qquad (11.56)$$

with radius of convergence r, i.e., (11.56) converges for $|z| < r$. (There is nothing special about taking the center $z_0 = 0$. It can be shifted to any point z_0 by replacing $z \to z - z_0$ in (11.56) and below.) We quote some theorems.

Theorem 11.3.3. *The power series (11.56) is unique, i.e., if two power series $\sum a_n z^n$ and $\sum b_n z^n$ are both convergent for $|z| < r$ and if they have the same sum for all these z, or even if they merely have the same sum at an infinite number of distinct points which cluster about 0 as a limit point, then the series are identical, i.e., $a_n = b_n$.*

Theorem 11.3.4. *The function $f(z)$ represented by (11.56) can also be developed in a unique power series about any other point z_1 in the interior of the circle of convergence (as shown in Figure 11.3),*

$$f(z) = \sum_{n=0}^{\infty} b_n (z - z_1)^n. \qquad (11.57)$$

The radius of convergence r_1 of this second series is at least $r_1 = r - |z_1|$. The coefficients b_n are found by setting $z = z_1 + z - z_1$ in (11.56), expanding in powers of $(z - z_1)$ and regrouping the terms.

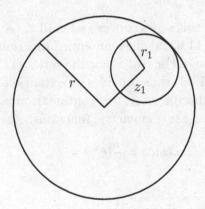

Figure 11.3: Convergence at point z_1 within circle of convergence of radius r

Theorem 11.3.5. *The function $f(z)$ represented by (11.56) is differentiable at every point z_1 inside its circle of convergence, and the derivative can be obtained by differentiating the series term-by-term:*

$$f'(z_1) = \sum_{n=1}^{\infty} n a_n z_1^{n-1}. \qquad (11.58)$$

Thus the function represented by (11.56) is analytic inside its circle of convergence.

Since the differentiated power series (11.58) has the same radius of convergence as the original power series (11.56), we can apply Theorem 11.3.5 over and over to obtain the kth derivative:

$$f^{(k)}(z_1) = \sum_{n=k}^{\infty} n(n-1)\dots(n-k+1)a_n z_1^{n-k}. \tag{11.59}$$

Evaluating this at $z_1 = 0$ we find

$$f^{(k)}(0) = k!\, a_k, \tag{11.60}$$

so that the original power series (11.56) is in fact a Taylor series:

$$f(z) = \sum_{n=0}^{\infty} \frac{f^{(n)}(0)}{n!} z^n. \tag{11.61}$$

Applying these same considerations to the series (11.57), we find

$$b_n = \frac{1}{n!} f^{(n)}(z_1). \tag{11.62}$$

To summarize, a power series has ideal properties inside its circle of convergence and purely formal manipulations will give us the correct answers. In particular, it defines an analytic function.

The exponential function is defined by the power series

$$e^z = \sum_{n=0}^{\infty} \frac{z^n}{n!} = 1 + z + \frac{z^2}{2!} + \frac{z^3}{3!} + \dots . \tag{11.63}$$

According to the ratio test, this series converges for all z, i.e., the radius of convergence is infinite, and the series (11.63) defines an entire function, i.e., a function analytic throughout the entire z plane. For real $z = x$ the series (11.63) reduces to the known result for e^x. The series (11.63) thus provides an extension or continuation of e^x into the complex plane. As we discuss later, this continuation is unique. Given (11.63) we can define the hyperbolic and trigonometric functions:

$$\cosh z = \frac{1}{2}(e^z + e^{-z}), \tag{11.64}$$

$$\sinh z = \frac{1}{2}(e^z - e^{-z}), \tag{11.65}$$

$$\cos z = \frac{1}{2}(e^{iz} + e^{-iz}), \tag{11.66}$$

$$\sin z = \frac{1}{2i}(e^{iz} - e^{-iz}), \tag{11.67}$$

$$\tan z = \frac{\sin z}{\cos z}. \tag{11.68}$$

The series for (11.64)-(11.67) converge throughout the whole z plane to define entire functions. Along the real axis they reduce to the usual series of real variable theory.

The radius of convergence of the series

$$\sum_{n=0}^{\infty} z^n \tag{11.69}$$

is $r = 1$ according to the ratio test. Taking the limit of the geometric series

$$1 + z + z^2 + \ldots + z^N = \frac{1 - z^{N+1}}{1 - z}, \tag{11.70}$$

we find, for $|z| < 1$,

$$\frac{1}{1 - z} = \sum_{n=0}^{\infty} z^n. \tag{11.71}$$

If we replace x by z in the well-known series for $\log(1 + x)$ we obtain

$$\log(1 + z) = \sum_{n=1}^{\infty} \frac{(-1)^{n-1}}{n} z^n = z - \frac{1}{2}z^2 + \frac{1}{3}z^3 - \frac{1}{4}z^4 + \ldots . \tag{11.72}$$

The radius of convergence of this series is $r = 1$, so in the region $|z| < 1$ it can be used to define the analytic function $\log(1 + z)$.

Similarly if we replace x by z in the binomial series we obtain

$$(1 + z)^\alpha = \sum_{n=0}^{\infty} \binom{\alpha}{n} z^n = 1 + \alpha z + \frac{\alpha(\alpha - 1)}{1 \cdot 2} z^2 + \frac{\alpha(\alpha - 1)(\alpha - 2)}{1 \cdot 2 \cdot 3} z^3 + \ldots . \tag{11.73}$$

The radius of convergence of this series is also $r = 1$ according to the ratio test, so the series (11.71) can be used to define the analytic function $(1 + z)^\alpha$ in the region $|z| < 1$. The formula (11.70) is a special case of (11.73).

The series

$$J_\alpha(z) = \left(\frac{z}{2}\right)^\alpha \sum_{n=0}^{\infty} \frac{(-1)^n}{n! \, \Gamma(n + \alpha + 1)} \left(\frac{z}{2}\right)^{2n} \tag{11.74}$$

for the Bessel function is obtained by solving Bessel's differential equation. According to the ratio test it converges for all z to define a function analytic throughout the entire z plane, all this for an arbitrary complex α.

Similarly, the hypergeometric series,

$$_2F_1(a, b; c; z) = \frac{\Gamma(c)}{\Gamma(a)\Gamma(b)} \sum_{n=0}^{\infty} \frac{\Gamma(a + n)\Gamma(b + n)}{\Gamma(c + n)} \frac{z^n}{n!} \tag{11.75}$$

$$= 1 + \frac{ab}{c} \frac{z}{1!} + \frac{a(a + 1)b(b + 1)}{c(c + 1)} \frac{z^2}{2!} + \ldots ,$$

obtained by solving the hypergeometric differential equation, converges for $|z| < 1$ to define an analytic function.

11.4 MULTIVALUED FUNCTIONS; CUTS; RIEMANN SHEETS

The functions discussed so far, i.e., rational functions and functions defined by power series, are single valued; for each value of z, the function has one definite value. A multivalued function $f(z)$ has two or more distinct values for each value of z. Important examples of multivalued functions are $\sqrt{z}$, $\sqrt[n]{z}$, $\sqrt{(z-a)(z-b)}$, $\log z$, z^α, $\sin^{-1} z$, and $\cos^{-1} z$. In order to apply the theorems of analytic function theory to such multivalued functions, one needs to pick out a single valued continuous function to deal with. The systematic way to do this is provided by the apparatus of cuts and Riemann sheets. We shall study this subject via the examples listed above; the extension to more general cases is conceptually clear.

We begin with the two valued function

$$w = \sqrt{z}. \tag{11.76}$$

With

$$z = re^{i\theta} \tag{11.77}$$

the two branches are

$$w_1 = r^{\frac{1}{2}} e^{i\frac{\theta}{2}} \tag{11.78}$$

and

$$w_2 = -r^{\frac{1}{2}} e^{i\frac{\theta}{2}} = r^{\frac{1}{2}} e^{i\frac{\theta+2\pi}{2}} \tag{11.79}$$

as shown in Figure 11.4

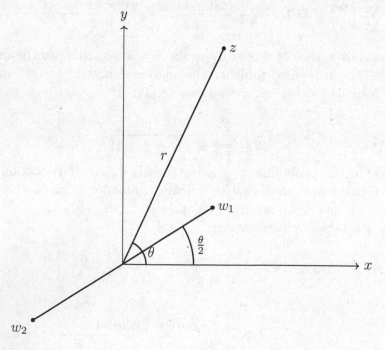

Figure 11.4: Two branches, w_1 and w_2 on the complex plane for $z = re^{i\theta}$

First, it is easy to check that the Cauchy Riemann equations are satisfied for the real and imaginary parts of one branch. Thus setting $w_1 = u + iv$ we find

$$u = r^{\frac{1}{2}} \cos \frac{\theta}{2}, \quad v = r^{\frac{1}{2}} \sin \frac{\theta}{2}, \tag{11.80}$$

where

$$r^{\frac{1}{2}} = \sqrt[4]{x^2 + y^2}, \quad \theta = \tan^{-1} \frac{y}{x}. \tag{11.81}$$

Computing the partial derivatives yields

$$\frac{\partial u}{\partial x} = \frac{\partial v}{\partial y} = \frac{1}{2r^{\frac{1}{2}}} \cos \frac{\theta}{2}, \tag{11.82}$$

$$\frac{\partial u}{\partial y} = -\frac{\partial v}{\partial x} = \frac{1}{2r^{\frac{1}{2}}} \sin \frac{\theta}{2}. \tag{11.83}$$

If we can manage to stay on one branch, $w = \sqrt{z}$ is thus an analytic function except at the point $z = 0$ where the partial derivatives become infinite in a nonunique way (depending on θ).

Let us now study the variation of w as z moves along a closed curve in the z plane, returning to its initial value. Start at some point $z = re^{i\theta}$ with the branch w_1, and let z move in any closed curve which goes <u>once around the point $z = 0$</u> as shown in Figure 11.5(a), so that $r \to r$, $\theta \to \theta + 2\pi$.

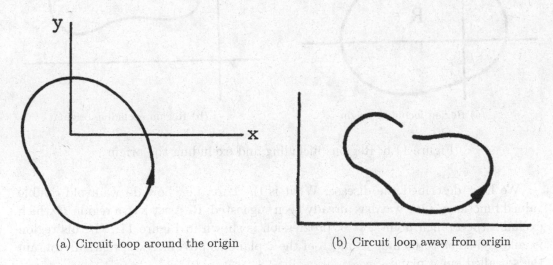

(a) Circuit loop around the origin   (b) Circuit loop away from origin

Figure 11.5: Circuit loops paths in complex plane

As z moves, w_1 changes continuously, and when z has returned to the original point, w_1 has changed to w_2. If z makes another circuit of the origin, $w(= w_2)$ continues to change continuously, and when z has returned to the starting point, w_2 has changed back to w_1. Further circuits of the origin lead to a repeat of the performance described above. We emphasize two points: (1) the continuous variation of w through <u>both</u> branches, with a continuous return to the original branch, (2) the special character of the point $z = 0$. Circuits around any other point, as shown in Figure 11.5(b) not including $z = 0$, lead to ordinary single valued behavior: $r \to r$, $\theta \to \theta$, so $w_1 \to w_1$ or $w_2 \to w_2$.

The point $z = 0$ is a singular point of the function $\sqrt{z}$. This particular kind of singularity is called a branch point and is very different from the poles considered earlier (11.23). The essential character of the branch point at $z = 0$ is that $\sqrt{z}$ is <u>unavoidably</u> double valued in any region including this point as an interior point, since if z just traverses a circuit around $z = 0$, $w = \sqrt{z}$ changes to $-w$ (illustrated in Figure 11.6(a)); if the branch point is inside the region, there is no way to separate one branch from the other. No other point has this character. In a region not including the point $z = 0$, as in Figure 11.6(b), it is possible to separate one branch of $\sqrt{z}$ from the other; as z moves around in such a region $w_1 = r^{1/2}e^{i\theta/2}$ does <u>not</u> change into $w_2 = -r^{1/2}e^{i\theta/2}$. (We note that the value of $\sqrt{z}$ at the branch point is not very significant. The value of the function is well defined and unique there, namely 0 for both branches. The partial derivatives (11.82) and (11.83) are, it is true, infinite in a nonunique way at $z = 0$, but for the function $z^{3/2} = zz^{1/2}$, which has a similar branch point, both the function and its derivative are well defined and unique at the branch point.)

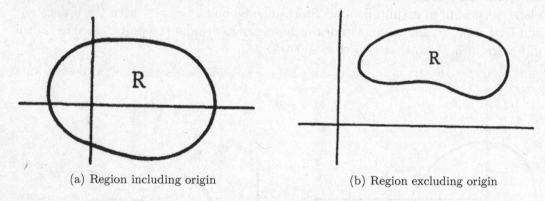

(a) Region including origin   (b) Region excluding origin

Figure 11.6: Regions including and excluding the origin

We have described the disease. What is the cure, i.e., how do we avoid double valued functions? One way has already been suggested. Restrict z to a region R which excludes the branch point $z = 0$, progression is shown in Figure 11.7. If this region is enlarged so as to include as much of the z plane as possible, we eventually obtain the so-called cut z plane.

The cut runs along some curve from the point $z = 0$ out to infinity in some direction. The shape of the curve and the direction it goes to infinity are not fixed; they are chosen for convenience for the problem at hand. First choose the negative real axis as the cut as in Figure 11.8. In this cut plane we define $\sqrt{z}$ to be

$$\sqrt{z} = r^{\frac{1}{2}}e^{i\frac{\theta}{2}}, \quad -\pi \le \theta < \pi. \tag{11.84}$$

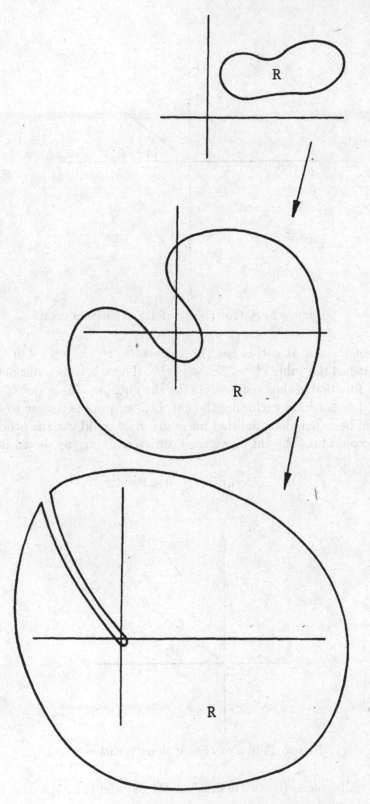

Figure 11.7: Progression of region enlarging yet avoiding branch point at $z = 0$

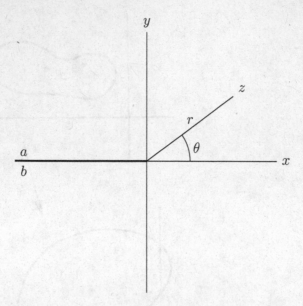

Figure 11.8: Cut choice along negative x axis

On the top of the cut at a this has the value $r^{1/2}e^{i\pi/2} = ir^{1/2}$. On the bottom of the cut at b it has the value $r^{1/2}e^{-i\pi/2} = -ir^{1/2}$. There is thus a discontinuity across the cut. The function is thus not analytic on the cut, but this is not of concern since z is restricted to a region excluding the cut, i.e., we promise never to cross the cut. The cut could be taken along another line and/or we could use the other branch. For instance, we could take the cut along the positive real axis, as shown in Figure 11.9, and define $\sqrt{z}$ by

$$\sqrt{z} = r^{\frac{1}{2}}e^{i\frac{\theta}{2}}, \quad 0 < \theta < 2\pi. \tag{11.85}$$

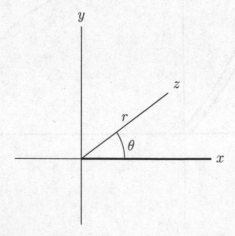

Figure 11.9: Cut choice along positive x axis

In the upper half z plane the two definitions (11.84) and (11.85) agree; in the bottom half-plane they differ by a minus sign.

In the method described above we obtained a well-defined single valued function by discarding one of the branches of a double valued function. An ingenious way of obtaining a continuous single valued function while keeping both branches was devised by Riemann. Imagine separating the z plane into two sheets connected along the cut – like two sheets of paper on top of each other sewn together along the cut.

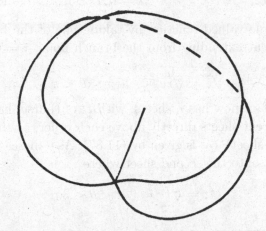

Figure 11.10: Riemann sheets connected along branch line

The independent variable z has the same value at the two points, one on each sheet, directly over one another. At points z on the top sheet, the function $\sqrt{z}$ has the values associated with one of its branches. At points z on the bottom sheet, $\sqrt{z}$ has the values associated with the other branch. Along the cut the two sheets are connected in such a way that z moves continuously from the top sheet to the bottom sheet while $\sqrt{z}$ changes continuously, see Figure 11.10. Suppose for example the cut is along the negative real axis. Then on the top sheet $\sqrt{z} = r^{1/2}e^{i\theta/2}$, $-\pi \leq \theta < \pi$. As θ increases through π, the point z goes down through the cut onto the bottom sheet where $\sqrt{z} = r^{1/2}e^{i\theta/2}$, $\pi < \theta < 3\pi$. As θ increases through 3π, z moves back up through the cut onto the top sheet. See Figure 11.11 for cross sectional view of the cut crossings. In the complete traversal of both sheets, θ increases by 4π, $\theta/2$ by 2π, and $\sqrt{z} = r^{1/2}e^{i\theta/2}$ is back to its starting value. Thus we see $\sqrt{z}$ is a continuous single valued analytic function on the two-sheeted z plane.

Figure 11.11: Cross section view of cut crossings

The circumstance that the two sheets intersect along the cut is not significant and arises from attempting to visualize the connections of the sheets in three-dimensional space. If you can do it, disregard the intersection of the sheets along the cut. This makes it easier to understand that the location of the cut is irrelevant for the two Riemann sheets.

We have used a lot of words to describe the cuts and Riemann sheets for $\sqrt{z}$.

Other examples are similar but more complicated. The function $\sqrt{z-a}$ has its branch point at $z = a$ but is otherwise identical to $\sqrt{z}$. The function $\sqrt[n]{z}$, $n = 3, 4, 5, \ldots$ has a branch point at $z = 0$ and is n valued. With $z = re^{i\theta}$

$$\sqrt[n]{z} = \sqrt[n]{r}\, e^{i\,\frac{\theta + 2\pi k}{n}}, \quad k = 0, 1, 2, 3, \ldots, n-1. \tag{11.86}$$

One can obtain a single valued function by taking one of the branches in the cut z plane, e.g., with the cut extending from the branch point $z = 0$ along the negative real axis:

$$\sqrt[n]{z} = \sqrt[n]{r}e^{i\frac{\theta}{n}}, \quad -\pi \le \theta < \pi. \tag{11.87}$$

The Riemann surface now has n sheets, with z, of course, having the same value at points on the different sheets directly above each other, as shown if Figure 11.12. On the top sheet (1st sheet) $\sqrt[n]{z}$ is given by (11.87). As θ increases through π, z goes down through the cut onto the second sheet where

$$\sqrt[n]{z} = \sqrt[n]{r}e^{i\frac{\theta}{n}}, \quad \pi < \theta < 3\pi. \tag{11.88}$$

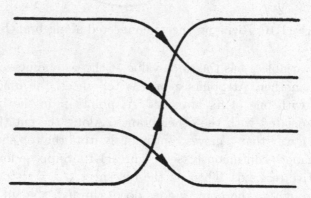

Figure 11.12: Cross section view of Riemann surface with multiple sheets

As θ increases through 3π, z goes down through the cut again onto the third sheet, etc. On the bottom sheet (n^{th} sheet)

$$\sqrt[n]{r} = \sqrt[n]{r}e^{i\frac{\theta}{n}}, \quad (2n-3)\pi \le \theta < (2n-1)\pi. \tag{11.89}$$

As θ increases through $(2n-1)\pi$, z moves through the cut back up to the top sheet and $\sqrt[n]{z}$ returns to its values (11.87) given on this sheet. The function $\sqrt[n]{z}$ is a continuous single valued analytic function on this n-sheeted Riemann surface.

Again the awkward intersection of the sheets along the cut arises only because we try to visualize the connections of the sheets in three-dimensional space and has no real significance.

The function $\sqrt{(z-a)(z-b)}$ has two branch points, one at $z = a$, on at $z = b$. The two cuts associated with these branch points can be oriented arbitrarily, running out to infinity (as illustrated in Figure 11.13), and a single valued branch of the function defined in the cut z plane by

$$\sqrt{(z-a)(z-b)} = \sqrt{r_1 r_2}\, e^{i\frac{1}{2}(\theta_1 + \theta_2)} \tag{11.90}$$

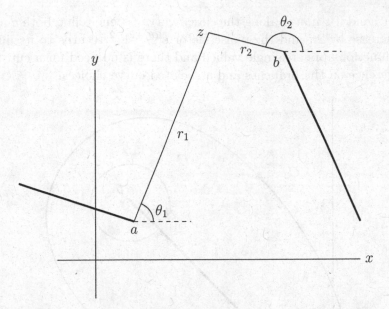

Figure 11.13: Two arbitrary cuts for two branch points

with θ_1 and θ_2 restricted so as not to cross the cuts. There is, however, a favored orientation of the cuts for this function, namely along the line joining a and b as shown in Figure 11.14. The two cuts then cancel, so to speak, on the extensions of this line beyond a or b, so that we only need a cut between a and b. To verify this statement notice that if z moves along a closed curve C_1 encircling a but not b, θ_1 increases by 2π, while θ_2 returns to its initial value, so that the phase factor $e^{i(\theta_1+\theta_2)/2}$ in (11.90) changes sign, corresponding to the discontinuity across the cut. Similarly if z moves along the closed curve C_2 encircling b but not a, θ_2 increases by 2π while θ_1 returns to its initial value so that again the phase factor $e^{i(\theta_1+\theta_2)/2}$ changes sign.

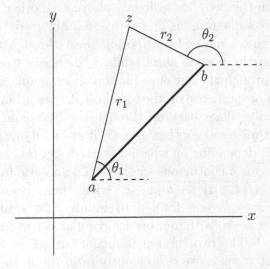

Figure 11.14: Cut along line between two branch points

On the other hand if z moves along the closed curve C_3 encircling both a and b, both θ_1 and θ_2 increase by 2π and the phase factor $e^{i(\theta_1+\theta_2)/2}$ returns to its initial value, so that the function appears single valued and there is no need for a cut. See Figure 11.15 for reference on the branches and associated curve choices.

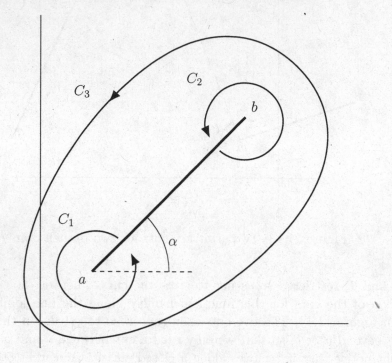

Figure 11.15: Curves around two branch points

The Riemann surface for $\sqrt{(z-a)(z-b)}$ has two sheets joined together along the cut, which can be taken between a and b. The independent variable z has the same value at points on the two sheets directly above each other. Following along the curve C_1 in the previous diagram, as θ_1 increases continuously through the value α corresponding to the cut, z moves from the top sheet through the cut down to the bottom sheet, stays on the bottom sheet while θ_1 increases from α to $\alpha + 2\pi$, and then moves back up through the cut onto the top sheet again. Similarly as z follows the curve C_2, it moves around once on the top sheet, moves around once again on the bottom sheet and finally comes back up through the cut again onto the top sheet. On the other hand if z moves along the curve C_3 it stays on one sheet, either the top one or the bottom one, depending on where it started. On this two-sheeted Riemann surface $\sqrt{(z-a)(z-b)}$ is a continuous single valued analytic function.

For the function $\sqrt{(z-a)(z-b)(z-c)}$ with three branch points we need two cuts, one from a to b, one from c to infinity, for example. The function $\sqrt{(z-a)(z-b)(z-c)(z-d)}$ with four branch points has two cuts, for example from a to b and from c to d. The two sheeted Riemann surface in this case has the interesting property that z can move continuously from the inside (z_0) to the outside (z_1) of a closed curve C without crossing the curve itself. The path is indicated in

Figure 11.16, where a solid line represents a path on the top sheet, a dotted line a path on the bottom sheet.

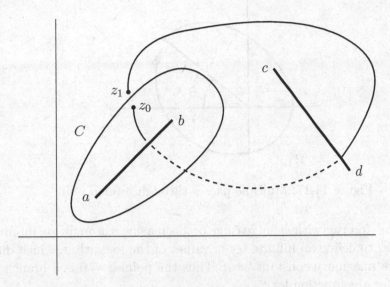

Figure 11.16: Curves for function with two cuts and four branch points

The function $\log z$ is the inverse of the exponential. Thus if

$$z = e^w, \tag{11.91}$$

then

$$w = \log z. \tag{11.92}$$

We can see right away that the logarithm is analytic. From the power series (11.63) for e^w we have

$$\frac{dz}{dw} = e^w = z \tag{11.93}$$

so that

$$\frac{dw}{dz} = \frac{d}{dz}\log z = \frac{1}{z} \tag{11.94}$$

and the existence and uniqueness of the derivative of the logarithm follows from the existence and uniqueness of the derivative of the exponential.

The logarithm is, however, multivalued and in fact has an infinite number of branches. In (11.91) we can write

$$z = re^{i\theta} = re^{i\theta + i2\pi n} = e^{\ln r + i\theta + i2\pi n}$$

where n is any integer, $n = 0, \pm 1, \pm 2, \ldots$, whence we see that

$$\log z = \ln r + i\theta + i2\pi n. \tag{11.95}$$

As z moves in a close curve around the point $z = 0$, θ increases by 2π and we move continuously from one branch of the logarithm to the next as shown in Figure 11.17.

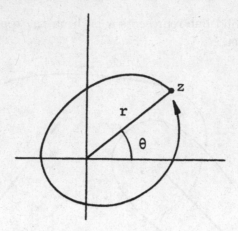

Figure 11.17: Multiple passes through branch point

This can go on forever with θ increasing or decreasing through an infinite set of multiples of 2π, to define an infinite set of values of the logarithm, which differ only by the additive imaginary constant $2\pi ni$. Thus the point $z = 0$ is a branch point of infinite order for the function $\log z$.

To obtain a single valued function we can make a cut in the z plane from the branch point $z = 0$ to infinity along, for example (as shown in Figure 11.18), the negative real axis and define $\log z$ by

$$\log z = \ln r + i\theta, \quad -\pi \leq \theta < \pi. \tag{11.96}$$

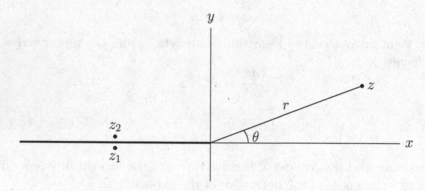

Figure 11.18: Cut along negative real axis for $\log z$

At the two points z_1 and z_2 on either side of the cut we then have

$$\log z_1 = \ln r - i\pi, \tag{11.97}$$
$$\log z_2 = \ln r + i\pi, \tag{11.98}$$

so that the discontinuity across the cut is constant:

$$\log z_2 - \log z_1 = 2\pi i. \tag{11.99}$$

To obtain a single valued function which includes all the branches of $\log z$, one must introduce a Riemann surface with an infinite number of sheets attached to each other at the branch point $z = 0$ – like the shaving made by a cutting tool with a blade rotating around a shaft of 0 diameter, as illustrated in Figure 11.19. As usual $z = re^{i\theta}$ has the same value at points directly under or over one another on the infinite set of sheets $[e^{i\theta} = e^{i(\theta + 2\pi n)}]$. As z circles around the branch point $z = 0$, θ increases or decreases continually and $\log z = \ln r + i\theta$ moves continuously through the whole range of its possible values.

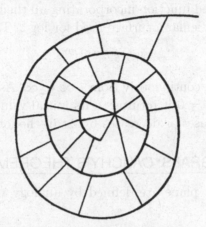

Figure 11.19: Infinite sheets attached at branch point

As a last example consider the function z^α, where α is some arbitrary complex number. Using $z = e^{\log z}$ we can define z^α by

$$z^\alpha = e^{\alpha \log z}. \tag{11.100}$$

We see that z^α is an analytic function, since the exponential is analytic and the $\log z$ is analytic. The multivalued character (11.95) of $\log z$ also determines the multivalued behavior of z^α:

$$z^\alpha = e^{\alpha(\ln r + i\theta + i2\pi n)}, \quad n = 0, \pm 1, \pm 2, \cdots. \tag{11.101}$$

For $\alpha = p/q$ a fractional number (p and q integers), $\alpha q = p = $ integer, so that there are only a finite set of distinct branches, $n = q$ giving the same value of z^α as $n = 0$, $n = q + 1$ the same value as $q = 1$, etc. The roots $\sqrt[n]{z}$ studied earlier fall into this class. If α is not a rational number, z^α has an infinite number of branches, like $\log z$. The point $z = 0$ is a branch point of infinite order. In the z plane cut from the branch point $z = 0$ along the negative real axis we can define a single valued function z^α by

$$z^\alpha = e^{\alpha(\ln r + i\theta)}, \quad -\pi \leq \theta < \pi. \tag{11.102}$$

As shown in Figure 11.20, this function is of course discontinuous across the cut:

$$z_1^\alpha = e^{\alpha(\ln r - i\pi)}, \tag{11.103}$$

$$z_2^\alpha = e^{\alpha(\ln r + i\pi)}. \tag{11.104}$$

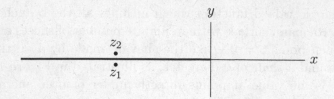

Figure 11.20: Points z_1 and z_2 near the cut on negative real axis

To obtain a single valued function incorporating all the branches of z^α we employ the same infinite-sheeted Riemann surface used for $\log z$. The formula

$$z^\alpha = e^{\alpha(\ln r + i\theta)} \tag{11.105}$$

then applies with θ varying continuously from $-\infty$ to ∞. As z moves in a closed path around the branch point $z = 0$ it returns to its initial value but on the next higher or lower sheet, and θ increases or decreases by 2π in the formula (11.105).

11.5 CONTOUR INTEGRALS; CAUCHY'S THEOREM

Contour integrals in the z plane are defined by analogy with the real integrals of calculus.

Definition 11.5.1. Subdivide the contour C into n segments in any manner by distributing points $z_0, z_1, \ldots, z_n$ along the contour, and choose points ζ_ν on C arbitrarily placed between the z_ν as shown in Figure 11.21. Consider then the sum

$$J_n = \sum_{\nu=1}^{n} (z_\nu - z_{\nu-1}) f(\zeta_\nu), \tag{11.106}$$

where $f(z)$ is a function to be integrated. If this sum approaches a limit J as $n \to \infty$ and the lengths of all path segments $\to 0$, i.e., if given an ϵ there exists a δ such that

$$|J_n - J| < \epsilon \tag{11.107}$$

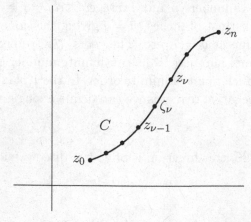

Figure 11.21: Contour of integration in complex plane

for all subdivisions such that the lengths of all path segments are less than δ, then the limit J is defined to be the integral:

$$J = \int_C f(z). \tag{11.108}$$

The integral can be shown to exist provided the contour C has certain reasonable smoothness properties[4] and provided $f(z)$ is continuous along the contour[5]. We note that $f(z)$ does <u>not</u> have to be analytic for the integral to exist.

If worse comes to worst, the contour integral (11.108) can be evaluated by reducing it to real integrals:

$$J = \int_C dz \, f(z)$$

$$= \int_C (dx + idy) \, [u(x,y) + iv(x,y)]$$

$$= \int_C dx \, u(x,y) - \int_C dy \, v(x,y) + i \left\{ \int_C dx \, v(x,y) + \int_C dy \, u(x,y) \right\}. \tag{11.109}$$

Along the contour, $y = y(x)$ or $x = x(y)$, so that these become ordinary one-dimensional integrals.

Integrals along circular contours are often easy to do. Suppose C is the circle $z - \zeta = re^{i\theta}$ as shown in Figure 11.22. Then $dz = ire^{i\theta}d\theta$, and if the contour is traversed in the counterclockwise sense, we have

$$\int_C dz \, f(z) = ir \int_0^{2\pi} d\theta \, e^{i\theta} f(\zeta + re^{i\theta}). \tag{11.110}$$

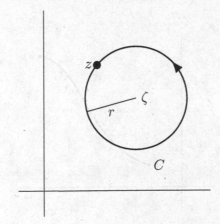

Figure 11.22: Circular contour in complex plane about point ζ

[4]It must be rectifiable, i.e., have a length, among other things. Knopp, <u>op. cit.</u>, Part I, p. 13.
[5]Knopp, <u>op. cit.</u>, Part I, p. 34.

An important example is the case $f(z) = (z - \zeta)^n$ where $n = 0, \pm 1, \pm 2, \ldots$ is a positive or negative integer:

$$\int_C dz \, (z - \zeta)^n = ir \int_0^{2\pi} d\theta \, e^{i\theta} r^n e^{in\theta}$$

$$= ir^{n+1} \int_0^{2\pi} d\theta e^{i(n+1)\theta}$$

$$= \begin{cases} 2\pi i, & n = -1 \\ 0, & n = 0, 1, \pm 2, \pm 3, \ldots \end{cases} . \tag{11.111}$$

If $f(z)$ is the derivative of some other function, i.e., if

$$f(z) = \frac{d}{dz} F(z), \tag{11.112}$$

then

$$\int_C dz \, f(z) = \int_C dF(z) = F(z_2) - F(z_1), \tag{11.113}$$

where z_1 and z_2 are the beginning and end of the interval, as an example see Figure 11.23. Note that (11.112) is not valid for all functions; in fact $F(z)$ must be analytic. For

$$f(z) = (z - \zeta)^n, \quad n = 0, \pm 1, \pm 2, \ldots, \tag{11.114}$$

we have

$$F(z) = \begin{cases} \dfrac{1}{n+1}(z - \zeta)^{n+1}, & n \neq -1 \\[2mm] \log(z - \zeta), & n = -1 \end{cases} . \tag{11.115}$$

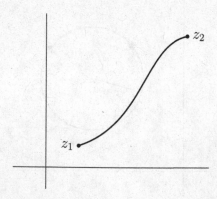

Figure 11.23: Path of integration from z_1 to z_2 in complex plane

If we integrate a function of the form (11.112) around a closed contour, (11.113) would seem to give zero, and this is indeed true provided $F(z)$ is single valued or we stay on the same branch of a multivalued function. Thus (11.113) and (11.115) easily explain the results (11.111) for $n \neq -1$. For the case $n = -1$ the function $F(z) = \log(z - \zeta)$. This multivalued function was discussed in some detail in Section 11.4 and we know that $\log(z - \zeta)$ increases by $2\pi i$ when z moves along a closed circuit around the branch point $z = \zeta$. Keeping this in mind we see that (11.113) also yields the result given in (11.111) for $n = -1$. We notice in passing that the derivation of (11.111) using (11.113) is independent of shape of the closed contour C. Thus generalizing (11.111) we find for any <u>closed</u> contour C

$$\int_C dz \, (z - \zeta)^n = 0, \quad n = 0, 1, \pm 2, \pm 3, \ldots \tag{11.116}$$

$$n \neq -1.$$

For $n = -1$

$$\int_C \frac{dz}{z - \zeta} = \left\{ \begin{array}{ll} 2\pi i, & \zeta \text{ inside } C \\ 0, & \zeta \text{ outside } C \end{array} \right. . \tag{11.117}$$

The contour integral $\int_C dz \, f(z)$ exists for $f(z)$ a continuous function. However, extremely interesting results are obtained if we restrict ourselves to the class of analytic functions $f(z)$ such that $f'(z)$ is uniquely defined. The fundamental theorem of the theory of functions of a complex variable is <u>Cauchy's Theorem</u>:

Theorem 11.5.1. *If $f(z)$ is analytic in a simply connected region R and if z_1 and z_2 are two interior points of R, then the integral*

$$\int_{z_1}^{z_2} dz \, f(z) \tag{11.118}$$

has the same value along every path of integration extending from z_1 to z_2 and lying entirely within R. Equivalently

$$\int_C dz \, f(z) = 0 \tag{11.119}$$

for C an arbitrary closed path lying within R (example shown in Figure 11.24).

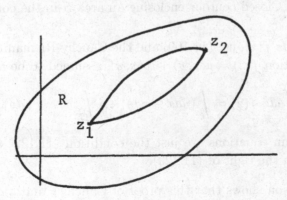

Figure 11.24: Example paths from z_1 to z_2 in a region R

The converse of Cauchy's theorem is <u>Morera's Theorem</u>:

Theorem 11.5.2. *If $f(z)$ is continuous in the simply connected region R and if*

$$\int_C dz\, f(z) = 0 \tag{11.120}$$

for every closed path C lying with R, then $f(z)$ is regular in R.

A simple but incomplete proof of these theorems is obtained from Stoke's theorem.

Proof. For a two-dimensional vector $\mathbf{A} = (A_x, A_y)$ whose components are functions only of x and y Stoke's theorem reads

$$\int_C (A_x dx + A_y dy) = \int_S dx\, dy \left[\frac{\partial A_y}{\partial x} - \frac{\partial A_x}{\partial y} \right], \tag{11.121}$$

where C is the closed contour enclosing the area S as shown in Figure 11.25. The necessary and sufficient condition that the contour integral on the left of (11.121) be zero for all closed contours C is

$$\frac{\partial A_y}{\partial x} - \frac{\partial A_x}{\partial y} = 0. \tag{11.122}$$

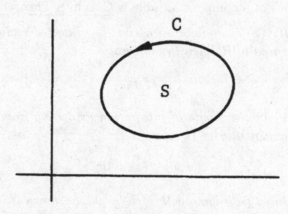

Figure 11.25: Closed contour enclosing an area S in the complex plane

To apply this to $\int_C dz\, f(z)$, use (11.109) and the Cauchy-Riemann equations (11.32), (11.33) for the function $f(z) = u(x,y) + iv(x,y)$, assumed to be analytic:

$$\int_C dz\, f(z) = \int_C (u\,dx - v\,dy) + i \int_C (v\,dx + u\,dy). \tag{11.123}$$

The Cauchy-Riemann equations are just the condition (11.122) applied to the two real line integrals on the right of (11.123). $\qquad\square$

Closer examination shows that this proof of Cauchy's and Morera's theorems is incomplete because it assumes the partial derivatives of u and v are continuous, i.e., $f'(z)$ is assumed not only to exist but also to be continuous. The more elaborate proofs

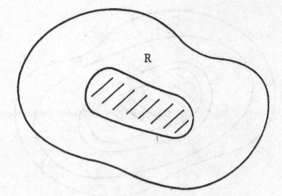

Figure 11.26: Annular region R with unknown behavior of function within shaded area

given in mathematical works[6] establish the theorems as stated above. It doesn't really matter because it turns out later in the development of the theory that an analytic function has derivatives at all orders so that $f'(z)$ must be continuous since $f''(z)$ exists.

An important generalization of Cauchy's theorem deals with a region which is not simply connected as in Figure 11.26. Suppose $f(z)$ is known to be analytic in an annular region R. Nothing is assumed about the behavior of $f(z)$ in the shaded region inside the annulus. Apply Cauchy's theorem to the two curves C_1 and C_2, each of which is separately contained in a simply connected region throughout which $f(z)$ is analytic as shown in Figure 11.27. Thus we have

$$\int_{C_1} dz \; f(z) = 0, \tag{11.124}$$

$$\int_{C_2} dz \; f(z) = 0. \tag{11.125}$$

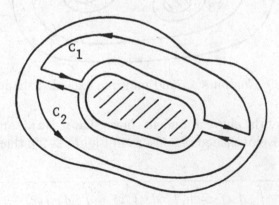

Figure 11.27: Separate curves in the annular region, R

---

[6]Knopp, ip. cit., Part I, Chap. 4, p. 47.

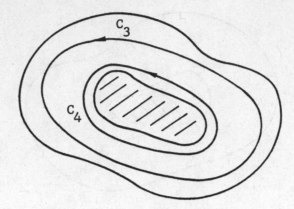

Figure 11.28: Combining contours leaves two concentric contours in annular region

Adding these together, cancelling the parts of the integrals which come from the contiguous parts of C_1 and C_2 running in opposite directions, and rewriting the result, we find

$$\int_{C_3} dz \, f(z) = \int_{C_4} dz \, f(z), \qquad (11.126)$$

where C_3 and C_4 are any two similarly directed contours inside the annular region as shown in Figure 11.28 where $f(z)$ is analytic. Note that the integrals (11.126) will not in general vanish if $f(z)$ has singularities inside the shaded region; of course, if $f(z)$ is analytic inside the shaded region as well as in the annular region surrounding it, Cauchy's theorem tells us that both sides of (11.126) will vanish.

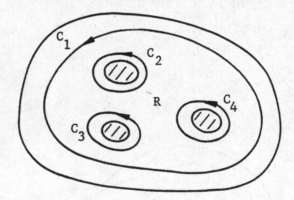

Figure 11.29: Contours around multiple excluded shaded regions

A generalization of the above argument shows that if $f(z)$ is analytic in an annular region R excluding several shaded regions, as in Figure 11.29, then we have a relation of the form

$$\int_{C_1} dz \, f(z) = \int_{C_2} dz \, f(z) + \int_{C_3} dz \, f(z) + \int_{C_4} dz \, f(z) + \dots . \qquad (11.127)$$

As we shall see, this is the basic relation which one uses to do integrals.

Let us return for a moment to the problem of calculating contour integrals. We see that if $f(z)$ is analytic throughout the region involved, the integral

$$f(z) = \int_{z_0}^{z} dz' \, f(z')$$
(11.128)

defines for all z a unique function $F(z)$ which is independent of the path of integration from z_0 to z (where an arbitrary path is shown in Figure 11.30). Then we have

$$F(z + \Delta z) = \int_{z_0}^{z+\Delta z} dz' \, f(z') = \int_{z_0}^{z} dz' \, f(z') + \int_{z}^{z+\Delta z} dz' \, f(z')$$
(11.129)

and

$$\frac{dF(z)}{dz} = \lim_{\Delta z \to 0} \frac{F(z + \Delta z) - F(z)}{\Delta z} = f(z),$$
(11.130)

so that $F(z)$ is analytic. Thus when $f(z)$ is analytic, the situation described in (11.112)ff. occurs and (11.113) holds:

$$\int_{z_1}^{z_2} dz' \, f(z') = F(z_2) - F(z_1).$$
(11.131)

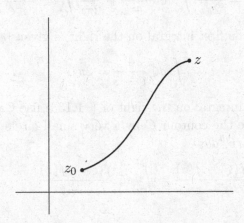

Figure 11.30: Integral from z_0 to z in complex plane

11.6 CAUCHY'S INTEGRAL FORMULA

The most important consequence of Cauchy's theorem is Cauchy's integral formula:

Theorem 11.6.1. *If $f(z)$ is analytic in a region R, C is a counterclockwise directed contour inside R, and z is a point inside C, as shown in Figure 11.31, then we have the formula for $f(z)$:*

$$\boxed{f(z) = \frac{1}{2\pi i} \int_C d\zeta \, \frac{f(\zeta)}{\zeta - z}.}$$
(11.132)

According to this remarkable result, if $f(z)$ is analytic, then its value at z is entirely determined by its values on the contour C. This is an extremely strong consequence of the assumption of unique differentiability for $f(z)$.

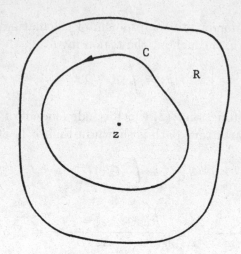

Figure 11.31: Counterclockwise contour, C, in region R

Proof. To prove the formula note that

$$\frac{1}{2\pi i}\int_C d\zeta\, \frac{f(\zeta)}{\zeta - z} = \frac{f(z)}{2\pi i}\int_C d\zeta\, \frac{1}{\zeta - z} + \frac{1}{2\pi i}\int_C d\zeta\, \frac{f(\zeta) - f(z)}{\zeta - z}. \tag{11.133}$$

According to (11.117) the first integral on the right is given by

$$\int_C d\zeta\, \frac{1}{\zeta - z} = 2\pi i. \tag{11.134}$$

To evaluate the second integral on the right of (11.133) use Cauchy's theorem in the form (11.126) to replace the contour C by a very small circle C' centered on z such that $\zeta - z = r e^{i\theta}$, $d\zeta = i r e^{i\theta} d\theta$:

$$\int_C d\zeta\, \frac{f(\zeta) - f(z)}{\zeta - z} = \int_{C'} d\zeta\, \frac{f(\zeta) - f(z)}{\zeta - z}$$

$$= i\int_0^{2\pi} d\theta\, \left[f(z + r e^{i\theta}) - f(z) \right]. \tag{11.135}$$

The radius r can be made arbitrarily small in this expression. Since $f(z)$ is assumed analytic, it is also continuous, so that $|f(z + r^e i\theta) - f(z)|$ can be made arbitrarily small by making r sufficiently small. Thus (11.135) is arbitrarily small and hence vanishes. Using this result and (11.134) in (11.133) establishes (11.132). ☐

The derivative of $f(z)$ can be obtained by differentiating (11.132) under the integral sign:

$$f'(z) = \frac{1}{2\pi i}\int_C d\zeta\, \frac{f(\zeta)}{(\zeta - z)^2}. \tag{11.136}$$

To justify the manipulation, use (11.132) and (11.136) to calculate

$$\frac{f(z + \Delta z) - f(z)}{\Delta z} - f'(z) = \frac{\Delta z}{2\pi i}\int_C d\zeta\, \frac{f(\zeta)}{(\zeta - z)^2(\zeta - z - \Delta z)}. \tag{11.137}$$

This can be made arbitrarily small by taking $|\Delta z|$ sufficiently small. In a similar way one easily shows that $f(z)$ possesses derivatives of all orders, given by the formula

$$f^{(n)}(z) = \frac{n!}{2\pi i} \int_C d\zeta \frac{f(\zeta)}{(\zeta - z)^{n+1}}. \tag{11.138}$$

Thus we have another remarkable result; we start with the assumption that $f(z)$ has a first derivative and find that in fact it has derivatives of all orders.

11.7 TAYLOR AND LAURENT EXPANSIONS

To proceed to the next step, we need the concept of uniform converge.

Definition 11.7.1. A convergent series of functions,

$$F(z) = \sum_{n=0}^{\infty} f_n(z), \tag{11.139}$$

is said to converge uniformly in a region R if given an arbitrary ϵ there exists an N_0 independent of z in R such that

$$\left| \sum_{n=0}^{N} f_n(z) - F(z) \right| < \epsilon \tag{11.140}$$

for $N \geq N_0$ for all z in R.

The important thing is that the same N_0 works for all z; we don't have to make N_0 bigger and bigger as z approaches some bad point. An obvious test for uniform convergence is the comparison test: if $\sum_n S_n$ is a convergent series of positive constants and $|f_n(z)| \leq S_n$ for z in the region R, then $\sum_n f_n(z)$ is uniformly convergent in R.

A uniformly convergent series of continuous or analytic functions has the useful property that it can be integrated term-by-term. We skip the proof that the sum function is continuous and therefore has an integral.[7] From (11.139) and (11.140) we find

$$\left| \int_C dz \, F(z) - \sum_{n=0}^{N} \int_C dz \, f_n(z) \right| = \left| \int_C dz \left\{ F(z) - \sum_{n=0}^{N} f_n(z) \right\} \right| < \epsilon L, \tag{11.141}$$

where L is the length of the contour. Since ϵL can be made arbitrarily small, we have

$$\int_C dz \, F(z) = \sum_{n=0}^{\infty} \int_C dz \, f_n(z). \tag{11.142}$$

We can use these results to expand Cauchy's integral formula (11.132) about a point z_0.

---

[7]Knopp. op. cit., Part I, p. 74.

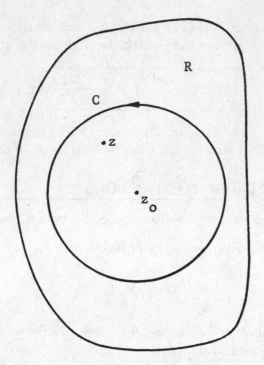

Figure 11.32: Circle centered at z_0 in a region, R

Proof. Take C to be a circle centered on z_0 and z a point inside C as in Figure 11.32. Then we can expand the denominator in (11.132) as follows:

$$\frac{1}{\zeta - z} = \frac{1}{\zeta - z_0 - (z - z_0)}$$

$$= \frac{1}{\zeta - z_0} \left[1 - \frac{z - z_0}{\zeta - z_0} \right]^{-1}$$

$$= \frac{1}{\zeta - z_0} \sum_{n=0}^{\infty} \left[\frac{z - z_0}{\zeta - z_0} \right]^n. \tag{11.143}$$

The geometric series here is convergent since $\left| \frac{z - z_0}{\zeta - z_0} \right| < 1$. Furthermore it is uniformly convergent according to the comparison test, since for z inside C, ζ anywhere on C, there is a number ρ such that $\left| \frac{z - z_0}{\zeta - z_0} \right| < \rho < 1$, and the series $1 + \rho + \rho^2 + \ldots = \sum_{n=0}^{\infty} \rho^n$ converges. Thus we are allowed to substitute (11.143) in (11.132) and integrate term-by-term. We find in this way

$$f(z) = \sum_{n=0}^{\infty} a_n (z - z_0)^n, \tag{11.144}$$

where

$$a_n = \frac{1}{2\pi i} \int_C d\zeta \, \frac{f(\zeta)}{(\zeta - z_0)^{n+1}}$$

$$= \frac{1}{n!} f^{(n)}(z_0),$$

(11.145)

according to (11.138). Thus (11.144) is just the Taylor series for $f(z)$ and we have shown that this Taylor series converges at least inside the largest circles which can be drawn about z_0 as center and remain inside the region R where $f(z)$ is known to be analytic, as illustrated in Figure 11.33. □

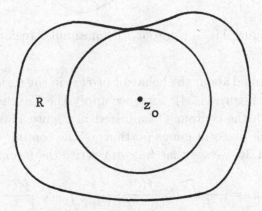

Figure 11.33: Convergence in largest circle centered about z_0 in region R

The results just presented form the converse to the theorems quoted without proof in Section 11.3 [see (11.58)] to the effect that a power series defines an analytic function inside its circle of convergence.

We can easily obtain a generalization of the Cauchy integral formula (11.132) which applies when $f(z)$ is known to be analytic in an annular region as in Figure 11.34.

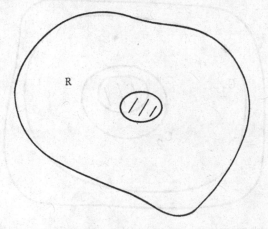

Figure 11.34: Annular region for an analytic function

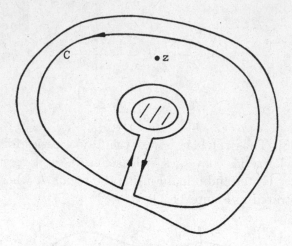

Figure 11.35: Contour within annular region

Proof. Nothing is assumed about the behavior of $f(z)$ in the region inside the annulus (the shaded region of Figure 11.34). First we apply the Cauchy integral formula in the form (11.132) with the contour C indicated in Figure 11.35. Noticing that the contributions along the two contiguous portions of the contour cancel, so C_1 and C_2 remain as in Figure 11.36, we see that we can rewrite the formula in the form

$$f(z) = \frac{1}{2\pi i} \int_{C_1} d\zeta \, \frac{f(\zeta)}{\zeta - z} - \frac{1}{2\pi i} \int_{C_2} d\zeta \, \frac{f(\zeta)}{\zeta - z}. \tag{11.146}$$

The minus sign in the second term arises from reversing the direction of the contour.

We can use (11.146) to establish a generalization of the Taylor series. Suppose we expand about a point z_0 located inside the forbidden region inside the annulus, taking C_1 and C_2 to be two circles centered about z_0 as represented in Figure 11.37. The first integral on the right of (11.146) can be expanded using the series (11.143) to lead to a series of the form (11.144). In the second integral on the right of (11.146)

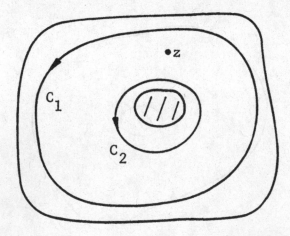

Figure 11.36: Remaining contours after removing contiguous portions

we expand the denominator in a different series:

$$\frac{1}{\zeta - z} = \frac{1}{\zeta - z_0 - (z - z_0)}$$

$$= \frac{1}{z - z_0}\left[1 - \frac{\zeta - z_0}{z - z_0}\right]^{-1}$$

$$= -\frac{1}{z - z_0}\sum_{n=0}^{\infty}\left[\frac{\zeta - z_0}{z - z_0}\right]^n. \qquad (11.147)$$

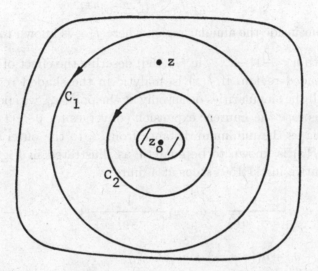

Figure 11.37: Forbidden region within annulus which includes a singularity

With ζ on C_2, $\left|\frac{\zeta - z_0}{z - z_0}\right| < \rho < 1$ so the series (11.147) is uniformly convergent and can be integrated term-by-term

$$-\frac{1}{2\pi i}\int_{C_2} d\zeta \, \frac{f(\zeta)}{\zeta - z} = \sum_{n=1}^{\infty} b_n \frac{1}{(z - z_0)^n}$$

$$= \sum_{n=-\infty}^{-1} a_n (z - z_0)^n, \qquad (11.148)$$

with b_n given by

$$b_n = a_{-n} = \frac{1}{2\pi i}\int_{C_2} d\zeta \, (\zeta - z_0)^{n-1} f(\zeta). \qquad (11.149)$$

Using the form (11.126) of Cauchy's theorem, we see the contour in (11.149) can be moved anywhere in the annular region where $f(z)$ is known to be analytic.

Combining our formulas for the two terms on the right side of (11.146) we obtained the Laurent expansion for a function $f(z)$ known to be analytic in an annular region

$$f(z) = \sum_{n=-\infty}^{\infty} a_n (z - z_0)^n$$

$$= a_0 + a_1(z - z_0) + a_2(z - z_0)^2 + \ldots$$

$$+ a_{-1} \frac{1}{z - z_0} + a_{-2} \frac{1}{(z - z_0)^2} + \ldots , \qquad (11.150)$$

where a_n is given by

$$a_n = \frac{1}{2\pi i} \int_C d\zeta \, \frac{f(\zeta)}{(\zeta - z_0)^{n+1}}, \qquad (11.151)$$

with C any contour inside the annular region where $f(z)$ is known to be analytic. □

The terms with $n = -1, -2, \ldots$ in (11.150) describe the effect of any singularities of $f(z)$ in the shaded region. If $f(z)$ is analytic in the shaded region, $a_n = 0$ for $n = -1, -2, \ldots$. If the singularities occur only at the point z_0, we speak of an isolated singularity. In this case the Laurent expansion converges for $0 < |z - z_0| < r$, where r is at least as big as the minimum distance from z_0 to the outer boundary of the region in which $f(z)$ is known to be analytic as illustrated in Figure 11.38. If the series with n negative in (11.150) stops at a finite N,

$$f(z) = a_{-N} \frac{1}{(z - z_0)^N} + a_{-N+1} \frac{1}{(z - z_0)^{N-1}} + \ldots + a_{-1} \frac{1}{(z - z_0)}$$

$$+ a_0 + a_1(z - z_0) + a_2(z - z_0)^2 + \ldots , \qquad (11.152)$$

we say $f(z)$ has a pole of order N at $z = z_0$. If the whole series with n negative is necessary $(N \to \infty)$, $f(z)$ is said to have an essential singularity at $z = z_0$.

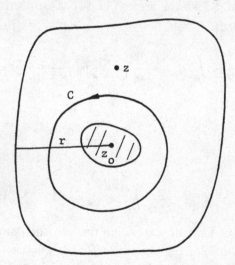

Figure 11.38: Convergence in a region outside a singularity, z_0

Finally we note that although the Laurent series is very general, it is tacitly assumed in its derivation that $f(z)$ is single valued in the annular region where the series converges. Thus the multivalued function $\log z$ <u>cannot</u> be expanded in an annular region surrounding the branch point at $z = 0$.

11.8   ANALYTIC CONTINUATION

As we have indicated on several occasions, an analytic function is a remarkable mathematical object. It is defined to be a function with a unique first derivative throughout a region. This definition is sufficiently broad to allow many interesting examples; most functions of interest in physics belong to this class. On the other hand, a function so defined has extraordinary properties. It turns out to have unique derivatives of all orders. Its real and imaginary parts satisfy Laplace's equation. Its line integral is independent of path. The values of the function at the points on a closed curve C determine its value at points inside C.

The principle of analytic continuation is the ultimate fabulous property of analytic functions. According to this principle an analytic function is uniquely determined everywhere in the complex plane by its values in any neighborhood, however small, of a point z_0. In fact, the values of the function along a path segment, however short, or even on an infinite set of discrete points with a limit point, suffice to uniquely determine the function everywhere – in these cases one must be given that the function is analytic in a region including the line segment or the infinite set of points. When we say everywhere we include that the type and location of all branch points is determined and that the analytic continuation onto all Riemann sheets associated with these branch points is determined. The only exception to these statements is the occurrence of a natural boundary line of singularities which separates off part of the complex plane.

The basic point here is contained in <u>the identity theorem for analytic functions</u>.

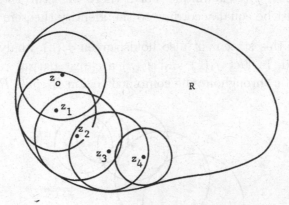

Figure 11.39: Coinciding points within circle neighborhoods as illustration of analytic continuation in a region

Theorem 11.8.1. *If two functions are analytic in a region R, and if they coincide in a neighborhood, however small, of a point z_0 of R, or only along a path segment,* *however small, terminating in z_0, or also only for an infinite number of distinct points with the limit point z_0, then the two functions are equal everywhere in R.*

Proof. Call the two functions $f_1(z)$ and $f_2(z)$. Use the expansion theorem (11.144) for analytic functions to expand the two functions in power series about the point z_0. These series will converge at least in the largest circle about z_0 which remains inside R. From the identity theorem for power series (Theorem 11.3.3; p. 283), the two power series must be identical so $f_1(z) = f_2(z)$ in the circle about z_0. Choose some point z_1 inside but near the boundary of this circle about z_0 and expand f_1 and f_2 in power series about z_1. Since $f_1(z) = f_2(z)$ near z_1 we conclude again that the power series will be identical so that $f_1(z) = f_2(z)$ inside the largest circle about z_1 which remains in R. Using a series of such circles, as illustrated in Figure 11.39, we conclude that $f_1(z) = f_2(z)$ throughout R. □

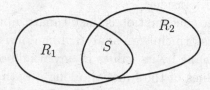

Figure 11.40: Subregion of overlapping regions

Lemma 11.8.2. *This theorem implies that an analytic continuation, if it exists at all, is unique.*

Proof. For suppose we have an analytic function $f_1(z)$ defined in a region R_1 and suppose R_2 is another region which overlaps R_1 in a subregion S as seen in Figure 11.40. If there exists a function $f_2(z)$, which is analytic in R_2 and equal to $f_1(z)$ in the common region S then $f_2(z)$ is unique. For if there were two such functions, $f_{2a}(z)$ and $f_{2b}(z)$, they must be equal according to the identity theorem just proved. □

The converse of this statement also holds: given $f_2(z)$, analytic in R_2, $f_1(z)$ is uniquely determined. In fact $f_1(z)$ and $f_2(z)$ are just partial representatives of a function $f(z)$ analytic throughout the combined region R_1 plus R_2.

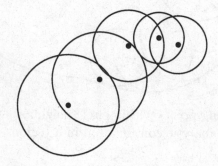

Figure 11.41: Circle chain method

The so-called circle-chain method employing power series, used in the proof of the identity theorem above, gives a method which <u>in principle</u> can be used to effect the analytic continuation. Start with the function defined by a power series in one circle. Use the values of the function so obtained to make a power series expansion about a point inside but near the edge of the circle. Etc., etc. The method is illustrated in Figure 11.41.

According to the results discussed above the analytic continuation obtained in this way is unique. The process can be pushed as far as possible by extending the circle-chain in all directions, the radii of the circles being, of course, the radii of convergence of the power series. Since wherever the function is analytic it can be expanded in a power series, these circles eventually reach to every nook and cranny of the complex plane where it is possible to analytically continue the function.

A singular point of a function is, according to the most general definition, a point ζ such that the function is not single valued and/or analytic in any small neighborhood of ζ. A function represented by a power series must have at least one such singular point on its circle of convergence. If it did not, the function would be analytic in a larger region, and since an analytic function can be expanded in a power series, it would have a larger circle of convergence.

Figure 11.42: Circle chain around a singularity

It is not possible to analytically continue through a singularity, but in general it will be possible to analytically continue around the singularity by the circle-chain method as illustrated in Figure 11.42. If one analytically continues from a given starting region to the same final region by two paths which differ by going around opposite sides of branch points, one may obtain different values, corresponding to the different branches of a multivalued function, or equivalently the different sheets of the Riemann surface.

An analytic continuation cannot be carried through only if one meets a continuous line of singularities separating one part of the complex plane from the rest. An example is provided by the power series

$$f(z) = \sum_{n=1}^{\infty} z^{n!}, \tag{11.153}$$

which has a radius of convergence $r = 1$. It can be shown[8] that as z approaches a point on the unit circle $z_0 = e^{2\pi i p/q}$, where p and q are any two integers, $|f(z)| \to \infty$. For this function the unit circle is a natural boundary.

Finally we want to point out that the circle-chain method is more a method in principle than a method to be used. Now that we know the analytic continuation is unique, any trick we can use to effect it is adequate.

An example of the sort of thing done in practice is provided by the gamma function. This function is usually defined in elementary courses for real $x > 0$ by Euler's formula

$$\Gamma(x) = \int_0^\infty dt \, e^{-t} t^{x-1}. \tag{11.154}$$

Direct integration gives $\Gamma(1) = 1$. Writing $t^{x-1} = \frac{1}{x}\frac{d}{dt}t^x$ and integrating by parts yields

$$\Gamma(x) = \frac{1}{x}\Gamma(x+1), \tag{11.155}$$

so that for integer $x = n$ we have

$$\Gamma(n+1) = n! \,. \tag{11.156}$$

The integral representation (11.154) does not converge for real $x \leq 0$.

We can analytically continue from the positive real axis to the right half complex plane with no effort at all. Since the integral converges for $\mathrm{Re}\, z > 0$ we have

$$\Gamma(z) = \int_0^\infty dt \, e^{-t} t^{z-1}, \quad z > 0. \tag{11.157}$$

The integral (11.157) diverges for $\mathrm{Re}\, z \leq 0$. We can however analytically continue to the whole complex plane by using a different integral representation due to Hankel which converges for all z:

$$\Gamma(z) = \frac{1}{2i \sin \pi z} \int_C dt \, e^t t^{z-1}, \tag{11.158}$$

where the contour C in the t plane is as indicated in Figure 11.43. If we keep the contour C away from $t = 0$, the integral converges for all z. In this formula the function t^{z-1} is defined with a cut along the negative real t axis:

$$t^{z-1} = e^{(z-1)\log t} = e^{(z-1)(\ln|t|+i\theta)}, \quad -\pi \leq \theta < \pi. \tag{11.159}$$

---

[8]Knopp. op. cit., Part I, p. 101.

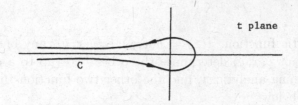

Figure 11.43: Contour in t plane avoiding $t = 0$

To show that (11.158) is the analytic continuation of (11.157) we transform (11.158) into (11.157) for $\operatorname{Re} z > 0$. According to Cauchy's theorem the contour C can be distorted as long as we do not cross any singularities. Let C hug the negative real t axis so that $\theta = -\pi$ for t coming in from $-\infty$ and $\theta = +\pi$ for t going back out to $-\infty$. Adding the contributions from the two sides of the contour, taking account of the minus sign due to reversal of the direction of integration, we find

$$e^{(z-1)(\ln|t|-i\pi)} - e^{(z-1)(\ln|t|+i\pi)} = -2i \sin \pi(z-1)e^{(z-1)\ln|t|}. \tag{11.160}$$

Letting $t' = -t = |t|$ for t on the negative axis we find easily that (11.158) becomes

$$\Gamma(z) = \int_0^\infty dt' \, e^{-t'} e^{(z-1)\ln t'}, \tag{11.161}$$

equivalent to (11.157).

The formula (11.158) thus provides one way of writing the unique analytic continuation of $\Gamma(z)$ throughout the entire z plane. Similarly the identity theorem proved at the beginning of this section tells us that since the analytic functions on the two sides of (11.155) are equal along the real axis they must be equal throughout the entire complex z plane:

$$\Gamma(z+1) = z\Gamma(z). \tag{11.162}$$

PROBLEMS

11.1 For an analytic function, $f(z) = u(x, y) + iv(x, y)$, any one of the functions $f(z)$, $u(x, y)$, or $v(x, y)$ determines the other two (up to a constant if u or v given). Assuming analyticity, find the other two functions from the one given for the cases below:

a) $u = x^2 - y^2 + x + 2$

b) $v = \sqrt{\frac{1}{2}\left(\sqrt{x^2 + y^2} - x\right)}$

c) $f = \tan^{-1} z$.

11.2 The Bernoulli numbers are defined to be the coefficients in the expansion

$$\frac{z}{e^z - 1} = \frac{1}{1 + \dfrac{z}{2!} + \dfrac{z^2}{3!} + \dfrac{z^3}{4!} + \ldots} = 1 + B_1 z + \frac{B_2}{2!}z^2 + \frac{B_3}{3!}z^3 + \ldots \quad .$$

a) Multiply up by the series in the denominator, obtain equations for the first few B_n, and show that $B_1 = -\frac{1}{2}$, $B_2 = \frac{1}{6}$, $B_3 = 0$, $B_4 = -\frac{1}{30}$, $B_5 = 0$, $B_6 = \frac{1}{42}$.

b) Show that $\dfrac{z}{e^z - 1} + \dfrac{z}{2}$ is an even function of z so that all B_n with n odd vanish after B_1.

c) Show that

$$z \cot z = iz + 1 + B_1(2iz) + \frac{B_2}{2!}(2iz)^2 + \ldots$$

$$= 1 - \frac{1}{3}z^2 - \frac{1}{45}z^4 - \frac{2}{945}z^6 + \ldots \quad .$$

11.3 Find the location and type of the singularities of the following functions. Introduce suitable cuts and discuss the Riemann sheet structure (number of sheets, topology):

a) $\dfrac{1}{4 + \sqrt{z^2 - 9}}$,

b) $\sqrt{4 + \sqrt{z^2 - 9}}$,

c) $\ln\left[5 + \sqrt{\dfrac{z + 1}{z - 1}}\right]$.

11.4 Show that

$$\arcsin z = \sin^{-1} z = \frac{1}{i}\log\left[iz + \sqrt{1 - z^2}\right].$$

Discuss in detail the branch points, cuts, and Riemann surface for this function.

11.5 Use Cauchy's integral formula with the contour a circle of radius r about z, as shown in Figure 11.44 to prove the <u>mean value theorem</u> for an analytic function $f(z)$:

$$f(z) = \frac{1}{2\pi} \int_0^{2\pi} d\theta \ f(z + re^{i\theta}).$$

Prove also the equivalent form

$$f(z) = \frac{1}{\pi R^2} \int_A dA \ f(z + re^{i\theta}).$$

Here the integral is over a circular area centered on z.

Figure 11.44: Contour of circle about z for Problem 11.5

11.6 From the second formula in Problem 10.5, show that

$$|f(z)| \leq \frac{1}{\pi R^2} \int_A dA \ \left| f(z + re^{i\theta}) \right| = \frac{1}{\pi R^2} \int_A dA \ |f(\zeta)|.$$

Use this to prove the <u>maximum modulus theorem</u>:

Theorem 11.8.3. *If $f(z)$ is analytic in a region and $|f(z)|$ attains a maximum value at some point z in the interior of the region, then $f(z)$ is identically constant throughout the region.*

An equivalent statement is that if $f(z)$ is analytic in a closed region R, then $|f(z)|$ attains its maximum value on the boundary of the region.

11.7 Use Cauchy's integral formula for the first derivative $f'(z)$ with the contour C a circle of radius r about z as in Figure 11.45 to show

$$|f'(z)| \leq \frac{M}{r},$$

where M is the maximum of $|f(z)|$ on C. Use this to prove <u>Liouville's theorem</u>:

Theorem 11.8.4. *a bounded entire function is necessarily a constant, i.e., a function which is analytic in the whole z plane and for which $|f(z)| \leq M$ everywhere is a constant.*

An equivalent statement is that the absolute value of a nonconstant entire function attains arbitrarily large values outside every circle.

Figure 11.45: Contour circle of radius r about z for Problem 11.7

11.8 Show that if $u(x, y)$ is a solution of Laplace's equation and if $v(x, y)$ is defined by the line integral

$$v(x, y) = \int_{(x_0, y_0)}^{(x,y)} \left[-\frac{\partial u}{\partial y} dx + \frac{\partial u}{\partial x} dy \right],$$

then $v(x, y)$ is independent of path and

$$f(z) = u(x, y) + iv(x, y)$$

is an analytic function.

11.9 Apply the Cauchy integral theorem to the two points

$$z_1 = r^{i\alpha}$$

and

$$z_2 = \frac{R^2}{r} e^{i\alpha},$$

one inside and one outside the contour $C = Re^{i\theta}$, as shown in Figure 11.46 to obtain for an analytic function $f(z)$

$$f(z) = \frac{1}{2\pi} \int_0^{2\pi} d\theta \, \frac{f(Re^{i\theta})}{1 - \frac{r}{R} e^{i(\alpha-\theta)}}$$

$$= \frac{1}{2\pi} \int_0^{2\pi} d\theta \, f(Re^{i\theta}) \frac{1 - \frac{r}{R} e^{-i(\alpha-\theta)}}{1 - 2\frac{r}{R} \cos(\alpha - \theta) + \frac{r^2}{R^2}},$$

$$0 = \frac{1}{2\pi} \int_0^{2\pi} d\theta \, \frac{f(Re^{i\theta})}{1 - \frac{R}{r} e^{i(\alpha-\theta)}}$$

$$= \frac{1}{2\pi} \int_0^{2\pi} d\theta \, f(Re^{i\theta}) \frac{\frac{r^2}{R^2} - \frac{r}{R} e^{-i(\alpha-\theta)}}{1 - 2\frac{r}{R} \cos(\alpha - \theta) + \frac{r^2}{R^2}},$$

so that

$$f(z) = \frac{1 - \frac{r^2}{R^2}}{2\pi} \int_0^{2\pi} d\theta \, \frac{f(Re^{i\theta})}{1 - 2\frac{r}{R} \cos(\alpha - \theta) + \frac{r^2}{R^2}}.$$

Take the real part to obtain the solution to Laplace's equation inside a circle in terms of the boundary values on the circle:

$$u(r, \alpha) = \frac{1 - \frac{r^2}{R^2}}{2\pi} \int_0^{2\pi} d\theta \, \frac{u(R, \theta)}{1 - 2\frac{r}{R}\cos(\alpha - \theta) + \frac{r^2}{R^2}}.$$

This is Poisson's solution of the Dirichlet problem for a circle.

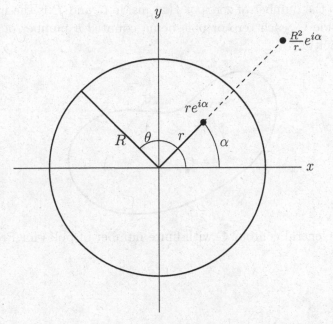

Figure 11.46: Points inside and outside a contour circle for Problem 11.9

11.10 Suppose $f(z)$ is analytic in the upper half-plane and on a portion of the real axis, as shown in Figure 11.47, and assume $f(z)$ real on this portion of real axis. Show that analytic continuation gives

$$f(z) = f^*(z^*)$$

for points z in the lower half-plane. This is the Schwarz reflection principle.

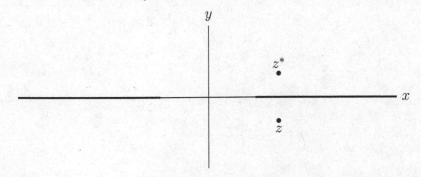

Figure 11.47: Restriction of $f(z)$ to being analytic on a portion of real axis and upper half-plane for Problem 11.10

11.11 Suppose the function $f(z)$ is <u>analytic</u> inside and on the closed curve C <u>except</u> that it has a <u>finite number of poles</u> inside C, as illustrated in Figure 11.48. Assume $f(z)$ does not vanish anywhere on C. Show that

$$\frac{1}{2\pi i}\int_C dz\, \frac{1}{f(z)}\frac{df(z)}{dz} = N - P,$$

where N is the number of zeros of $f(z)$ inside C and P is the number of poles of $f(z)$ inside C, each zero or pole being counted a number of times equal to its multiplicity.

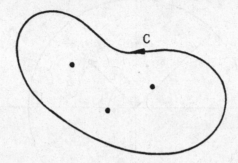

Figure 11.48: General contour C with finite number of poles for Problem 11.11

CHAPTER 12

Evaluation of Integrals

12.1 INTRODUCTION

Analytic function theory provides some powerful methods for calculating definite integrals. We discuss first the basic result known as the residue theorem and then a number of examples of its application.

12.2 THE RESIDUE THEOREM

The fundamental relation has already been exhibited in (11.127). Let us specialize to the case where $f(z)$ is analytic except for certain poles at the points z_i. The poles can be of finite or infinite order (essential singularities). Branch points are not allowed in the region under consideration since the function must be single valued in this region. We have

$$\int_C dz\, f(z) = \sum_i \int_{C_i} dz\, f(z_i)$$

$$= 2\pi i \sum_i \mathrm{Res}(f, z_i). \tag{12.1}$$

The residue of $f(z)$ at the singularity at z_i is defined by

$$\boxed{\mathrm{Res}(f, z_i) = \frac{1}{2\pi i} \int_{C_i} dz\, f(z),} \tag{12.2}$$

where C_i is a counterclockwise oriented contour enclosing the pole at z_i and only that pole as illustrated in Figure 12.1.

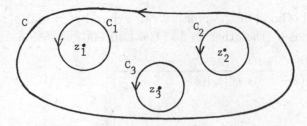

Figure 12.1: Contour with poles within contours

There are several ways to evaluate the residues. From (11.150), (11.151) we see that for a pole of any order at the point z_i the residue is the coefficient a_{-1} in the Laurent expansion of $f(z)$ about the point z_i. Thus if by any method we can find the coefficient a_{-1} in the expansion

$$f(z) = a_0 + a_1(z - z_i) + a_2(z - z_i) + \ldots$$

$$+ a_{-1}\frac{1}{z - z_i} + a_{-2}\frac{1}{(z - z_i)^2} + \ldots \quad , \tag{12.3}$$

we have

$$\text{Res}(f, z_i) = a_{-1} \tag{12.4}$$

for the residue which appears in (12.1).

If the function $f(z)$ has a simple pole of order one at z_i, the Laurent series about z_i has the form

$$f(z) = \frac{a_{-1}}{(z - z_i)} + a_0 + a_1(z - z_i) + a_2(z - z_i)^2 + \ldots \quad , \tag{12.5}$$

and we see that

$$\text{Res}(f, z_i) = a_{-1} = \lim_{z \to z_i}[(z - z_i)f(z)]. \tag{12.6}$$

For a pole of order n we have a similar but more complicated formula:

$$f(z) = \frac{a_{-n}}{(z - z_i)^n} + \frac{a_{-n+1}}{(z - z_i)^{n-1}} + \ldots + \frac{a_{-1}}{z - z_i}$$

$$+ a_0 + a_1(z - z_i) + \ldots \quad , \tag{12.7}$$

$$\text{Res}(f, z_i) = a_{-1} = \frac{1}{(n - 1)!} \lim_{z \to z_i}\left[\frac{d^{n-1}}{dz^{n-1}}\{(z - z_i)^n f(z)\}\right]. \tag{12.8}$$

A useful, easily remembered rule for a <u>simple pole</u> $(n - 1)$ is obtained by the following argument. Suppose $f(z)$ is of the form

$$f(z) = \frac{p(z)}{q(z)}, \tag{12.9}$$

where near the pole

$$p(z) \quad \text{analytic near } z_i,$$
$$q(z) \sim c(z - z_i). \tag{12.10}$$

Then in the neighborhood of the pole we can expand $p(z)$ and $q(z)$ in Taylor series and divide one series by the other to find the Laurent expansion

$$f(z) = \frac{p(z_i) + p'(z_i)(z - z_i) + \ldots}{q'(z_i)(z - z_i) + q''(z_i)(z - z_i)^2 + \ldots} \tag{12.11}$$

$$= \frac{p(z_i)}{q'(z_i)}\frac{1}{z - z_i} + \frac{p(z_i)}{q'(z_i)}\left[\frac{p'(z_i)}{p(z_i)} - \frac{q''(z_i)}{q'(z_i)}\right] + O[(z - z_i)] + \ldots \quad . \tag{12.12}$$

We see that

$$\text{Res}(f, z_i) = a_{-1} = \frac{p(z_i)}{q'(z_i)}. \tag{12.13}$$

Thus the residue of (12.9) is obtained by replacing the factor $q(z)$, which contains the simple zero, by its derivative and then evaluating everything at z_i, the pole position.

We now give some examples which illustrate how the residue theorem is used in practice.

12.3 RATIONAL FUNCTIONS $(-\infty, \infty)$

A simple example is

$$I_1 = \int_{-\infty}^{\infty} dx \, \frac{1}{x^2 + a^2}. \tag{12.14}$$

To employ the residue theorem, we need a closed contour as shown in Figure 12.2. We can add on the contribution from the semicircular arc C_2 with $R \to \infty$ since this contribution vanishes:

$$\int_{C_2} dz \, \frac{1}{z^2 + a^2} = \int_0^\pi \frac{R i e^{i\theta} d\theta}{R^2 e^{2i\theta} + a^2} \xrightarrow{R \to \infty} 0. \tag{12.15}$$

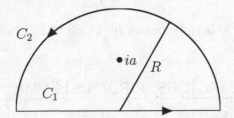

Figure 12.2: Semicircle closed contour of radius, R, with pole, ia

Thus we find

$$I_1 = \int_{C_1} dx \, \frac{1}{x^2 + a^2} = \int_{C_1 + C_2} \frac{dz}{z^2 + a^2}$$

$$= 2\pi i \text{Res}\left[\frac{1}{z^2 + a^2}, ia\right] = \frac{\pi}{a}, \quad \text{Re}\, a > 0. \tag{12.16}$$

The integral

$$I_n = \int_{-\infty}^{\infty} \frac{dx}{(x^2 + a^2)^n}, \quad n = 1, 2, 3, \dots \quad , \tag{12.17}$$

can be evaluated by using the formula (12.8) for an nth order pole or by differentiating (12.16) $n-1$ times under the integral sign:

$$I_n = \frac{(-1)^{n-1}}{(n-1)!} \frac{d^{n-1}}{d(a^2)^{n-1}} \int_{-\infty}^{\infty} \frac{dx}{x^2 + a^2}$$

$$= \frac{(-1)^{n-1}}{2^{n-1}(n-1)!} \left[\frac{1}{a}\frac{d}{da}\right]^n \left(\frac{\pi}{a}\right). \tag{12.18}$$

Variations on the above technique will work for an integral

$$I = \int_{-\infty}^{\infty} dx \, f(x) \tag{12.19}$$

with $f(x)$ a rational function, i.e.,

$$f(x) = \frac{\sum\limits_{n=0}^{N} a_n x^n}{\sum\limits_{m=0}^{M} b_m x^m}. \tag{12.20}$$

The semicircular contour at $R \to \infty$ gives a vanishing contribution as long as $M \geq N + 2$, which is necessary for (12.19) to converge. Of course, for a high order polynomial in the denominator of (12.20) it may not be easy to locate the poles, i.e., the zeros of the denominator.

Suppose now the limits on the integral are $(0, \infty)$. If the integral is an even function, i.e., $f(x) = f(-x)$, we have

$$\int_0^{\infty} dx \, f(x) = \frac{1}{2} \int_{-\infty}^{\infty} dx \, f(x), \tag{12.21}$$

and we can proceed as above. For cases in which $f(x)$ is not an even function see Section 12.5.

12.4 EXPONENTIAL FACTORS; JORDAN'S LEMMA

Consider an integral of the type

$$I = \int_{-\infty}^{\infty} dx \, e^{iax} g(x). \tag{12.22}$$

with a a real number. In order to employ the residue theorem we need a closed contour so we try to add in a semicircular contour of radius $R \to \infty$ as seen in Figure 12.3. The integral along this contour C_2 is

$$I_2 = \int_{C_2} dz \, e^{iaz} g(z) = \int_0^{\pi} iRe^{i\theta} \, d\theta \, e^{iaR\cos\theta - aR\sin\theta} g(Re^{i\theta}). \tag{12.23}$$

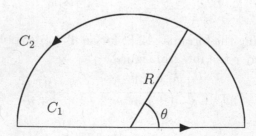

Figure 12.3: Semicircle contour of radius R without pole

Under what conditions will this integral vanish in the limit $R \to \infty$? If $a > 0$, the exponential factor $e^{-aR\sin\theta}$ approaches zero very rapidly except in a narrow angular range where $|\sin\theta| \leq C/R$, i.e., $\theta < C/R$ or $\theta > \pi - C/R$. Keeping the factor $Rd\theta$ in mind, we see that the integral (12.23) will certainly vanish provided $g(Re^{i\theta}) \to 0$ as $R \to \infty$ for all θ. Formally, if $|g(Re^{i\theta})| < G(R)$ and $a > 0$,

$$|I_2| \leq R \int_0^\pi d\theta \, e^{-aR\sin\theta} G(R)$$

$$\leq 2RG(R) \int_0^{\pi/2} d\theta \, e^{-2aR\theta/\pi}$$

$$= \frac{\pi}{a} G(R)(1 - e^{-aR}), \tag{12.24}$$

since $\sin\theta > 2\theta/\pi$ for $0 \leq \theta \leq \pi/2$. For a negative we must use a semicircular contour in the lower half-plane; the contour in the upper half-plane leads to a result which diverges exponentially as $R \to \infty$. Thus we have established <u>Jordan's lemma</u> (see Figures 12.4 and 12.5):

Lemma 12.4.1. *If* $|g(z)| = |g(Re^{i\theta})| < G(R) \xrightarrow[R\to\infty]{} 0,$

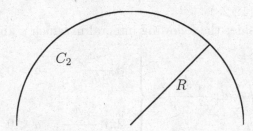

Figure 12.4: Jordan's Lemma: contour for upper half-plane

$$\int_{C_2} dz \, e^{iaz} g(z) \xrightarrow[R\to\infty]{} 0, \quad a > 0, \tag{12.25}$$

and

$$\int_{C_3} dz \, e^{iaz} g(z) \xrightarrow[R\to\infty]{} 0, \quad a < 0. \tag{12.26}$$

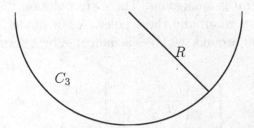

Figure 12.5: Jordan's Lemma: contour for lower half-plane

We can use these results to complete the contour and thus do the integral (12.22) by the residue theorem. If $g(z)$ is analytic except for simple poles of order one at points z_k, and $|g(z)| \to 0$ as $|z| \to \infty$, we find

$$a > 0 : \quad \int_{-\infty}^{\infty} dx \, e^{iax} g(x) = \int_{C_1 + C_2} dz \, e^{iaz} g(z)$$

$$= 2\pi i \sum_{\text{Im } z_k > 0} e^{iaz_k} \text{Res}(g, z_k), \tag{12.27}$$

$$a < 0 : \quad \int_{-\infty}^{\infty} dx \, e^{iax} g(x) = \int_{C_1 + C_3} dz \, e^{iaz} g(z)$$

$$= -2\pi i \sum_{\text{Im } z_k < 0} e^{iaz_k} \text{Res}(g, z_k). \tag{12.28}$$

For the two cases we obtain the residues in the upper or lower half z plane. The minus sign in (12.28) arises because the contour is traversed in the clockwise rather than the counterclockwise direction. The function $g(z)$ can have branch points and cuts only in the unused half of the z plane: the lower half z plane for (12.27), the upper half z plane for (12.28).

As an example consider the following integral in which a and b are real numbers:

$$\int_{-\infty}^{\infty} dx \frac{e^{iax}}{x^2 + b^2} = \begin{cases} 2\pi i \dfrac{e^{-ab}}{2ib}, & a > 0 \\[3mm] -2\pi i \dfrac{e^{ab}}{2(-ib)}, & a < 0 \end{cases}$$

$$= \frac{\pi}{b} e^{-|a|b}. \tag{12.29}$$

A very important integral differs from (12.29) only by a minus sign in the denominator:

$$\int_{-\infty}^{\infty} dx \frac{e^{iax}}{x^2 - b^2}. \tag{12.30}$$

As it stands, this integral is ambiguous. There are poles on the real axis at $x = \pm b$. We must specify how to go around these poles. A few of the many possibilities are given below; the contour around the poles is indicated by a symbol under the integral sign:

$$\int_{-\infty}^{\infty} dx \frac{e^{iax}}{x^2 - b^2} = \begin{cases} 2\pi i \left[\dfrac{e^{iab}}{2b} + \dfrac{e^{-iab}}{-2b} \right] = -\dfrac{2\pi}{b} \sin ab, & a > 0 \\[3mm] 0, & a < 0. \end{cases} \tag{12.31}$$

$$\int_{-\infty}^{\infty} dx \frac{e^{iax}}{x^2 - b^2} = \begin{cases} 0, & a > 0 \\ -2\pi i \left[\dfrac{e^{iab}}{2b} + \dfrac{e^{-iab}}{-2b}\right] = \dfrac{2\pi}{b}\sin ab, & a < 0. \end{cases} \qquad (12.32)$$

$$\int_{-\infty}^{\infty} dx \frac{e^{iax}}{x^2 - b^2} = \begin{cases} 2\pi i \dfrac{e^{-iab}}{-2b} = -i\dfrac{\pi}{b}e^{-iab}, & a > 0 \\ -2\pi i \dfrac{e^{iab}}{2b} = -i\dfrac{\pi}{b}e^{iab}, & a < 0 \end{cases} \qquad (12.33)$$

Integrals involving $\sin ax$ or $\cos ax$ can be done by writing these functions in terms of e^{iax} and e^{-iax} or by using $\cos ax = \operatorname{Re} e^{iax}$, $\sin ax = \operatorname{Im} e^{iax}$. Thus for example

$$\int_0^{\infty} dx \; \frac{\cos ax}{x^2 + b^2} = \frac{1}{2}\int_{-\infty}^{\infty} dx \; \frac{\cos ax}{x^2 + b^2}$$

$$= \frac{1}{2}\operatorname{Re}\int_{-\infty}^{\infty} dx \; \frac{e^{iax}}{x^2 + b^2}$$

$$= \frac{\pi}{2b}e^{-|a|b}, \qquad (12.34)$$

where we have used (12.29).

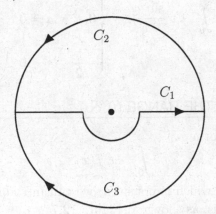

Figure 12.6: Combined contours for semicircle with displaced contour around pole on axis

Sometimes one must be careful in replacing sines or cosines by exponentials. Thus

$$\int_0^{\infty} dx \; \frac{\sin ax}{x} = \frac{1}{2}\int_{-\infty}^{\infty} dx \; \frac{\sin ax}{x} \qquad (12.35)$$

is convergent while

$$\int_0^{\infty} dx \; \frac{e^{iax}}{x} \qquad (12.36)$$

diverges near $x = 0$. Since $(\sin az)/z$ is analytic near $z = 0$, we can use Cauchy's theorem to displace the contour away from $z = 0$ in (12.35) before replacing $\sin ax$ by exponentials. See Figure 12.6 for illustration. With a positive we find

$$\int_0^\infty dx \, \frac{\sin ax}{x} = \frac{1}{2} \int_{C_1} dz \, \frac{\sin az}{z}$$

$$= \frac{1}{4i} \int_{C_1} dz \left\{ \frac{e^{iaz}}{z} - \frac{e^{-iaz}}{z} \right\}$$

$$= \frac{1}{4i} \int_{C_1+C_2} dz \, \frac{e^{iaz}}{z} - \frac{1}{4i} \int_{C_1+C_3} dz \, \frac{e^{-iaz}}{z}$$

or

$$\int_0^\infty dx \, \frac{\sin ax}{x} = \frac{1}{4i} \left[2\pi i \frac{1}{1} - 0 \right]$$

$$= \frac{\pi}{2}. \tag{12.37}$$

Since $\sin ax$ is an odd function of a and $\sin 0 = 0$ we thus find

$$\int_0^\infty dx \, \frac{\sin ax}{x} = \begin{cases} \dfrac{\pi}{2}, & a > 0 \\[2mm] 0, & a = 0 \; . \\[2mm] -\dfrac{\pi}{2}, & a < 0 \end{cases} \tag{12.38}$$

12.5 INTEGRALS ON THE RANGE $(0, \infty)$

To do an integral of the form

$$\int_0^\infty dx \, f(x) \tag{12.39}$$

involving a function $f(x)$ which is not even, we can introduce $\log z$. Taking the cut along the positive real axis, as shown in Figure 12.7

$$\log z = \ln r + i\theta, \quad 0 \le \theta < 2\pi. \tag{12.40}$$

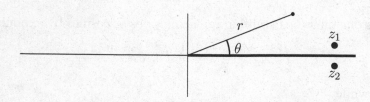

Figure 12.7: Cut along positive real axis

The discontinuity of $\log z$ across the cut is

$$\log z_1 - \log z_2 = -2\pi i. \tag{12.41}$$

Thus if $f(z)$ is analytic along the real axis, and provided the small semicircle around the origin gives a vanishing contribution, as shown in Figure 12.8, we have

$$\int_0^\infty dx \; f(x) = -\frac{1}{2\pi i} \int_{C_1} dz \; f(z) \log z. \tag{12.42}$$

If furthermore $f(z) \to 0$ for $|z| \to \infty$ as fast as $|z|^{-1-\epsilon}$, $\epsilon > 0$, we can complete the contour by adding in a vanishing contribution from C_2. If, then, $f(z)$ has only simple poles of order one at points z_k, with residues $\mathrm{Res}(f, z_k)$, we find

$$\int_0^\infty dx \; f(x) = -\sum_k \mathrm{Res}(f, z_k) \log z_k. \tag{12.43}$$

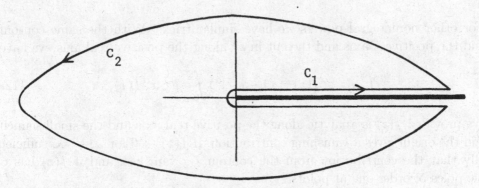

Figure 12.8: Contour avoiding positive x axis

For example with $n = 2, 3, 4, 5, \ldots$

$$\int_0^\infty dx \; \frac{1}{1+x^n} = -\sum_k \frac{\log z_k}{n z_k^{n-1}}, \quad z_k = n^{th} \text{ roots of } (-1). \tag{12.44}$$

For the case $n = 3$, $z_1 = e^{i\pi/3}$, $z_2 = e^{i\pi}$, $z_3 = e^{i5\pi/3}$,

$$\int_0^\infty dx \; \frac{1}{1+x^3} = -\left[\frac{i\pi/3}{3e^{2i\pi/3}} + \frac{i\pi}{3e^{2i\pi}} + \frac{5i\pi/3}{3e^{10i\pi/3}} \right]$$

$$= -\frac{i\pi}{3}\left[\frac{1}{3}\left(-\frac{1}{2} - i\frac{\sqrt{3}}{2}\right) + 1 + \frac{5}{3}\left(-\frac{1}{2} + i\frac{\sqrt{3}}{2}\right) \right]$$

$$= \frac{2\pi\sqrt{3}}{9}. \tag{12.45}$$

A similar device is useful for

$$\int_0^\infty dx \ \sqrt{x} f(x). \tag{12.46}$$

Take the cut for $\sqrt{z}$ along the positive axis. Then with z_1 and z_2, as indicated in Figure 12.7, $\sqrt{z_2} = -\sqrt{x} = -\sqrt{z_1}$. Assuming $f(z)$ is analytic along the positive real axis and provided the small semicircle around the origin gives a vanishing contribution,

$$\int_0^\infty dx \ \sqrt{x} f(x) = \frac{1}{2} \int_{C_1} dz \ \sqrt{z} f(z), \tag{12.47}$$

with C_1 the contour around the positive real axis in Figure 12.8. If furthermore $f(z)$ goes to zero sufficiently rapidly as $|z| \to \infty$ so that the contribution from the contour C_2 vanishes and if $f(z)$ has only simple poles of order one at points z_k, we find

$$\int_0^\infty dx \ \sqrt{x} f(x) = i\pi \sum_k \sqrt{z_k} \ \text{Res}(f, z_k). \tag{12.48}$$

For other nonintegral powers we have similar tricks. With the same contour C around the positive x axis and the cut in z^α along the positive real axis, we have

$$\int_{C_1} dz \ z^\alpha f(z) = (1 - e^{2\pi i \alpha}) \int_0^\infty dx \ x^\alpha f(x), \tag{12.49}$$

always provided $f(z)$ is analytic along the positive real axis and the small semicircle around the origin gives a vanishing contribution. If $f(z) \to 0$ for $|z| \to \infty$ sufficiently rapidly that the contribution from the contour C_2 vanishes, and if $f(z)$ has only simple poles of order one at points z_k,

$$\int_0^\infty dx \ x^\alpha f(x) = \frac{2\pi i}{1 - e^{2\pi i \alpha}} \sum_k z_k^\alpha \ \text{Res}(f, z_k). \tag{12.50}$$

Thus, for example, we find

$$\int_0^\infty dt \ \frac{t^{z-1}}{1+t} = \frac{2\pi i}{1 - e^{2\pi i(z-1)}} e^{i\pi(z-1)}$$

$$= \frac{-\pi}{\sin \pi(z-1)}$$

$$= \frac{\pi}{\sin \pi z} \tag{12.51}$$

for $0 < \text{Re } z < 1$, so that the integral converges.

12.6 ANGULAR INTEGRALS

Integrals with respect to θ of functions of $\sin\theta$, $\cos\theta$ can often be done by making the substitution

$$e^{i\theta} = z,$$

$$\cos\theta = \frac{1}{2}\left(z + \frac{1}{z}\right),$$

$$\sin\theta = \frac{1}{2i}\left(z - \frac{1}{z}\right),$$

$$d\theta = \frac{dz}{iz}. \tag{12.52}$$

If $f(\cos\theta, \sin\theta)$ is a rational function of $\cos\theta$ and $\sin\theta$, the above substitution will transform the integral

$$\int d\theta\, f(\cos\theta, \sin\theta) = \int dz\, R(z) \tag{12.53}$$

into the integral with respect to z of a rational function of z. Such integrals can be done by the usual partial fractions decomposition technique of elementary calculus.

If the limits on the θ integral are $(0, 2\pi)$, the corresponding contour in the z plane is the unit circle, and with such a contour we can use the residue theorem. For example we have

$$\int_0^{2\pi} d\theta\, \frac{1}{a + b\cos\theta} = \int_C \frac{dz}{iz}\, \frac{1}{a + \frac{b}{2}\left(z + \frac{1}{z}\right)}$$

$$= \frac{2}{ib}\int_C \frac{dz}{z^2 + \frac{2az}{b} + 1}, \tag{12.54}$$

where C is the unit circle. Assume $a > |b| > 0$, so that $a + b\cos\theta$ has no zeros in the region of integration. The denominator in the integral on the right side of (12.54) has zeros at

$$z_1 = -\frac{a}{b} + \sqrt{\frac{a^2}{b^2} - 1}\,, \quad z_2 = -\frac{a}{b} - \sqrt{\frac{a^2}{b^2} - 1}\,, \tag{12.55}$$

which are such that $z_1 z_2 = 1$, so that one zero is inside the unit circle, the other outside. Using the residue theorem we find, for $b > 0$, z_1 inside the unit circle,

$$\int_0^{2\pi} d\theta\, \frac{1}{a + b\cos\theta} = \frac{2}{ib}2\pi i\, \frac{1}{2z_1 + \frac{2a}{b}} = \frac{2\pi}{\sqrt{a^2 - b^2}}\,. \tag{12.56}$$

The case $b < 0$ leads to the same result.

12.7 TRANSFORMING THE CONTOUR

Occasionally an integral can be transformed to a more convenient form by using Cauchy's theorem to move the contour. Consider the integrals

$$F_1 = \int_0^\infty dx \, \cos ax^2$$

$$F_2 = \int_0^\infty dx \, \sin ax^2,$$ (12.57)

which form the real and imaginary parts of

$$F = F_1 + iF_2 = \int_0^\infty dx \, e^{iax^2}.$$ (12.58)

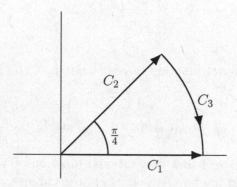

Figure 12.9: Transforming positive x axis, C_1, to a preferred contour, $C_2 + C_3$

Since e^{iaz^2} is analytic in the entire z plane we can use Cauchy's theorem to write, with reference to Figure 12.9 for the contours,

$$F = \int_{C_1} dz \, e^{iaz^2} = \int_{C_2} dz \, e^{iaz^2} + \int_{C_3} dz \, e^{iaz^2}.$$ (12.59)

Along the contour C_3 at large R we have

$$e^{iaz^2} = e^{iaR^2(\cos^2\theta - \sin^2\theta) - 2aR^2\cos\theta\sin\theta}.$$ (12.60)

This goes exponentially to zero as $R \to \infty$ except in a small angular range $\theta < c/R^2$ near $\theta = 0$. Since this angular range $\propto 1/R^2$ and $dz = iRe^{i\theta}d\theta$, we see that the contribution from the contour C_3 vanishes in the limit $R \to \infty$. Thus we find

$$F_1 + iF_2 = \int_{C_2} dz \, e^{iaz^2} = e^{i\pi/4} \int_0^\infty dr \, e^{-ar^2}$$

$$= \frac{1+i}{\sqrt{2}} \frac{1}{2} \sqrt{\frac{\pi}{a}}$$ (12.61)

or

$$\int_0^\infty dz \, \cos ax^2 = \int_0^\infty dx \, \sin ax^2 = \frac{1}{2} \sqrt{\frac{\pi}{2a}}.$$ (12.62)

In this and the preceding sections we have shown some ways in which Cauchy's theorem and the residue theorem can be used to do integrals. Evidently there are many tricks; the main limitation is generally the ingenuity of the practitioner.

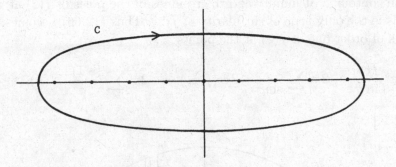

Figure 12.10: Contour including the poles of $1/\sin \pi z$

12.8 PARTIAL FRACTION AND PRODUCT EXPANSIONS

A slightly different application of the residue theorem leads to a method for calculating certain infinite sums. The function $\sin \pi z$ has an infinite series of zeros

$$\sin \pi z = 0 \quad \text{at} \quad z = 0, \pm 1, \pm 2, \dots , \tag{12.63}$$

so that its reciprocal has an infinite series of poles:

$$\frac{1}{\sin \pi z} \quad \text{has } 1^{st} \text{ order poles at } z = 0, \pm 1, \pm 2, \dots . \tag{12.64}$$

The residues are given by

$$\text{Res}\left[\frac{1}{\sin \pi z}, n\right] = \frac{1}{\pi \cos \pi n} = \frac{(-1)^n}{\pi}. \tag{12.65}$$

Thus if $f(z)$ is analytic near the integers $z = n$,

$$\int_C dz \, \frac{f(z)}{\sin \pi z} = -2\pi i \sum_n \frac{(-1)^n}{\pi} \, f(n), \tag{12.66}$$

where C is a contour surrounding the poles of $(\sin \pi z)^{-1}$ in the clockwise sense as shown in Figure 12.10. Note that if $f(z)$ has poles on or near the real axis we would indent the contour C so as to exclude them as in Figure 12.11.

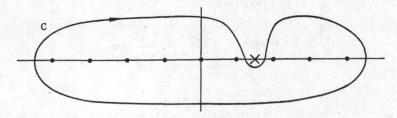

Figure 12.11: Contour choice to exclude a pole on or near the real axis

We now use Cauchy's theorem to push the contour out to infinity, carefully going around any poles z_k of $f(z)$ as shown in Figure 12.12. The contributions from contiguous portions of the contour traversed in opposite directions cancel, and we are left with the contour Γ at infinity and the residues of the poles of $f(z)$ at z_k, assuming that this is the only type of singularity of $f(z)$. Thus (12.66) becomes, assuming simple poles of order one of $f(z)$ at the points z_k,

$$\int_{\Gamma} dz \, \frac{f(z)}{\sin \pi z} + 2\pi i \sum_{z_k} \frac{1}{\sin \pi z_k} \mathrm{Res}(f, z_k) = -2\pi i \sum_{n} \frac{(-1)^n}{\pi} \, f(n). \qquad (12.67)$$

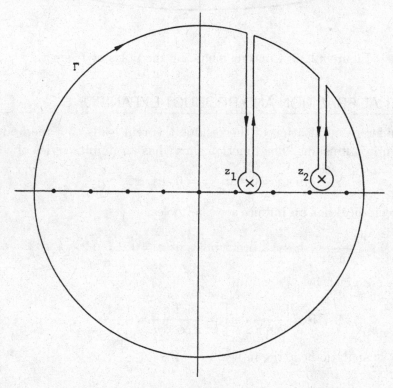

Figure 12.12: Contour, Γ, expanding to infinity while avoiding poles z_1 and z_2

If $f(z)/\sin \pi z \to 0$ at infinity sufficiently rapidly, the integral around the contour Γ at infinity $\to 0$. This is not a very restrictive requirement since $(\sin \pi z)^{-1} \to 0$ exponentially except near the real axis. However, it must be checked for each case. Assuming then $\int_{\Gamma} \to 0$, (12.67) becomes

$$\sum_{n=-\infty}^{\infty} (-1)^n f(n) = -\pi \sum_{z_k} \frac{1}{\sin \pi z_k} \mathrm{Res}(f, z_k). \qquad (12.68)$$

As an example take

$$f(z) = \frac{1}{z - \zeta}, \quad z_k = \zeta, \quad \mathrm{Res}(f, z_k) = 1, \qquad (12.69)$$

to obtain

$$\sum_{n=-\infty}^{\infty} (-1)^n \frac{1}{n-\zeta} = -\frac{\pi}{\sin \pi \zeta} \tag{12.70}$$

or equivalently

$$\frac{1}{\sin \pi \zeta} = \frac{1}{\pi \zeta} - \frac{2\zeta}{\pi} \sum_{n=1}^{\infty} \frac{(-1)^n}{n^2 - \zeta^2}. \tag{12.71}$$

This is the partial fraction expansion of $\csc \pi \zeta$.

To cancel the factor $(-1)^n$ in the sum on the left of (12.68) one can include a factor $\cos \pi z$ in $f(z)$. Consider, for example, the case

$$f(z) = \frac{\cos \pi z}{z - \zeta}. \tag{12.72}$$

For this case one might worry that the integral over the contour at infinity,

$$\int_\Gamma dz \, \frac{f(z)}{\sin \pi z} = \int_\Gamma dz \, \frac{1}{z - \zeta} \frac{\cos \pi z}{\sin \pi z}, \tag{12.73}$$

might not vanish. However, for $|z| \to \infty$ we have

$$\frac{\cos \pi z}{\sin \pi z} = i \frac{e^{i\pi z} + e^{-i\pi z}}{e^{i\pi z} - e^{-i\pi z}} \longrightarrow \begin{cases} -i, & 0 < \theta < \pi \\ +i, & \pi < \theta < 2\pi \end{cases} \tag{12.74}$$

as $r \to \infty$ at fixed θ in $z = re^{i\theta}$. Thus the contribution to (12.73) from the upper half z plane cancels the contribution from the lower half z plane. The formula (12.68) with $f(z)$ given by (12.72) thus yields

$$\sum_{n=-\infty}^{\infty} \frac{1}{n-\zeta} = -\pi \frac{\cos \pi \zeta}{\sin \pi \zeta}, \tag{12.75}$$

or equivalently

$$\cot \pi \zeta = \frac{1}{\pi \zeta} - \frac{2\zeta}{\pi} \sum_{n=1}^{\infty} \frac{1}{n^2 - \zeta^2}, \tag{12.76}$$

which is the partial fraction expansion of $\cot \pi \zeta$.

A less daring way to derive this result is to take

$$f(z) = \frac{\cos \pi z}{(z - \alpha)(z - \beta)} \tag{12.77}$$

in (12.68); then there is no doubt that the contribution from the contour integral at $|z| \to \infty$ vanishes. We find

$$\sum_{n=-\infty}^{\infty} \frac{1}{(n-\alpha)(n-\beta)} = \frac{1}{\alpha\beta} + \sum_{n=1}^{\infty} \frac{2(n^2 + \alpha\beta)}{(n^2 - \alpha^2)(n^2 - \beta^2)}$$

$$= -\pi \frac{\cot \pi \alpha - \cot \pi \beta}{\alpha - \beta}. \tag{12.78}$$

Expanding in the limit $\beta \to 0$ gives (12.76) with ζ replaced by α.

We can derive an infinite product expansion from (12.76) by writing it as

$$\frac{1}{\pi} \frac{d}{d\zeta} \log \left[\frac{\sin \pi \zeta}{\pi z} \right] = \frac{1}{\pi} \sum_{n=1}^{\infty} \frac{d}{d\zeta} \log(\zeta^2 - n^2), \tag{12.79}$$

integrating from $\zeta = 0$ to $\zeta = z$,

$$\log \left[\frac{\sin \pi z}{\pi z} \right] = \sum_{n=1}^{\infty} \log \left[1 - \frac{z^2}{n^2} \right], \tag{12.80}$$

and exponentiating,

$$\sin \pi z = \pi z \prod_{n=1}^{\infty} \left[1 - \frac{z^2}{n^2} \right]. \tag{12.81}$$

In addition to its utility in partial fraction expansions such as (12.76), the device described in this section is important in the part of high energy physics called Regge pole theory. The partial wave expansion for a scattering problem

$$f(\theta) = \sum_{l=0}^{\infty} (2l + 1) f_l P_l(\cos \theta) \tag{12.82}$$

is rewritten as an integral in the l plane, as shown in Figure 12.13, with the aid of the so-called Sommerfeld-Watson transformation:

$$f(\theta) = \frac{-\pi}{2\pi i} \int_{C_1} dl \, \frac{(2l + 1)}{\sin \pi l} f_l P_l(-\cos \theta). \tag{12.83}$$

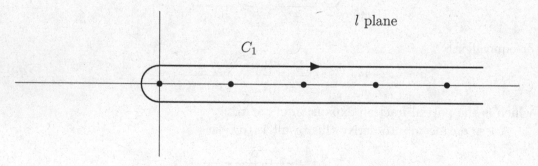

Figure 12.13: Contour, C_1, in complex l plane

Notice that the property $P_l(\cos \theta) = (-1)^l P_l(-\cos \theta)$ of the Legendre polynomials has been used to absorb the factor $(-1)^l$ from the residue of $(\sin \pi l)^{-1}$. The contour C_1 is now deformed so as to go up parallel to the imaginary axis, as shown in Figure 12.14. If f_l regarded as a function of l has poles at $l = \alpha_1, \alpha_2, \ldots$ with residues $\beta_1, \beta_2, \ldots$, we find

$$f(\theta) = -\pi \sum_k \frac{2\alpha_k + 1}{\sin \pi \alpha_k} \beta_k P_{\alpha_k}(-\cos \theta) - \frac{\pi}{2\pi i} \int_{C_2} dl \, \frac{2l + 1}{\sin \pi l} f_l P_l(-\cos \theta). \tag{12.84}$$

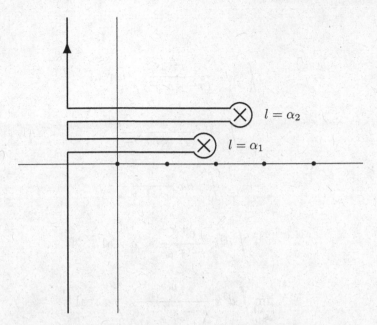

Figure 12.14: Deformed contour, C_1, parallel to imaginary axis avoiding poles

Under certain conditions the background integral along the contour C_2 can be neglected, leaving only the contributions of the Regge poles at $l = \alpha_k$ as illustrated in Figure 12.15.

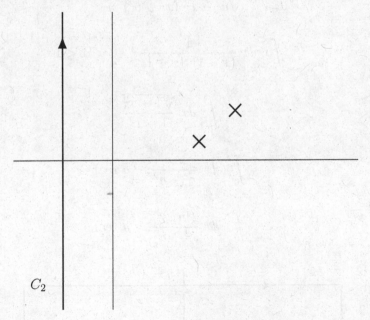

Figure 12.15: Contour, C_2, remains with sum of the residue poles

PROBLEMS

12.1
$$\int_0^\infty dx\, \frac{x^2}{(x^2+a^2)^2}, \quad a \text{ real}$$

12.2
$$\int_0^\infty dx\, \frac{1}{x^4+5x^2+6}$$

12.3
$$\int_0^\infty dx\, \frac{1}{x^4+1}$$

12.4
$$\int d^3k\, \frac{e^{i\mathbf{k}\cdot\mathbf{r}}}{k^2+a^2}, \quad a \text{ real}$$

12.5
$$\lim_{\substack{\epsilon\to 0 \\ \epsilon>0}} \int d^3k\, \frac{e^{i\mathbf{k}\cdot\mathbf{r}}}{k^2-a^2-i\epsilon}, \quad a \text{ real}$$

12.6
$$\int_0^\infty dx\, \frac{\sin x}{x(x^2+a^2)}, \quad a \text{ real}$$

12.7
$$\int_{-a}^a dx\, \frac{1}{\sqrt{a^2-x^2}\,(x^2+b^2)}, \quad a,b \text{ real}$$

12.8
$$\int_1^\infty dx\, \frac{1}{(x-1)^\alpha(x+1)}, \quad 0<\alpha<1$$

12.9
$$\int_{-1}^1 dx\, \frac{\sqrt{1-x^2}}{1+x^2}$$

12.10
$$\int_0^\infty dx\, \frac{\ln x}{x^2+a^2}, \quad a \text{ real}$$

12.11
$$\int_0^\infty dx\, \frac{(\ln x)^2}{x^2+a^2}, \quad a \text{ real}$$

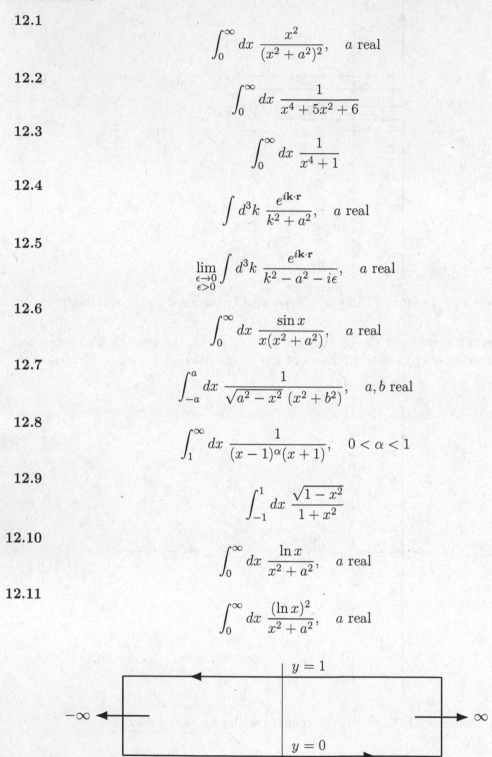

Figure 12.16: Contour to use for Problem 12.12

12.12 Use contour in Figure 12.16 to solve

$$\int_0^\infty dx \, \frac{\cosh \alpha x}{\cosh \pi x}, \quad -\pi < \alpha < \pi$$

12.13 Evaluate

$$\sum_{n=1}^\infty \frac{1}{n^2 + a^2}, \quad a \text{ real}$$

12.14 Show that

$$\frac{1}{\sin^2 z} = \sum_{n=-\infty}^\infty \frac{1}{(z - n\pi)^2}$$

12.15 Use (12.71) and (12.76) to evaluate the following numerical series

$$\frac{\pi}{4} = 1 - \frac{1}{3} + \frac{1}{5} - \frac{1}{7} + \frac{1}{9} - \cdots$$

$$\frac{\pi^2}{12} = 1 - \frac{1}{2^2} + \frac{1}{3^2} - \frac{1}{4^2} + \cdots$$

$$\frac{\pi^2}{6} = 1 + \frac{1}{2^2} + \frac{1}{3^2} + \frac{1}{4^2} + \cdots$$

$$\frac{\pi^2}{8} = 1 + \frac{1}{3^2} + \frac{1}{5^2} + \frac{1}{7^2} + \cdots$$

CHAPTER 13

Dispersion Relations

13.1 INTRODUCTION

A dispersion relation is a formula for the real part of an analytic function as an integral over its imaginary part. It is an application of Cauchy's integral formula (11.132). The name dispersion relation is due to physicists and will not be found in the mathematical literature. Similar problems have been studied by the mathematicians under names such as Hilbert transforms, the Riemann-Hilbert problem, the Wiener-Hopf method and Carleman's method. The physicist's name arises from the original applications of the technique by Kramers and Krönig to the study of the index of refraction describing the dispersion of light waves. More recently further refinements of the method have been employed extensively in elementary particle theory as an alternative to quantum field theory and perturbation expansions.

In this chapter we study a few of the basic mathematical questions without any indication of the physical significance of the complex variable z or the analytic function $f(z)$. Suffice it to say that it is often possible by analytic continuation to attach significance to such things as a complex energy plane or a complex wave number plane or a complex angular momentum plane and also to the corresponding amplitudes for physical processes as analytic functions of these complex variables.

A dispersion relation is an identity satisfied by the real and imaginary parts of a wide class of analytic functions. If some further relation is specified between the real and imaginary parts of the analytic function, the dispersion relation can be converted into a certain type of singular integral equation. In the last section of this chapter we study some methods for solving such equations.

13.2 PLEMELJ FORMULAS; DIRAC'S FORMULA

Consider an integral of the type

$$F(z) = \int dx' \, \frac{f(x')}{x' - z}. \tag{13.1}$$

We are deliberately vague about the limits on the integral. For our purposes here we suppose the integration is along all or part of the real axis, e.g.,

$$\int_a^b dx' \, \frac{f(x')}{x' - z} \quad \text{or} \quad \int_a^\infty dx' \, \frac{f(x')}{x' - z} \quad \text{or} \quad \int_{-\infty}^{-b} dx' \, \frac{f(x')}{x' - z} + \int_a^\infty dx' \, \frac{f(x')}{x' - z}. \tag{13.2}$$

We can treat the various cases together by taking $f(x')$ equal to zero along certain portions of the real axis.

An integral of the type (13.1) appears in the Cauchy integral formula (11.132) except that there we have a closed contour, not necessarily along the real axis. We shall restrict ourselves in this chapter to integrals along the real axis. However, our considerations can obviously be generalized to integrals along other arcs or closed contours in the complex plane.

Of particular interest are the values of the integral (13.1) for z on or near the real axis:

$$F(x) = \int dx' \, \frac{f(x')}{x' - x}. \tag{13.3}$$

As it stands this integral is ambiguous, like the examples (12.30)-(12.33). We must specify how one goes around the pole at $x' = x$ on the real axis. We can approach from above,

$$F_+(x) = \lim_{\substack{\epsilon \to 0 \\ \epsilon > 0}} F(x + i\epsilon) = \lim_{\substack{\epsilon \to 0 \\ \epsilon > 0}} \int dx' \, \frac{f(x')}{x' - x - i\epsilon}. \tag{13.4}$$

or below,

$$F_-(x) = \lim_{\substack{\epsilon \to 0 \\ \epsilon > 0}} F(x - i\epsilon) = \lim_{\substack{\epsilon \to 0 \\ \epsilon > 0}} \int dx' \, \frac{f(x')}{x' - x + i\epsilon}. \tag{13.5}$$

Another possibility is the Cauchy principle value integral designated by the letter $\mathcal{P}$ and defined by

$$\mathcal{P} \int_a^b dx' \, \frac{f(x')}{x' - x} = \lim_{\substack{\epsilon \to 0 \\ \epsilon > 0}} \left[\int_a^{x-\epsilon} dx' \, \frac{f(x')}{x' - x} + \int_{x+\epsilon}^b dx' \, \frac{f(x')}{x' - x} \right]. \tag{13.6}$$

If $f(x')$ is sufficiently smooth near $x' = x$, the contributions from opposite sides of the singular point will cancel in (13.6) because of the change in sign of $x' - x$, and the limit will exist.

If we assume that $f(z')$ is analytic near the real axis, we can use Cauchy's theorem to deform the contours in (13.4), (13.5) with a semicircular indentation of radius $\epsilon \to 0$ with the singularity above the contour as shown in Figure 13.1. We can then set $z = x$ with no difficulty. Thus we find

$$F_+(x) = \int_{C_1} dz' \, \frac{f(z')}{z' - x}$$

$$= \lim_{\substack{\epsilon \to 0 \\ \epsilon > 0}} \left[\int^{x-\epsilon} dx' \, \frac{f(x')}{x' - x} + \int_{x+\epsilon} dx' \, \frac{f(x')}{x' - x} + i \int_{-\pi}^0 d\theta \, f(x + \epsilon e^{i\theta}) \right]$$

$$= \mathcal{P} \int dx' \, \frac{f(x')}{x' - x} + i\pi f(x), \tag{13.7}$$

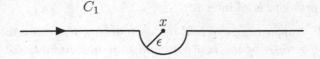

Figure 13.1: Singularity above contour

and similarly, with the singularity below the indented contour as shown in Figure 13.2,

$$F_-(x) = \int_{C_2} dz' \, \frac{f(z')}{z' - x}$$

$$= \lim_{\substack{\epsilon \to 0 \\ \epsilon > 0}} \left[\int^{x-\epsilon} dx' \, \frac{f(x')}{x' - x} + \int_{x+\epsilon} dx' \, \frac{f(x')}{x' - x} + i \int_\pi^0 d\theta \, f(x + \epsilon e^{i\theta}) \right]$$

$$= \mathcal{P} \int dx' \, \frac{f(x')}{x' - x} - i\pi f(x). \tag{13.8}$$

Figure 13.2: Singularity below contour

The results (13.7), (13.8) are called the Plemelj formulas. Physicists usually remember the formulas by writing them in the form

$$\boxed{\int dx' \, \frac{f(x')}{x' - x \mp i\epsilon} = \mathcal{P} \int dx' \, \frac{f(x')}{x' - x} \pm i\pi f(x)} \tag{13.9}$$

and noting that since $f(x')$ is an arbitrary function, this implies

$$\boxed{\frac{1}{x' - x \mp i\epsilon} = \mathcal{P} \frac{1}{x' - x} \pm i\pi\delta(x - x'),} \tag{13.10}$$

which is sometimes called Dirac's formula.

13.3 DISCONTINUITY PROBLEM

The function $F(z)$ given by the integral (13.1) is an analytic function of z except along those portions of the real axis where $f(x)$ is nonvanishing. $F(z)$ has a cut along these portions of the real axis, and the discontinuity across this cut is given by (13.7), (13.8):

$$F_+(x) - F_-(x) = 2\pi i f(x). \tag{13.11}$$

The converse problem is of interest.

Theorem 13.3.1. *Suppose we are told that a function $F(z)$ is analytic except for a cut along certain portions of the real axis and that the discontinuity across this cut is given by a specified $f(x)$ according to (13.11). Can we find $F(z)$? Evidently one function satisfying these requirements is*

$$F(z) = \int dx' \, \frac{f(x')}{x' - z}.$$ (13.12)

Proof. If there is another function $G(z)$ satisfying the same requirements, then $F(z) - G(z)$ is an entire function, i.e., a function analytic throughout the entire z plane. So the general solution of the problem is

$$G(z) = \int dx' \, \frac{f(x')}{x' - z} + E(z),$$ (13.13)

with $E(z)$ an arbitrary entire function.

To restrict the solution further, one can impose limitations on how fast $E(z) \to \infty$ as $z \to \infty$. An entire function can be expanded in a Taylor series

$$E(z) = \sum_{n=0}^{\infty} a_n z^n,$$ (13.14)

which converges for all z. The coefficients are given by the formula (11.145):

$$a_n = \frac{1}{2\pi i} \int_C d\zeta \, \frac{E(\zeta)}{\zeta^{n+1}}.$$ (13.15)

Taking C to be a circle of radius R, we see that

$$|a_n| \leq \frac{E_{\max}(R)}{R^n},$$ (13.16)

where $E_{\max}(R)$ is the maximum value of $|E(z)|$ on the circle of radius R. If we specify that $E(z)$ go to infinity no faster than $|z|^N$, i.e.,

$$|E(z)| \leq C \, |z|^N,$$ (13.17)

then we have

$$a_n \leq C R^{N-n}.$$ (13.18)

Since R is arbitrary this implies

$$a_n = 0, \quad n > N.$$ (13.19)

Thus the limitation (13.17) on the growth rate of $E(z)$ implies that $E(z)$ is a polynomial:

$$E(z) = \sum_{n=0}^{N} a_n z^n.$$ (13.20)

In particular, if $E(z)$ is to remain finite as $|z| \to \infty$, $E(z)$ can only be a constant, independent of z.

The function given by the integral (13.12) vanishes as $z \to \infty$. If we impose the boundary condition $F(z) \xrightarrow[z \to \infty]{} 0$, then $E(z) = 0$ and (13.12) is the unique solution of the discontinuity problem. $\square$

13.4 DISPERSION RELATIONS; SPECTRAL REPRESENTATIONS

Consider a function $f(z)$ which has the properties

1) $f(z)$ is analytic in the upper half z plane,

2) $f(z) \to 0$ as $|z| \to \infty$ in the upper half z plane.

Apply Cauchy's integral formula (11.132) to the function $f(z)$ with z a point in the upper half-plane and C a contour in the upper half-plane, as shown in Figure 13.3:

$$f(z) = \frac{1}{2\pi i} \int_C dz' \, \frac{f(z')}{z' - z}. \tag{13.21}$$

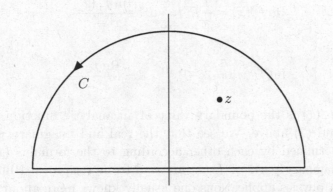

Figure 13.3: Contour in upper half-plane

Now expand the contour until it consists of the real axis plus a semicircular arc at infinity. In view of the property (2) above, the contribution from the semicircular arc will vanish, and we are left with

$$f(z) = \frac{1}{2\pi i} \int_{-\infty}^{\infty} dx' \, \frac{f_+(x')}{x' - z}. \tag{13.22}$$

In this formula

$$f_+(x') \equiv \lim_{\substack{\epsilon \to 0 \\ \epsilon > 0}} f(x' + i\epsilon) \tag{13.23}$$

is the boundary value of the analytic function $f(z)$ as the real axis is approached from above. This is evidently what is needed in (13.22), since we start with a contour above the real axis and push it down onto the real axis from above.

We now let z approach the real axis from above: $z \to x + i\epsilon$, $\epsilon \to 0$, $\epsilon > 0$. Then (13.22) becomes

$$f_+(x) = \frac{1}{2\pi i} \int_{-\infty}^{\infty} dx' \, \frac{f_+(x')}{x' - x - i\epsilon}, \tag{13.24}$$

where the limit symbol $\lim_{\substack{\epsilon \to 0 \\ \epsilon > 0}}$ is suppressed but understood. We can now use Dirac's formula (13.10) to evaluate (13.24):

$$f_+(x) = \frac{1}{2\pi i} \int_{-\infty}^{\infty} dx' \, f_+(x') \left[\frac{\mathcal{P}}{x' - x} + i\pi\delta(x' - x) \right]$$

$$= \frac{1}{2\pi i} \mathcal{P} \int_{-\infty}^{\infty} dx' \, \frac{f_+(x')}{x' - x} + \frac{1}{2} f_+(x), \tag{13.25}$$

or equivalently

$$f_+(x) = \frac{1}{\pi i} \mathcal{P} \int_{-\infty}^{\infty} dx' \, \frac{f_+(x')}{x' - x}. \tag{13.26}$$

Taking the real and imaginary parts of (13.26) we find

$$\operatorname{Re} f_+(x) = \frac{1}{\pi} \mathcal{P} \int_{-\infty}^{\infty} dx' \, \frac{\operatorname{Im} f_+(x')}{x' - x}, \tag{13.27}$$

$$\operatorname{Im} f_+(x) = -\frac{1}{\pi} \mathcal{P} \int_{-\infty}^{\infty} dx' \, \frac{\operatorname{Re} f_+(x')}{x' - x}. \tag{13.28}$$

We recall that $f_+(x)$ is the boundary value of an analytic function $f(z)$ satisfying properties (1) and (2) above. We see that the real and imaginary parts of such a function are determined by each other according to the formulas (13.27), (13.28). The real and imaginary parts of $f_+(x)$ are said to be reciprocal Hilbert transforms of each other. In physics applications one usually knows more about $\operatorname{Im} f_+(x)$ than $\operatorname{Re} f_+(x)$, so one concentrates on the relation (13.27), which is a formula for the real part $\operatorname{Re} f_+(x)$, given the imaginary part $\operatorname{Im} f_+(x')$ for all x'. In physics applications (13.27) would be called a dispersion relation for $f_+(x)$.

A related formula for $f_+(x)$ is obtained as follows:

$$f_+(x) = \operatorname{Re} f_+(x) + i \operatorname{Im} f_+(x)$$

$$= \frac{1}{\pi} \mathcal{P} \int_{-\infty}^{\infty} dx' \, \frac{\operatorname{Im} f_+(x')}{x' - x} + i \operatorname{Im} f_+(x)$$

$$= \frac{1}{\pi} \int_{-\infty}^{\infty} dx' \, \operatorname{Im} f_+(x') \left[\frac{\mathcal{P}}{x' - x} + i\pi\delta(x' - x) \right]$$

$$\boxed{f_+(x) = \frac{1}{\pi} \int_{-\infty}^{\infty} dx' \, \frac{\operatorname{Im} f_+(x')}{x' - x - i\epsilon},} \tag{13.29}$$

where again the limit as $\epsilon \to 0$, $\epsilon > 0$ is understood. The formula (13.29) would be called a spectral representation for the function $f_+(x)$. The point is that one can find the whole function $f_+(x)$ from just its imaginary part. If some other considerations give us information about the imaginary part, this may be very useful.

From the last formula (13.29) we can analytically continue back up into the upper half z plane by the simple expedient of replacing $x + i\epsilon \to z$:

$$f(z) = \frac{1}{\pi} \int_{-\infty}^{\infty} dx' \, \frac{\mathrm{Im}\, f_+(x')}{x' - z}. \tag{13.30}$$

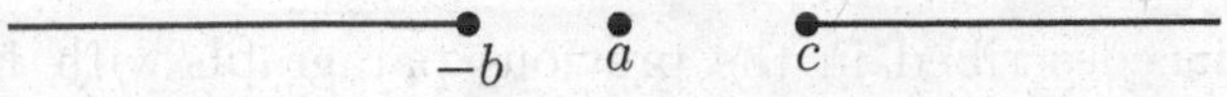

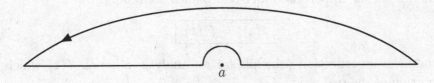

Figure 13.4: Pole a along x axis for function defined for x between $-b$ and c

In many applications $\mathrm{Im}\, f_+(x')$ vanishes along part of the real axis, and there may be poles along this part. Thus suppose $\mathrm{Im}\, f_+(x')$ vanishes in the interval $-b < x' < c$, and suppose there is a pole at $x' = a$ as shown in Figure 13.4. We can retrace the derivation from (13.21) to (13.30) using a contour with a semicircular indentation of radius $r \to 0$ around the pole at $z = a$, as shown in Figure 13.5. Near the pole, assumed to be of first-order, at $z = a$, we see from the Laurent expansion (11.150) that

$$f(z') \xrightarrow[z' \to a]{} a_{-1} \frac{1}{z' - a} = \frac{\mathrm{Res}(f, a)}{z' - a}, \tag{13.31}$$

so that the contribution from the small semicircle to the right side of (13.21) is

$$\frac{1}{2\pi i} \int_{\frown} dz' \, \frac{f(z')}{z' - z} = \frac{\mathrm{Res}(f, a)}{a - z} \frac{1}{2\pi i} \int_{\frown} dz' \, \frac{1}{z' - a}$$

$$= \frac{1}{2} \frac{\mathrm{Res}(f, a)}{z - a}. \tag{13.32}$$

Figure 13.5: Contour with semicircular indentation about a pole

We assume that the residue $\mathrm{Res}(f, a)$ is real so that (13.32) makes no contribution to $\mathrm{Im}\, f_+(x)$ along the real axis. Tracing through the added contribution of (13.32), we find in place of (13.30),

$$f(z) = \frac{\mathrm{Res}(f, a)}{z - a} + \frac{1}{\pi} \int_{-\infty}^{-b} dx' \, \frac{\mathrm{Im}\, f_+(x')}{x' - z} + \frac{1}{\pi} \int_{c}^{\infty} dx' \, \frac{\mathrm{Im}\, f_+(x')}{x' - z}, \tag{13.33}$$

for the situation in which $\mathrm{Im}\, f_+(x')$ vanishes between $x' = -b$ and $x' = c$ and $f(z)$ has a pole at $z = a$, $-b < a < c$. Another way to say it is to notice that for $z \to x + i\epsilon$

$$\frac{\mathrm{Res}(f, a)}{z - a} \to \frac{\mathrm{Res}(f, a)}{x + i\epsilon - a} = \mathrm{Res}(f, a) \left[\frac{\mathcal{P}}{x - a} - i\pi \delta(x - a) \right] \tag{13.34}$$

has an imaginary part proportional to $\delta(x - a)$,

$$\text{Im} \frac{\text{Res}(f, a)}{x' + i\epsilon - a} = -\pi \text{Res}(f, a)\delta(x' - a). \tag{13.35}$$

Substituting this imaginary part in (13.30) leads to the pole term explicitly displayed in (13.33).

For the situation described in the previous paragraph with $\text{Im} f_+(x) = 0$ for $-b < x < c$, except possibly for isolated poles, we can analytically continue from the upper half z plane through the gap and use (13.33) to define $f(z)$ throughout the entire z plane. As usual, this analytic continuation procedure leads to a unique result. We see that the resulting function has cuts along the portions of the real axis from $-\infty < x < -b$ and $c < x < \infty$. The discontinuity across the cut, as shown in Figure 13.6, is given by

$$f(x + i\epsilon) - f(x - i\epsilon) = \frac{1}{\pi} \int dx' \, \text{Im} \, f_+(x') \left[\frac{1}{x' - x - i\epsilon} - \frac{1}{x' - x + i\epsilon} \right]. \tag{13.36}$$

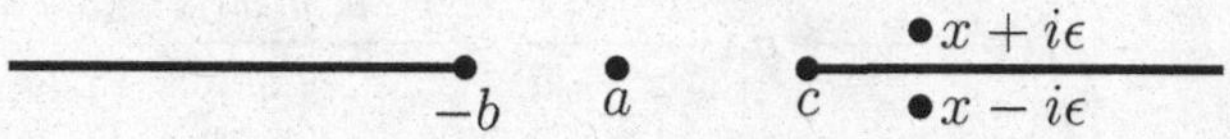

Figure 13.6: Singularity at a with cuts on axis for less than $-b$ and greater than c

Using Dirac's formula (13.10) for $(x' - x \mp i\epsilon)^{-1}$ we find

$$f(x + i\epsilon) - f(x - i\epsilon) = 2i \, \text{Im} \, f_+(x). \tag{13.37}$$

It is also clear from (13.33) that the values of the function $f(z)$ at conjugate points z and z^* in the upper and lower half-planes are related by

$$\boxed{f(z^*) = f^*(z).} \tag{13.38}$$

This is the Schwarz reflection principle for an analytic function $f(z)$ which is real along a portion of the real axis.

Sometimes the formula (13.33) is obtained by a slightly different argument. Suppose $f(z)$ is a function with the properties

1) $f(z)$ is analytic except for cuts on the intervals $(-\infty, -b)$, (c, ∞) and a pole at $z = a$.

2) $f(z) \to 0$ at $|z| \to \infty$.

3) $f(z)$ is real along a portion of the real axis, $-b < x < c$.

The Schwarz reflection principle is easily deduced by analytic continuation from the assumption (3). An analytic function can be expanded in a Taylor series

$$f(z) = \sum_{n=0}^{\infty} a_n (z - z_0)^n \tag{13.39}$$

with the coefficients given by

$$a_n = \frac{1}{n!} \frac{d^n f(z)}{dz^n}\bigg|_{z=z_0}. \qquad (13.40)$$

Expanding first about points z_0 along the real axis on the interval where $f(z)$ is real, and using real increments $\Delta z = \Delta x$ to calculate the derivatives (since for an analytic function we can use any increment), we see that the a_n will be real for these z_0 so that $f(z^*) = f^*(z)$ in the circles of convergence of such a power series. We can now compare Taylor series about a point z_1 and its complex conjugate z_1^*, both points z_1 and z_1^* being inside the region where we have already shown that $f(z^*) = f^*(z)$, as illustrated in Figure 13.7:

$$f(z) = \sum_n b_n(z - z_1)^n, \qquad (13.41)$$

$$f(z) = \sum_n c_n(z - z_1^*)^n. \qquad (13.42)$$

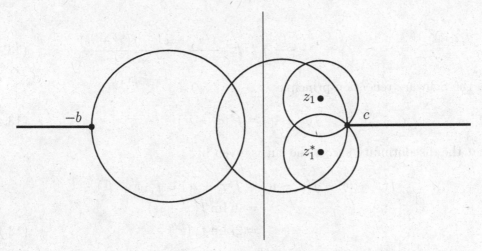

Figure 13.7: Region of convergence for points z_1 and z_1^*

Using the formulas like (13.40) for b_n, c_n we find

$$c_n = \frac{1}{n!} \frac{d^n f(z_1^*)}{dz_1^{*n}} = \left[\frac{1}{n!} \frac{d^n f(z_1)}{dz_1^n}\right]^* = b_n^*, \qquad (13.43)$$

so that

$$f(z^*) = \sum_n b_n^*(z^* - z_1^*)^n = f^*(z) \qquad (13.44)$$

in the extended regions where the series (13.41), (13.42) converge. Continuing the analytic continuation with the circle-chain method we find the Schwarz reflection principle for all conjugate points z, z^*.

We can now apply Cauchy's integral formula to $f(z)$ with a contour C going

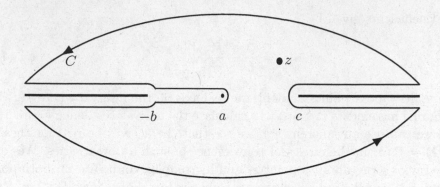

Figure 13.8: Contour going around poles and cuts

around the poles and cuts, as shown in Figure 13.8. Using condition (2) to discard the portions of the contour at infinity we find

$$f(z) = \frac{\text{Res}(f,a)}{z-a} + \frac{1}{2\pi i} \int_{-\infty}^{-b} dx' \, \frac{f(x'+i\epsilon) - f(x'-i\epsilon)}{x'-z}$$

$$+ \frac{1}{2\pi i} \int_{c}^{\infty} dx' \, \frac{f(x'+i\epsilon) - f(x'-i\epsilon)}{x'-z}. \tag{13.45}$$

From the Schwarz reflection principle

$$f(x'-i\epsilon) = f^*(x'+i\epsilon), \tag{13.46}$$

so that the discontinuity across the cut is given by

$$\begin{aligned}
f(x'+i\epsilon) - f(x'-i\epsilon) &= f(x'+i\epsilon) - f^*(x'+i\epsilon) \\
&= 2i \, \text{Im} \, f(x'+i\epsilon) \\
&= 2i \, \text{Im} \, f_+(x'), \tag{13.47}
\end{aligned}$$

and we find

$$\boxed{f(z) = \frac{\text{Res}(f,a)}{z-a} + \frac{1}{\pi} \int_{-\infty}^{-b} dx' \, \frac{\text{Im} \, f_+(x')}{x'-z} + \frac{1}{\pi} \int_{c}^{\infty} dx' \, \frac{\text{Im} \, f_+(x')}{x'-z}} \tag{13.48}$$

in agreement with (13.33).

Of course all our results are just special cases of the discontinuity problem discussed in Section 13.3. Just as there, we obtain a unique result only by assuming that $f(z) \to 0$ as $z \to \infty$. Even if the function $f(z)$ of interest does not vanish at infinity, it may be polynomial bounded, i.e., $f(z)/z^n \to 0$ as $z \to \infty$ for some integer n. For such a case we can apply dispersion theory to the new function $f(z)/z^n$ at the expense of generating the nth order pole at $z = 0$. [Of course we can divide by $(z-a)^n$ and put the nth order pole at a or divide by $(z-a)(z-b)\dots$ and split up the nth order pole into simple poles.] Consider the simplest case of this type, $f(z) \to$ const.

as $z \to \infty$ so that $f(z)/z \to 0$ as $z \to \infty$. Assume that $f(z)$ has no poles in the gap between the two cuts. The residue of $f(z)/z$ at $z = 0$ is then $f(0)$. Applied to this case the representation (13.48) becomes

$$\frac{f(z)}{z} = \frac{f(0)}{z} + \frac{1}{\pi} \int_{-\infty}^{-b} dx' \, \frac{\mathrm{Im}\, f_+(x')}{x'(x'-z)} + \frac{1}{\pi} \int_c^\infty dx' \, \frac{\mathrm{Im}\, f_+(x')}{x'(x'-z)}, \qquad (13.49)$$

or

$$f(z) = f(0) + \frac{z}{\pi} \int_{-\infty}^{b} dx' \, \frac{\mathrm{Im}\, f_+(x')}{x'(x'-z)} + \frac{z}{\pi} \int_c^\infty dx' \, \frac{\mathrm{Im}\, f_+(x')}{x'(x'-z)}. \qquad (13.50)$$

This is called a subtracted dispersion relation because it can be obtained by the procedure (illegitimate since the integrals do not converge) of writing down the unsubtracted dispersion relation

$$f(z) = \frac{1}{\pi} \int_{-\infty}^{-b} dx' \, \frac{\mathrm{Im}\, f_+(x')}{x'-z} + \frac{1}{\pi} \int_c^\infty dx' \, \frac{\mathrm{Im}\, f_+(x')}{x'-z} \qquad (13.51)$$

and subtracting the same relation evaluated at $z = 0$,

$$f(0) = \frac{1}{\pi} \int_{-\infty}^{-b} dx' \, \frac{\mathrm{Im}\, f_+(x')}{x'} + \frac{1}{\pi} \int_c^\infty dx' \, \frac{\mathrm{Im}\, f_+(x')}{x'}, \qquad (13.52)$$

using

$$\frac{1}{x'-z} - \frac{1}{x'} = \frac{z}{x'(x'-z)}. \qquad (13.53)$$

In order to employ (13.50) one must know $f(0)$ as well as $\mathrm{Im}\, f_+(x')$ along the cuts. If two subtractions are necessary in order to achieve convergent integrals, two constants will be introduced, etc.

As a final point we can consider the possibility of analytically continuing through the cuts on the intervals $(-\infty, -b)$, (c, ∞) onto another sheet of the Riemann surface for the analytic function $f(z)$. For a general discontinuity function $\mathrm{Im}\, f_+(x')$ this may be very difficult. For an important subclass of problems $\mathrm{Im}\, f_+(x')$ is itself an analytic function $g(z')$ evaluated on the real axis:

$$\mathrm{Im}\, f_+(x') = g(z') \Big|_{z'=x'}. \qquad (13.54)$$

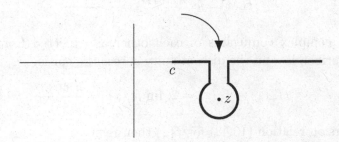

Figure 13.9: Distorting the contour

In this case we can analytically continue through the cut by using Cauchy's theorem to distort the contour for the $z' = x'$ integration: move the contour out of the way and let z follow. Thus suppose we wish to analytically continue down through the right-hand cut in (13.33) for a case in which there is only a right-hand cut along which (13.54) holds, as shown in Figure 13.9:

$$f(z) = \frac{1}{\pi} \int_c^\infty dx' \, \frac{g(x')}{x' - z}$$

$$= \frac{1}{\pi} \int_C dz' \, \frac{g(z')}{z' - z}. \tag{13.55}$$

Assuming $g(z')$ has no singularities in the region considered, we can use the residue theorem to evaluate the contribution from the pole at $z' = z$, obtaining

$$f(z) = 2ig(z) + \frac{1}{\pi} \int_c^\infty dz' \, \frac{g(x')}{x' - z} \tag{13.56}$$

for the function $f(z)$ on the next sheet of the Riemann surface.

13.5 EXAMPLES

1) Consider the function

$$f(z) = \frac{1}{(c - z)^\alpha}, \quad 0 < \alpha < 1, \tag{13.57}$$

on the branch such that $f(z)$ is real for $z = x$ real and $x < c$, refer to Figure 13.10. Thus there is a cut along the real z axis from $z = c$ to $z = +\infty$ and in the cut plane we have

$$f(z) = e^{-\alpha \log(c-z)} = e^{-\alpha[\ln r + i\theta]}, \quad -\pi \leq \theta < \pi. \tag{13.58}$$

On the top and bottom of the cut we find

$$f_+(x) = e^{-\alpha[\ln r - i\pi]} = \frac{e^{i\pi\alpha}}{(x - c)^\alpha}, \tag{13.59}$$

$$f_-(x) = e^{-\alpha[\ln r + i\pi]} = \frac{e^{-i\pi\alpha}}{(x - c)^\alpha}. \tag{13.60}$$

These are complex conjugates of each other as expected from the Schwarz reflection principle, and the discontinuity across the cut is

$$f_+(x) - f_-(x) = 2i \, \text{Im} \, f_+(x) = \frac{2i \sin \pi\alpha}{(x - c)^\alpha}. \tag{13.61}$$

The dispersion relation (13.48) for $f(z)$ then reads

$$\frac{1}{(c - z)^\alpha} = \frac{\sin \pi\alpha}{\pi} \int_c^\infty dx' \, \frac{1}{(x' - c)^\alpha (x' - z)}. \tag{13.62}$$

This holds for any z in the cut plane. Taking z at the two positions $z = x \pm i\epsilon$, just above and below the cut, and adding, we find

$$\frac{1}{2}\left[f_+(x) + f_-(x)\right] = \frac{\cos \pi\alpha}{(x-c)^\alpha}$$

$$= \frac{\sin \pi\alpha}{2\pi} \int_c^\infty dx' \, \frac{1}{(x'-c)^\alpha} \left[\frac{1}{x'-x-i\epsilon} + \frac{1}{x'-x+i\epsilon}\right].$$
(13.63)

Using Dirac's formula (13.10) we see that the integral on the right reduces to the principal value integral and so find

$$\mathcal{P} \int_c^\infty dx' \, \frac{1}{(x'-c)^\alpha(x'-x)} = \frac{\pi \cot \pi\alpha}{(x-c)^\alpha}, \quad x > c.$$
(13.64)

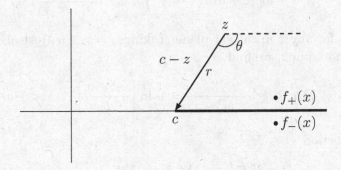

Figure 13.10: Diagram to use with evaluating Example (1) – equation (13.57)

2) For the function

$$f(z) = \frac{1}{z}\log(z+b),$$
(13.65)

we can take the cut along the negative real axis $(-\infty, -b)$, as shown in Figure 13.11, and pick the branch with $f(z)$ real for real $z = x > -b$:

$$\log(z+b) = \ln r + i\theta, \quad -\pi \le \theta < \pi.$$
(13.66)

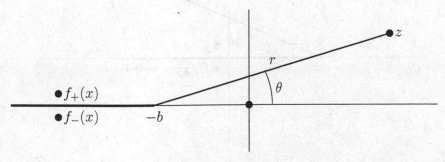

Figure 13.11: Cut along negative real axis for $x < -b$ for Example (2)

On the top and bottom of the cut we find

$$f_+(x) = \frac{1}{x} \left[\ln r + i\pi \right], \tag{13.67}$$

$$f_-(x) = \frac{1}{x} \left[\ln r - i\pi \right], \tag{13.68}$$

so that the discontinuity across the cut is

$$f_+(x) - f_-(x) = 2i \, \text{Im} \, f_+(x) = \frac{2\pi i}{x}. \tag{13.69}$$

The function also has a pole at $z = 0$ with residue $\ln b$. The dispersion relation for $f(z)$ is then

$$\frac{1}{z} \log(z + b) = \frac{\ln b}{z} + \int_{-\infty}^{-b} dx' \, \frac{1}{x'(x' - z)}. \tag{13.70}$$

This holds for any z in the cut plane. Taking $z = x \pm i\epsilon$, just above and below the cut, and adding, we find

$$P \int_{-\infty}^{-b} dx' \, \frac{1}{x'(x' - x)} = \frac{1}{x} \ln \left[\frac{-b - x}{b} \right], \quad x < -b. \tag{13.71}$$

3) For the function

$$f(z) = \frac{(z - 1)^{\alpha - 1}}{(z + 1)^{\alpha}} = \frac{1}{z - 1} \left[\frac{z - 1}{z + 1} \right]^{\alpha}, \quad 0 < \alpha < 1, \tag{13.72}$$

we can choose the cut along the real axis between $x = -1$ and $x = +1$ and choose the branch so that $f(z)$ is real and positive for real $z = x > 1$, as shown in Figure 13.12:

$$\left[\frac{z - 1}{z + 1} \right]^{\alpha} = \left[\frac{r_1}{r_2} \right]^{\alpha} e^{i\alpha(\theta_1 - \theta_2)}. \tag{13.73}$$

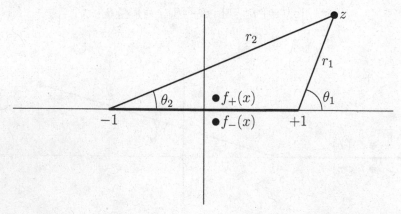

Figure 13.12: Cut along real axis between -1 and +1 for Example (3)

Then we find

$$f_+(x) = -\frac{r_1^{\alpha-1}}{r_2^{\alpha}} e^{+i\pi\alpha}, \tag{13.74}$$

$$f_-(x) = -\frac{r_1^{\alpha-1}}{r_2^{\alpha}} e^{-i\pi\alpha}, \tag{13.75}$$

and the discontinuity across the cut is given by

$$f_+(x) - f_-(x) = 2i \operatorname{Im} f_+(x) = -2i \frac{r_1^{\alpha-1}}{r_2^{\alpha}} \sin \pi\alpha. \tag{13.76}$$

The dispersion relation is

$$\frac{(z-1)^{\alpha-1}}{(z+1)^{\alpha}} = -\frac{\sin \pi\alpha}{\pi} \int_{-1}^{1} dx' \frac{(1-x')^{\alpha-1}}{(1+x')^{\alpha}} \frac{1}{x'-z}. \tag{13.77}$$

This holds for all z in the cut plane. Setting $z = x \pm i\epsilon$, just above and below the cut, and adding, we find

$$\mathcal{P} \int_{-1}^{1} dx' \frac{(1-x')^{\alpha-1}}{(1+x')^{\alpha}} \frac{1}{x'-x} = \pi \frac{(1-x)^{\alpha-1}}{(1+x)^{\alpha}} \cot \pi\alpha. \tag{13.78}$$

4) The second solution of Legendre's equation for $l = 0, 1, 2, \ldots$ can be written in the form

$$Q_l(z) = \frac{1}{2} \int_{-1}^{1} dt \frac{P_l(t)}{z-t}, \tag{13.79}$$

in terms of the Legendre polynomial $P_l(t)$, which is the first solution (see Section 3.4). The formula (13.79), which has the form of a dispersion integral, holds for all z in the complex plane cut on the interval $(-1, 1)$. For $z = x \pm i\epsilon$, $-1 < x < 1$, just above and below the cut, we find

$$Q_l(x + i\epsilon) = \frac{1}{2} \int_{-1}^{+1} dt \frac{P_l(t)}{x+i\epsilon-t} = \frac{1}{2} \mathcal{P} \int_{-1}^{+1} dt \frac{P_l(t)}{x-t} - \frac{i\pi}{2} P_l(x), \tag{13.80}$$

$$Q_l(x - i\epsilon) = \frac{1}{2} \int_{-1}^{+1} dt \frac{P_l(t)}{x-i\epsilon-t} = \frac{1}{2} \mathcal{P} \int_{-1}^{+1} dt \frac{P_l(t)}{x-t} + \frac{i\pi}{2} P_l(x), \tag{13.81}$$

so that the discontinuity across the cut is

$$Q_l(x + i\epsilon) - Q_l(x - i\epsilon) = -i\pi P_l(x). \tag{13.82}$$

5) The function

$$f(z) = \frac{1}{\pi} \int_{-\infty}^{\infty} dx' \frac{e^{-x'^2}}{x'-z} \tag{13.83}$$

arises continually in plasma physics. We start with z in the upper half z plane. Moving z down onto the real axis we find

$$f_+(x) = \frac{1}{\pi} \int_{-\infty}^{\infty} dx' \frac{e^{-x'^2}}{x' - x - i\epsilon}$$

$$= \frac{1}{\pi} \int_C dz' \frac{e^{-z'^2}}{z' - x}. \tag{13.84}$$

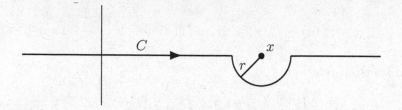

Figure 13.13: Contour deformed around a point x on the real axis

Here we have used Cauchy's theorem to make an indentation in the contour for the z' integration around the point x, as shown in Figure 13.13. Note that we have

$$f_+(0) = \frac{1}{\pi} \int_{-\pi}^{0} ire^{i\theta} d\theta \frac{e^{-z'^2}}{re^{i\theta}} = i, \tag{13.85}$$

since the contributions from the straight portions of the contour cancel for $z = 0$ and the radius r of the circle can be made arbitrarily small.

We can show that $f_+(x)$ is related to an error function of imaginary argument by deriving a differential equation for it. From the formula (13.84)

$$\frac{df_+(x)}{dx} = \frac{1}{\pi} \int_{-\infty}^{\infty} dx' \, e^{-x'^2} \frac{d}{dx} \frac{1}{x' - x - i\epsilon}$$

$$= -\frac{1}{\pi} \int_{-\infty}^{\infty} dx' \, e^{-x'^2} \frac{d}{dx'} \frac{1}{x' - x - i\epsilon}$$

$$= +\frac{1}{\pi} \int_{-\infty}^{\infty} dx' \frac{1}{x' - x - i\epsilon} \frac{d}{dx'} e^{-x'^2}$$

$$= -\frac{1}{\pi} \int_{-\infty}^{\infty} dx' \, e^{-ix'^2} \frac{2x'}{x' - x - i\epsilon}$$

$$= -\frac{2}{\pi} \int_{-\infty}^{\infty} dx' \, e^{-x'^2} \left[1 + \frac{x}{x' - x - i\epsilon} \right]$$

$$= -\frac{2}{\sqrt{\pi}} - 2x f_+(x). \tag{13.86}$$

After multiplication by the factor e^{x^2} this differential equation can be put in the form

$$\frac{d}{dx}\left\{e^{x^2}f_+(x)\right\} = -\frac{2}{\sqrt{\pi}}e^{x^2}.$$ (13.87)

Integrating from 0 to x and using the initial value (13.85) this becomes

$$e^{x^2}f_+(x) - i = -\frac{2}{\sqrt{\pi}}\int_0^x dx'\ e^{x'^2} = \frac{2i}{\sqrt{\pi}}\int_0^{i\pi} dy\ e^{-y^2},$$ (13.88)

i.e.,

$$f_+(x) = ie^{-x^2}[1 + \text{erf}(ix)],$$ (13.89)

where

$$\text{erf}(z) = \frac{2}{\sqrt{\pi}}\int_0^z dy\ e^{-y^2}.$$ (13.90)

13.6 INTEGRAL EQUATIONS WITH CAUCHY KERNELS

A dispersion relation is an identity satisfied by a wide class of analytic functions. If some other consideration, e.g., in the physics underlying the mathematics problem, relates the imaginary part under the integral sign to the function itself, the dispersion relation is converted into an integral equation, which can be solved to yield a more or less unique answer. Suppose, for example, that in the dispersion relation (13.48) we take the left-hand cut as a given function $g(x)$ and assume that $\text{Im} f_+(x') = h(x')f_+(x')$ along the right-hand cut, with $h(x')$ a given function. Evaluating (13.48) at $z = x + i\epsilon$ and suppressing the $(+)$ notation, i.e., $f(x) = f_+(x)$, we find an integral equation for $f(x)$ of the form

$$f(x) = g(x) + \frac{1}{\pi}\int_a^\infty dx'\ \frac{h(x')f(x')}{x' - x - i\epsilon}.$$ (13.91)

This is a singular integral equation, but it has a simplifying feature, namely that the entire x dependence of the kernel is given explicitly in the Cauchy denominator $(x' - x - i\epsilon)$; thus $h(x')$ depends on x' but not on x. This enables us to solve the integral equations by a trick devised by the mathematicians Hilbert and Carleman and applied in high energy physics by the physicist Omnès.[1]

To solve (13.91) we convert it into a discontinuity problem of the type discussed in Section 13.3. Consider the integral term in (13.91), analytically continued into the complex plane; define

$$F(z) = \frac{1}{2\pi i}\int_a^\infty dx'\ \frac{h(x')f(x')}{x' - z}.$$ (13.92)

Thus we have

$$\frac{1}{\pi}\int_a^\infty dx'\ \frac{h(x')f(x')}{x' - x - i\epsilon} = 2iF_+(x)$$ (13.93)

[1]The definitive text on the subject is N. I. Muskhelishvili, <u>Singular Integral Equations</u> (Nordoff NV, Groningen, Netherlands, 1953). A short survey is given by Carrier, Krook, and Pearson, <u>op. cit.</u>, p. 422. The application to high energy physics can be found in Omnès' paper, Nuovo Cimento, <u>8</u>, 316 (1958).

and

$$F_+(x) - F_-(x) = \frac{1}{2\pi i} \int_a^\infty dx' \, h(x')f(x') \left[\frac{1}{x'-x-i\epsilon} - \frac{1}{x'-x+i\epsilon} \right]$$

$$= h(x)f(x). \tag{13.94}$$

Using (13.93) and (13.94) we can thus rewrite (13.91) as a relation between $F_+(x)$ and $F_-(x)$:

$$\frac{1}{h(x)}[F_+(x) - F_-(x)] = g(x) + 2iF_+(x) \tag{13.95}$$

or

$$F_+(x)[1 - 2ih(x)] - F_-(x) = g(x)h(x). \tag{13.96}$$

This is a linear relation between $F_+(x)$ on the top of the cut and the $F_-(X)$ on the bottom. We want to somehow get rid of the factor $[1 - 2ih(x)]$ so as to transform it into a formula for the discontinuity across the cut. To this end define a function $\Omega(z)$ which satisfies a relation like (13.96) but with the right-hand side set equal to zero:

$$\Omega_+(x)[1 - 2ih(x)] - \Omega_-(x) = 0. \tag{13.97}$$

Provided $[1 - 2ih(x)] \neq 0$ along the cut (a, ∞), this can be converted into a discontinuity relation by taking logarithms,

$$\log \Omega_+(x) - \log \Omega_-(x) = -\log[1 - 2ih(x)], \tag{13.98}$$

and so has a solution

$$\log \Omega(z) = -\frac{1}{2\pi i} \int_a^\infty dx' \, \frac{\log[1 - 2ih(x')]}{x' - z}, \tag{13.99}$$

$$\Omega(z) = \exp\left(-\frac{1}{2\pi i} \int_a^\infty dx' \, \frac{\log[1 - 2ih(x')]}{x' - z} \right). \tag{13.100}$$

Evidently we better have $h(x') \xrightarrow[x' \to \infty]{} 0$ so that the integral (13.99) will converge. With

$$\Omega_+(x) = \exp\left(-\frac{1}{2\pi i} \int_a^\infty dx' \, \frac{\log[1 - 2ih(x')]}{x' - x - i\epsilon} \right), \tag{13.101}$$

$$\Omega_-(x) = \exp\left(-\frac{1}{2\pi i} \int_a^\infty dx' \, \frac{\log[1 - 2ih(x')]}{x' - x + i\epsilon} \right), \tag{13.102}$$

we can check directly the relation

$$\frac{\Omega_-(x)}{\Omega_+(x)} = 1 - 2ih(x). \tag{13.103}$$

If it is useful later to secure convergent integrals, we can evidently multiply our solution (13.100) by any single valued function with no cut, which will then cancel in the ratio (13.103) and the defining equation (13.97).

The function $\Omega(z)$ provides a solution of (13.96) when $g(x) = 0$. When $g(x) \neq 0$ we can use $\Omega(z)$ to convert (13.96) to a discontinuity problem. Thus substituting (13.103) for $1 - 2ih(x)$ into (13.96) we find

$$\frac{F_+(x)}{\Omega_+(x)} - \frac{F_-(x)}{\Omega_-(x)} = \frac{g(x)h(x)}{\Omega_-(x)}. \tag{13.104}$$

This is the discontinuity problem for $F(z)/\Omega(z)$ and we find as solution

$$\frac{F(z)}{\Omega(z)} = \frac{1}{2\pi i} \int_a^\infty dx' \frac{g(x')h(x')}{\Omega_-(x')} \frac{1}{x' - z} + E(z), \tag{13.105}$$

where $E(z)$ is a single valued function with no cut and at most singularities at $z = a$ and $z = \infty$.[2] Thus we have

$$F_+(x) = \Omega_+(x) \left[\frac{1}{2\pi i} \int_a^\infty dx' \frac{g(x')h(x')}{\Omega_-(x')} \frac{1}{x' - x - i\epsilon} + E(x) \right], \tag{13.106}$$

and from (13.91) the solution of the integral equation is

$$f(x) = g(x) + 2iF_+(x). \tag{13.107}$$

Although we have solved a slightly more general problem with a more general $h(x)$ in (13.91), for the case of most interest in physics we have

$$f(x) = f_+(x) = f_-^*(x) \tag{13.108}$$

and

$$h(x)f_+(x) = \operatorname{Im} f_+(x) = \frac{1}{2i}[f_+(x) - f_-(x)], \tag{13.109}$$

so that

$$h(x) = \frac{1}{2i} \left[1 - \frac{f_-(x)}{f_+(x)} \right]$$

$$= \frac{1}{2i} \left[1 - \frac{f_-(x)}{f_-^*(x)} \right] \tag{13.110}$$

has the form

$$h(x) = \frac{1}{2i} \left[1 - e^{-2i\delta(x)} \right]$$

$$= e^{-i\delta(x)} \sin \delta(x), \tag{13.111}$$

---

[2]See the paper by Omnès, op. cit., where the form of $E(z)$ is related to the behavior of $h(x)$ at $x = a$ and $x = \infty$. In important cases $E(z) = 0$.

where $\delta(x)$ is a real function, assumed known. For this case (13.101) and (13.102) give

$$\Omega_+(x) = e^{\rho(x)+i\delta(x)}, \tag{13.112}$$

$$\Omega_-(x) = e^{\rho(x)-i\delta(x)}, \tag{13.113}$$

with

$$\rho(x) = \frac{1}{\pi}\mathcal{P}\int_a^\infty dx' \, \frac{\delta(x')}{x'-x}. \tag{13.114}$$

Substituting these expressions in (13.106), (13.107), we find as solution of the integral equation

$$f(x) = e^{i\delta(x)}\times$$

$$\left[g(x)\cos\delta(x) + e^{\rho(x)}\frac{1}{\pi}\mathcal{P}\int_a^\infty dx' \, g(x')e^{-\rho(x')}\sin\delta(x')\frac{1}{x'-x} + e^{\rho(x)}E'(x)\right], \tag{13.115}$$

where the $E'(x) = 2iE(x)$ must be real for (13.109) to be satisfied.

Variations on the technique described here can be applied to numerous problems, including integral equations with kernels proportional to $(x-x')^\alpha$ and $\log(x'-x)$.[3]

---

[3]See Carrier, Krook, and Pearson, op. cit., p. 426 ff.

PROBLEMS

13.1 A function $F(z)$ is specified in the following way:

a) $F(z)$ is analytic except for a cut along the imaginary axis from $z = i$ to $z = i\infty$, as shown in Figure 13.14.

b) $F(z) \to 0$ as $z \to \infty$.

c) The discontinuity across the cut is given by

$$F(iy + \epsilon) - F(iy - \epsilon) = \frac{1}{y^2}, \quad \epsilon = \text{positive infinitesimal.}$$

Find the function $F(z)$. Specify precisely the desired branches on any multivalued functions.

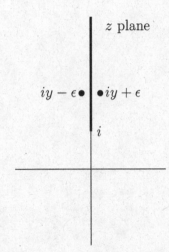

Figure 13.14: Branch cut along imaginary axis from $+i$ to ∞ for Problem 13.1

13.2 a) Express the function

$$f(z) = \frac{1}{\sqrt{z^2 - 1}}$$

as a dispersion integral over an appropriate cut.

b) Evaluate the principal value integral

$$I = \mathcal{P} \int_{-1}^{+1} dx' \, \frac{1}{x' - x} \, \frac{1}{\sqrt{1 - x'^2}}, \quad -1 < x < 1.$$

13.3 Find the solution of the singular integral equation

$$\mathcal{P} \int_{-1}^{+1} dt \, \frac{f(t)}{t - x} = 1.$$

CHAPTER 14

Special Functions

14.1 INTRODUCTION

Special functions of a complex variable is a vast subject exhaustively studied by the mathematicians of the last century. In the very brief treatment here we can only study a few of the most important points. A comprehensive treatment is given by Whittaker and Watson and briefer treatments by Copson also Carrier, Krook and Pearson.[1] Many results are listed without proof in mathematical handbooks.[2] These handbooks have become invaluable references for physicists and others not too interested in proofs.

14.2 THE GAMMA FUNCTION

We have already established some of the principal results for this function at the end of Section 11.8:

$$\Gamma(z) = \int_0^\infty dt\ e^{-t} t^{z-1}, \quad \operatorname{Re} z > 0 \tag{14.1}$$

$$\Gamma(n+1) = n!, \quad n = 1, 2, 3, \dots, \tag{14.2}$$

$$\Gamma(z) = \frac{1}{2i\sin\pi z} \int_C dt\ e^t t^{z-1}, \quad t^{z-1} = e^{(z-1)(\ln|t|+i\theta)}, \quad -\pi < \theta < \pi, \tag{14.3}$$

$$\Gamma(z+1) = z\Gamma(z) \tag{14.4}$$

---

[1] References are given in Section 11.1.

[2] W. Magnus, F. Oberhettinger and R. P. Soni, Formulas and Theorems for the Special Functions of Mathematical Physics (Springer-Verlag, New York, 1966).

E. Jahnke and F. Emde, Tables of Functions (Dover Publications, New York, 1945).

M. Abramowitz and I. Stegun, Handbook of Mathematical Functions, National Bureau of Standards Applied Mathematics Series. 55.

A. Erdélyi, W. Magnus, F. Oberhettinger, and F. G. Tricomi, Bateman Manuscript Project: Higher Transcendental Functions, Vols. I-III, Tables of Integral Transforms, Vols. I and II (McGraw-Hill, New York, 1953).

The integral representation (14.3), valid for all z, teaches us that $\Gamma(z)$ has simple poles at $z = -n = 0, -1, -2, \ldots$. Near $z = -n$, $\sin \pi z = (-1)^n \sin \pi(z + n) \simeq (-1)^n \pi(z + n)$. At $z = -n$ the contour integral in (14.3) can be collapsed to a closed contour around the origin, as shown in Figure 14.1, since t^{z-1} is single-valued; we find

$$\int_C dt \, e^t t^{-n-1} = \frac{1}{n!} 2\pi i. \tag{14.5}$$

Thus near $z = -n$, $n = 0, 1, 2, \ldots$, we have

$$\Gamma(z) \simeq \frac{(-1)^n}{n!} \frac{1}{z + n}, \tag{14.6}$$

i.e., the residue of $\Gamma(z)$ at $z = -n$ is $(-1)^n/n!$.

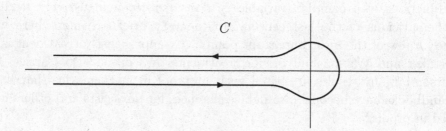

Figure 14.1: Contour looped around the origin

Some useful formulas are obtained by first introducing a related function, the beta function, defined by

$$B(u, v) = \int_0^1 dt \, t^{u-1}(1 - t)^{v-1}. \tag{14.7}$$

In defining t^{u-1} and $(1 - t)^{v-1}$, $\log t$ and $\log(1 - t)$ are chosen real on the interval $0 \le t \le 1$. Like the definition (14.1) of the gamma function, this representation is valid only for $\operatorname{Re} u > 0$, $\operatorname{Re} v > 0$. Setting $t = \sin^2 \theta$, (14.7) becomes

$$B(u, v) = 2 \int_0^{\pi/2} d\theta \, (\sin \theta)^{2u-1}(\cos \theta)^{2v-1}. \tag{14.8}$$

Setting $t = \tau/(1 + \tau)$, (14.7) is transformed to

$$B(u, v) = \int_0^\infty d\tau \frac{\tau^{u-1}}{(1 + \tau)^{u+v}}. \tag{14.9}$$

The beta function can be related to gamma functions by computing the product $\Gamma(u)\Gamma(v)$, using the expression (14.1) with $t \to x^2$ or y^2, and transforming to polar coordinates, $x = r\cos\theta$, $y = r\sin\theta$:

$$\Gamma(u)\Gamma(v) = 2\int_0^\infty dx\ e^{-x^2}x^{2u-1} \times 2\int_0^\infty dy\ e^{-y^2}y^{2v-1}$$

$$= 4\int_0^\infty r\,dr\ e^{-r^2}r^{2(u+v-1)}\int_0^{\pi/2} d\theta\ (\cos\theta)^{2u-1}(\sin\theta)^{2v-1}$$

$$= \Gamma(u+v)B(u,v), \tag{14.10}$$

where we used the expression (14.8). Thus we find

$$\boxed{B(u,v) = \frac{\Gamma(u)\Gamma(v)}{\Gamma(u+v)},} \tag{14.11}$$

and since the beta function is given in terms of gamma functions we do not study it further.

Setting $u = z$, $v = 1 - z$ in (14.11) and using the expression (14.9) for the beta function and the integral (12.51) we obtain the important identity

$$\Gamma(z)\Gamma(1-z) = \Gamma(1)B(z,1-z)$$

$$= \int_0^\infty d\tau\ \frac{\tau^{z-1}}{1+\tau}$$

$$\boxed{\Gamma(z)\Gamma(1-z) = \frac{\pi}{\sin\pi z}.} \tag{14.12}$$

Although our derivation of this formula is valid only for $\mathrm{Re}\,z > 0$, $\mathrm{Re}(1-z) > 0$, i.e., $0 < \mathrm{Re}\,z < 1$, the result can be extended throughout the entire complex plane by analytic continuation.

The formula (14.11) can also be used to find the partial fraction expansion of $\Gamma'(z)/\Gamma(z)$. Set $u = h$, $v = z - h$ in (14.11) and use (14.7) for $B(u,v)$:

$$\frac{\Gamma(z-h)\Gamma(h)}{\Gamma(z)} = \int_0^1 dt\ t^{h-1}(1-t)^{z-h-1}$$

$$= \frac{1}{h} + \int_0^1 dt\ \left[(1-t)^{z-h-1} - 1\right]t^{h-1}, \quad 0 < h < \mathrm{Re}\,z. \tag{14.13}$$

Consider now the limit $h \to 0$. In this limit the recursion relation (14.4) and the Taylor series for $\Gamma(h+1)$ give

$$\Gamma(h) = \frac{1}{h}\Gamma(h+1) = \frac{1}{h}\left[\Gamma(1) + h\Gamma'(1) + \ldots\right]$$

$$= \frac{1}{h} - \gamma + \ldots, \tag{14.14}$$

where

$$\gamma = -\Gamma'(1) \tag{14.15}$$

is Euler's constant [see (4.32)]. Similarly, the Taylor series for $\Gamma(z - h)$ gives

$$\frac{\Gamma(z - h)}{\Gamma(z)} = 1 - h\frac{\Gamma'(z)}{\Gamma(z)} + \ldots . \tag{14.16}$$

Substituting (14.14) and (14.16) in (14.13) and comparing coefficients of power of h, in particular the coefficient of h^0, we find

$$\frac{\Gamma'(z)}{\Gamma(z)} = -\gamma + \int_0^1 \frac{dt}{t}\left[1 - (1-t)^{z-1}\right]. \tag{14.17}$$

We can use the expansion

$$\frac{1}{t} = \frac{1}{1 - (1-t)} = \sum_{n=0}^{\infty}(1-t)^n, \tag{14.18}$$

and do the integral (14.17) term-by-term to obtain

$$\frac{\Gamma'(z)}{\Gamma(z)} = -\gamma + \sum_{n=0}^{\infty}\left[\frac{1}{n+1} - \frac{1}{n+z}\right]$$

$$= -\frac{1}{z} - \gamma + \sum_{n=1}^{\infty}\left[\frac{1}{n} - \frac{1}{n+z}\right]. \tag{14.19}$$

This is the partial fraction expansion of $\Gamma'(z)/\Gamma(z)$. Rewriting it in the form

$$\frac{d}{dz}\log[z\Gamma(z)] = \frac{d}{dz}\left[-\gamma z + \sum_{n=1}^{\infty}\left\{\frac{z}{n} - \log\left(1 + \frac{z}{n}\right)\right\}\right], \tag{14.20}$$

integrating from 0 to z and exponentiating, we find

$$\frac{1}{\Gamma(z)} = ze^{\gamma z}\prod_{n=1}^{\infty}\left\{\left(1 + \frac{z}{n}\right)e^{-\frac{z}{n}}\right\}, \tag{14.21}$$

an infinite product expansion for the gamma function. Setting $z = 1$ in (14.21) and taking logarithms we find another expression for Euler's constant γ:

$$\gamma = -\Gamma'(1) = \lim_{n\to\infty}\left(1 + \frac{1}{2} + \frac{1}{3} + \ldots + \frac{1}{n} - \ln n\right). \tag{14.22}$$

14.3   ASYMPTOTIC EXPANSIONS; STIRLING'S FORMULA

An asymptotic expansion gives an approximate expression for a function $f(z)$ for large $|z|$ in terms of elementary functions. Technically speaking, $f(z)$ has an asymptotic expansion

$$f(z) \sim a_1\phi_1(z) + a_2\phi_2(z) + a_3\phi_3(z) + \ldots \tag{14.23}$$

if for fixed n and

$$S_n(z) = \sum_{j=1}^{n} a_j \phi_j(z), \tag{14.24}$$

we have

$$\left| \frac{f(z) - S_n(z)}{\phi_n(z)} \right| \xrightarrow[z \to \infty]{} 0. \tag{14.25}$$

Note that this is different from the definition of convergence, which has to do with how well $f(z)$ is approximated at fixed z and increasing n, rather than fixed n and increasing $|z|$. Thus the series on the right side of (14.23) may not converge to $f(z)$, so we use the symbol $\sim$ rather than $=$ to indicate the relation between the two sides of (14.23). In practice asymptotic expansions usually have the form

$$f(z) \sim p(z) \left[a_0 + \frac{a_1}{z} + \frac{a_2}{z^2} + \ldots \right], \tag{14.26}$$

and the criterion (14.25) becomes

$$\left| \frac{\frac{f(z)}{p(z)} - \sum_{j=0}^{n} \frac{a_j}{z^j}}{\frac{a_n}{z^n}} \right| \xrightarrow[z \to \infty]{} 0. \tag{14.27}$$

If a function $f(z)$ is such that $f(z)/p(z)$ is analytic at $z = \infty$, i.e., if $f\left(\frac{1}{z}\right)/p\left(\frac{1}{z}\right)$ is analytic near $z = 0$, then there exists a convergent power series

$$\frac{f(z)}{p(z)} = a_0 + \frac{a_1}{z} + \frac{a_2}{z^2} + \ldots \tag{14.28}$$

valid for $|z| > R$, and this provides an asymptotic expansion since

$$\left| \frac{\frac{f(z)}{p(z)} - \sum_{j=0}^{n} \frac{a_j}{z^j}}{\frac{a_n}{z^n}} \right| = \left| \frac{a_{n+1}}{a_n} \frac{1}{z} + \frac{a_{n+2}}{a_n} \frac{1}{z^2} + \ldots \right| \tag{14.29}$$

can be made arbitrarily small for large z. A simple example is the expansion

$$e^{1/z} = 1 + \frac{1}{z} + \frac{1}{2!} \frac{1}{z^2} + \frac{1}{3!} \frac{1}{z^3} + \ldots . \tag{14.30}$$

An example of an asymptotic expansion which is not a convergent power series is provided by the exponential integral function, denoted by $Ei(x)$,

$$\boxed{Ei(x) = \int_x^\infty dt \, \frac{e^{-t}}{t}} \tag{14.31}$$

for $x \to \infty$. Repeated integration by parts gives

$$Ei(x) = \frac{e^{-x}}{x} - \int_x^\infty dt \, \frac{e^{-t}}{t^2}$$

$$= \frac{e^{-x}}{x} - \frac{e^{-x}}{x^2} + 2 \int_x^\infty dt \, \frac{e^{-t}}{t^3}$$

$$\vdots$$

or

$$Ei(x) = e^{-x} \left[\frac{1}{x} - \frac{1}{x^2} + \frac{2!}{x^3} - \frac{3!}{x^4} + \ldots + \frac{(-1)^{n-1}(n-1)!}{x^n} \right] + (-1)^n n! \int_x^\infty dt \, \frac{e^{-t}}{t^{n+1}}.$$

$$(14.32)$$

The remainder term is simple to approximate:

$$\left| (-1)^n n! \int_x^\infty dt \, \frac{e^{-t}}{t^{n+1}} \right| < \frac{n!}{x^{n+1}} \int_x^\infty dt \, e^{-t} = \frac{n! \, e^{-x}}{x^{n+1}}. \qquad (14.33)$$

For this case $p(x) = e^{-x}$ and the quantity on the left side of (14.27) is

$$\left| \frac{(-1)^n n! \int_x^\infty dt \, \frac{e^{-t}}{t^{n+1}}}{e^{-x} \frac{(-1)^{n-1}(n-1)!}{x^n}} \right| < \frac{n}{x} \xrightarrow[x \to \infty]{} 0, \qquad (14.34)$$

so that we have the asymptotic expansion

$$\boxed{Ei(x) \sim e^{-x} \left[\frac{1}{x} - \frac{1}{x^2} + \frac{2!}{x^3} - \frac{3!}{x^4} + \ldots + \frac{(1)^{n-1}(n-1)!}{x^n} \ldots \right].} \qquad (14.35)$$

On the other hand, the ratio test shows that the series on the right side of (14.35) diverges for any x, no matter how large.

Asymptotic expansions are very useful in practical calculations and much effort has gone into finding them in cases not so simple as the exponential integral. Such expansions are given in the mathematical handbooks listed in Section 14.1.

An important case is Stirling's approximation for $\Gamma(z)$ for large $|z|$. In order to derive this, we shall employ a general method known as the method of steepest descents, which will also be applied later to the Bessel function. The method can be applied to an integral representation, such as (14.1) for $\Gamma(z)$, and consists of expanding the integrand about the point where it is a maximum so as to obtain a series in z^{-1} with coefficients which are Gaussian integrals.

Consider first real $z = x$, large and positive. From (14.4) and (14.1) we find

$$\Gamma(x) = \frac{1}{x}\Gamma(x+1) = \frac{1}{x} \int_0^\infty dt \, e^{-t} t^x = \frac{1}{x} \int_0^\infty dt \, e^{-t + x \ln t}. \qquad (14.36)$$

We can remove part of the x dependence with the substitution $t = xu$:

$$\Gamma(x) = x^x \int_0^\infty du\, e^{-x(u - \ln u)}. \tag{14.37}$$

The integrand has a maximum at $u = 1$, where $u - \ln u$ has its minimum. Expanding $u - \ln u$ in a Taylor series about this point we find

$$u - \ln u = 1 + \frac{1}{2}(u - 1)^2 - \frac{1}{3}(u - 1)^3 + \frac{1}{4}(u - 1)^4 + \ldots \tag{14.38}$$

and hence

$$\Gamma(x) \sim x^x \int_0^\infty du\, \exp\left\{ -x\left[1 + \frac{1}{2}(u - 1)^2 - \frac{1}{3}(u - 1)^3 + \frac{1}{4}(u - 1)^4 + \ldots \right] \right\}. \tag{14.39}$$

Letting $v = \sqrt{x}\,(u - 1)$ this becomes

$$\Gamma(x) \sim \frac{x^x e^{-x}}{\sqrt{x}} \int_{-\sqrt{x}}^\infty dv\, e^{-v^2/2}\, \exp\left\{ \frac{1}{3\sqrt{x}}v^3 - \frac{1}{4x}v^4 + \frac{1}{5x^{\frac{3}{2}}}v^5 - \frac{1}{6x^2}v^6 + \ldots \right\}. \tag{14.40}$$

For very large x we can expand the exponentials involving inverse powers of $\sqrt{x}$ and replace the lower limit on the integral by $-\infty$:

$$\Gamma(x) \sim \frac{x^x e^{-x}}{\sqrt{x}} \int_{-\infty}^\infty dv\, e^{-v^2/2} \left[1 + \left\{ \frac{1}{3\sqrt{x}}v^3 - \frac{1}{4x}v^4 + \frac{1}{5x^{3/2}}v^5 - \frac{1}{6x^2}v^6 + \ldots \right\} \right.$$

$$\left. + \frac{1}{2!}\left\{ \frac{1}{3\sqrt{x}}v^3 - \frac{1}{4x}v^4 + \ldots \right\}^2 + \frac{1}{3!}\left\{ \frac{1}{3\sqrt{x}}v^3 + \ldots \right\}^3 + \ldots \right]. \tag{14.41}$$

Using

$$\int_{-\infty}^\infty dv\, e^{-v^2/2}\, v^n = \begin{cases} \sqrt{2\pi} & , \quad n = 0 \\ \sqrt{2\pi} \cdot 1 \cdot 3 \cdot 5 \cdot \ldots \cdot (n - 1) & , \quad n = 2, 4, \ldots, \\ 0 & , \quad n \text{ odd} \end{cases} \tag{14.42}$$

one finds

$$\Gamma(x) \sim \sqrt{\frac{2\pi}{x}}\, x^x e^{-x} \left[1 + \frac{1}{12x} + \frac{1}{288x^2} + \ldots \right]. \tag{14.43}$$

We note that the series (14.38) converges only in the region $|u - 1| < 1$. For large x this is expected to be the most important region by far because $u = 1$ is the place where the integrand is largest and because of the x in the exponent in (14.39). This is most clear in the form (14.40) where the inverse powers of x appear explicitly. In any event the approximations made are good for large x, and we may expect the series (14.43) to be an asymptotic expansion rather than a convergent power series.

Another important point concerns the region of validity of the expansion (14.43). The manipulations above were carried through for large positive real x. Since the integral expression (14.36) with which we started is valid for complex z such that

Re $z > 0$, it is fairly easy to see the (14.43) also holds for such z. A slight variation on the derivation given above[3] shows, moreover, that the formula holds for the whole z plane excluding a wedge about the negative real axis where the gamma function has poles. Thus for $z = re^{i\theta}$ we have

$$\Gamma(z) \sim \sqrt{\frac{2\pi}{z}} \, z^z e^{-z} \left[1 + \frac{1}{12z} + \frac{1}{288z^2} + \cdots \right] \tag{14.44}$$

for large r and $-\pi + \delta \leq \theta \leq \pi - \delta$, where δ is an arbitrary positive number.

14.4   THE HYPERGEOMETRIC FUNCTION

This function $F(a, b; c; z)$ is a very general one and depends on three complex parameters a, b, c as well as the complex variable z. Many functions can be written as hypergeometric functions with special choices of a, b, c, and z. Despite its generality it has some well-defined properties with respect to its analytic continuation which are sometimes useful in practice. We present here a very brief treatment meant to introduce the reader to the function so that he will have some appreciation of it in case he should meet it again. Mostly we shall quote a few results without proof. A good pedagogical treatment is given by Copson.[4] Many results are given without proof in the mathematical handbooks listed in the footnote of Section 14.1.

Hypergeometric functions provide the solution of the most general second-order, linear differential equation with three regular singular points. We write this equation in the form

$$\frac{d^2 w}{dz^2} + p(z) \frac{dw}{dz} + q(z) w = 0 \tag{14.45}$$

and recall[5] that for ξ to be a regular singular point of the equation, $p(z)$ and/or $q(z)$ are to be singular at ξ but $(z - \xi) p(z)$ and $(z - \xi)^2 q(z)$ are to be regular, i.e., analytic. With three regular singular points, ξ, η, ζ, the functions

$$P(z) = (z - \xi)(z - \eta)(z - \zeta) p(z) \tag{14.46}$$

$$Q(z) = (z - \xi)^2 (z - \eta)^2 (z - \zeta)^2 q(z) \tag{14.47}$$

are then entire functions, i.e., analytic throughout the z plane. We moreover require that the point at infinity be regular. This means that if we replace $z \to 1/z$ in (14.45), the new differential equation, which turns out to be

$$\frac{d^2 w}{dz^2} + \left[\frac{2}{z} - \frac{p(1/z)}{z^2} \right] \frac{dw}{dz} + \frac{q(1/z)}{z^4} w = 0, \tag{14.48}$$

must have $z = 0$ as an ordinary point. Thus $[2z - p(1/z)]/z^2$ and $q(1/z)/z^4$ must be regular at $z = 0$, or $2z - z^2 p(z)$ and $z^4 q(z)$ must be regular at $z \to \infty$, i.e.,

---

[3]See, e.g., Copson, op. cit., p. 222.
[4]Copson, op. cit., Chap. X, p. 233.
[5]See Sections 2.6 and 2.7. The theory was developed there for real variables. The same theory holds in the complex plan: just replace x by z.

go to finite constants as $z \to \infty$. Substituting these results in (14.46), (14.47) and using the theorem of Section 13.3 that a polynomial-bounded entire function is a polynomial, we see that $P(z)$ and $Q(z)$ must be quadratic polynomials in z and that the coefficient of z^2 in $P(z)$ is 2. Substituting these polynomials in (14.46), (14.47) and working out the partial fraction expansion of $p(z)$ and $q(z)$, we must then obtain formulas of the type

$$p(z) = \frac{A}{z - \xi} + \frac{B}{z - \eta} + \frac{C}{z - \zeta} \tag{14.49}$$

with

$$A + B + C = 2 \tag{14.50}$$

and

$$q(z) = \frac{1}{(z - \xi)(z - \eta)(z - \zeta)} \left[\frac{D}{z - \xi} + \frac{E}{z - \eta} + \frac{F}{z - \zeta} \right]. \tag{14.51}$$

The requirement that the differential equation have only three regular singular points thus restricts the form of $p(z)$ and $q(z)$ to (14.49) and (14.51), respectively, depending on six constants A, B, C, D, E, F related by (14.50). If we attempt a power series solution about one of the singular points, e.g., ξ,

$$w(z) = (z - \xi)^\alpha \sum_{n=0}^{\infty} a_n (z - \xi)^n, \tag{14.52}$$

we find as indicial equation for α

$$\alpha(\alpha - 1) + A\alpha + \frac{D}{(\xi - \eta)(\xi - \zeta)} = 0. \tag{14.53}$$

If the two solutions of this quadratic equation are α and α', then we have

$$\alpha + \alpha' = 1 - A \tag{14.54}$$

and

$$\alpha\alpha' = \frac{D}{(\xi - \eta)(\xi - \zeta)}, \tag{14.55}$$

or

$$A = 1 - \alpha - \alpha', \quad D = (\xi - \eta)(\xi - \zeta)\alpha\alpha'. \tag{14.56}$$

In a similar way we can express the constants B, E and C, F in terms of the roots of the indicial equation for series solution about the other two singular points, η and ζ:

$$B = 1 - \beta - \beta', \qquad E = (\eta - \zeta)(\eta - \xi)\beta\beta', \tag{14.57}$$

$$C = 1 - \gamma - \gamma', \qquad F = (\zeta - \xi)(\zeta - \eta)\gamma\gamma'. \tag{14.58}$$

In terms of this notation the differential equation assumes the form

$$\frac{d^2w(z)}{dz^2} + \left\{ \frac{1-\alpha-\alpha'}{z-\xi} + \frac{1-\beta-\beta'}{z-\eta} + \frac{1-\gamma-\gamma'}{z-\zeta} \right\} \frac{dw(z)}{dz}$$

$$- \left\{ \frac{\alpha\alpha'}{(z-\xi)(\eta-\zeta)} + \frac{\beta\beta'}{(z-\eta)(\zeta-\xi)} + \frac{\gamma\gamma'}{(z-\zeta)(\xi-\eta)} \right\} \frac{(\xi-\eta)(\eta-\zeta)(\zeta-\xi)}{(z-\xi)(z-\eta)(z-\zeta)} w(z)$$

$$= 0, \tag{14.59}$$

and the condition (14.50) becomes

$$\alpha + \alpha' + \beta + \beta' + \gamma + \gamma' = 1. \tag{14.60}$$

A solution of the equation (14.59) is designated by the Riemann symbol

$$\boxed{w = P \left\{ \begin{array}{ccc} \xi & \eta & \zeta \\ \alpha & \beta & \gamma & z \\ \alpha' & \beta' & \gamma' \end{array} \right\}.} \tag{14.61}$$

Note that this is not any particular solution, but just a solution of the differential equation with the parameters listed.

The interesting point is that the rather general equation (14.60) can be reduced to a standard form by using certain simple transformations. Thus by substitution in the differential equation one can show

$$\left(\frac{z-\xi}{z-\eta} \right)^p \left(\frac{z-\zeta}{z-\eta} \right)^q P \left\{ \begin{array}{ccc} \xi & \eta & \zeta \\ \alpha & \beta & \gamma & z \\ \alpha' & \beta' & \gamma' \end{array} \right\} = P \left\{ \begin{array}{ccc} \xi & \eta & \zeta \\ \alpha+p & \beta-p-q & \gamma+q & z \\ \alpha'+p & \beta'-p-q & \gamma'+q \end{array} \right\},$$
$$\tag{14.62}$$

which means that if $w(z)$ is a solution of the differential equation (14.59) with parameters $\xi, \eta, \zeta, \alpha, \beta, \gamma, \alpha', \beta', \gamma'$, then $\left(\frac{z-\xi}{z-\eta}\right)^p \left(\frac{z-\zeta}{z-\eta}\right)^q w(z)$ is a solution of the equation with parameters $\xi, \eta, \zeta, \alpha+p, \beta-p-q, \gamma+q, \alpha'+p, \beta'-p-q, \gamma'+q$. Similarly one finds

$$P \left\{ \begin{array}{ccc} \xi & \eta & \zeta \\ \alpha & \beta & \gamma & z \\ \alpha' & \beta' & \gamma' \end{array} \right\} = P \left\{ \begin{array}{ccc} \xi_1 & \eta_1 & \zeta_1 \\ \alpha & \beta & \gamma & z_1 \\ \alpha' & \beta' & \gamma' \end{array} \right\}, \tag{14.63}$$

where z and z_1, ξ and ξ_1, η and η_1, and ζ and ζ_1 are related by a bilinear transformation of the type

$$z_1 = \frac{Kz+L}{Mz+N}, \quad \xi_1 = \frac{K\xi+L}{M\xi+N}, \quad \eta_1 = \frac{K\eta+L}{M\eta+N}, \quad \zeta_1 = \frac{K\zeta+L}{M\zeta+N}, \tag{14.64}$$

with K, L, M, N, arbitrary complex constants. The formula (14.63) says that if $w(z)$ is a solution of the differential equation with parameters $\xi, \eta, \zeta, \ldots$, then $w\left(\frac{Kz+L}{Mz+N}\right)$ is a solution of the equation with parameters $\xi_1, \eta_1, \zeta_1, \ldots$.

We can use these transformations to reduce the differential equation (14.59) to a standard form. First we use (14.63) to move the location of the three regular singular points from ξ, η, ζ to 0, ∞, 1:

$$P \left\{ \begin{matrix} \xi & \eta & \zeta \\ \alpha & \beta & \gamma & z \\ \alpha' & \beta' & \gamma' \end{matrix} \right\} = P \left\{ \begin{matrix} 0 & \infty & 1 \\ \alpha & \beta & \gamma & \dfrac{(z-\xi)(\eta-\zeta)}{(z-\eta)(\xi-\zeta)} \\ \alpha' & \beta' & \gamma' \end{matrix} \right\}. \tag{14.65}$$

Next we apply (14.62) to the right side of (14.65). For the case with $\eta = \infty$ we need a special form of (14.62). With η infinite $z - \eta$ is a constant. Multiplying through by this constant to the power $p + q$, we find

$$z^p (1-z)^q\, P \left\{ \begin{matrix} 0 & \infty & 1 \\ \alpha & \beta & \gamma & z \\ \alpha' & \beta' & \gamma' \end{matrix} \right\} = P \left\{ \begin{matrix} 0 & \infty & 1 \\ \alpha+p & \beta-p-q & \gamma+q & z \\ \alpha'+p & \beta'-p-q & \gamma'+q \end{matrix} \right\}. \tag{14.66}$$

Applying this to the right side of (14.65) we obtain

$$P \left\{ \begin{matrix} \xi & \eta & \zeta \\ \alpha & \beta & \gamma & z \\ \alpha' & \beta' & \gamma' \end{matrix} \right\} = t^\alpha (1-t)^\gamma\, P \left\{ \begin{matrix} 0 & \infty & 1 \\ 0 & \alpha+\beta+\gamma & 0 & t \\ \alpha'-\alpha & \alpha+\beta'+\gamma & \gamma'-\gamma \end{matrix} \right\}, \tag{14.67}$$

with $t = \dfrac{(z-\xi)(\eta-\zeta)}{(z-\eta)(\xi-\zeta)}$. Thus the solutions of the general equation (14.59) are obtained in terms of functions of the type

$$P \left\{ \begin{matrix} 0 & \infty & 1 \\ 0 & a & 0 & t \\ 1-c & b & c-a-b \end{matrix} \right\}. \tag{14.68}$$

Putting these values $\xi = 0$, $\eta = \infty$, $\zeta = 1$, $\alpha = 0$, $\alpha' = 1 - c$, $\beta = a$, $\beta' = b$, $\gamma = 0$, $\gamma' = c - a - b$ in (14.59), the differential equation reduces to the so-called hypergeometric equation:

$$t(1-t)w''(t) + \{c - (a+b+1)t\}\, w'(t) - abw(t) = 0. \tag{14.69}$$

The two solutions of the hypergeometric equation (14.69), obtained by series expansion about the point $t = 0$, are the hypergeometric function

$$F(a,b;c;t) = 1 + \frac{ab}{c}\frac{t}{1!} + \frac{a(a+1)b(b+1)}{c(c+1)}\frac{t^2}{2!} + \cdots$$

or

$$\boxed{F(a,b;c;t) = \sum_{n=0}^{\infty} \frac{\Gamma(a+n)}{\Gamma(a)}\frac{\Gamma(b+n)}{\Gamma(b)}\frac{\Gamma(c)}{\Gamma(c+n)}\frac{t^n}{n!}} \tag{14.70}$$

and

$$t^{1-c}F(1+a-c, 1+b-c; 2-c; t). \tag{14.71}$$

The series (14.70) converges to define an analytic function for $|t| < 1$. It can be defined by an analytic continuation outside this region. We return to this question shortly.

In (14.67) we have indicated one set of transformations which reduces the general differential equation with three regular singular points to the hypergeometric equation. There are, however, 24 different ways in which this can be done, $3! = 6$ ways corresponding to interchange of the locations ξ, η, ζ of the singularities, and for each of these 4 possibilities corresponding to interchange of α, α' and β, β'. The original differential equation is, of course, invariant with respect to these interchanges, but they lead to different results on the right side (14.65). If the function on the right side of (14.67) is taken to be the hypergeometric function, we obtain in this way 24 different solutions of the differential equation (14.59):

$$w_1(z) = \left(\frac{z-\xi}{z-\eta}\right)^{\alpha} \left(\frac{z-\zeta}{z-\eta}\right)^{\gamma} F\left[\alpha + \beta + \gamma, \alpha + \beta' + \gamma; 1 + \alpha - \alpha'; \frac{(\zeta - \eta)(z - \xi)}{(\zeta - \xi)(z - \eta)}\right],$$

$$\vdots$$

$$w_{12}(z) = \left(\frac{z-\zeta}{z-\xi}\right)^{\gamma'} \left(\frac{z-\eta}{z-\xi}\right)^{\beta'} \times$$

$$F\left[\gamma' + \alpha + \beta', \gamma' + \alpha' + \beta'; 1 + \gamma' - \gamma; \frac{(\eta - \xi)(z - \zeta)}{(\eta - \zeta)(z - \xi)}\right],$$

$$\vdots$$

$$w_{24}(z) = \left(\frac{z-\eta}{z-\xi}\right)^{\beta'} \left(\frac{z-\zeta}{z-\xi}\right)^{\gamma'} \times$$

$$F\left[\beta' + \alpha + \gamma', \beta' + \alpha' + \gamma', 1 + \beta' - \beta, \frac{(\zeta - \xi)(z - \eta)}{(\zeta - \eta)(z - \xi)}\right]. \tag{14.72}$$

The complete list of 24 solutions is given, for example, by Magnus, Oberhettinger and Soni.[6] While these solutions are different from each other, they are not independent. There is a linear relation between any three of the 24 solutions.

This same game can be played with the hypergeometric equation itself, since it is, after all, a special case of the general equation (14.59). Thus to obtain the second solution (14.71) of the hypergeometric equation, start by interchanging α, α' and then use (14.66) to reduce the equation to standard form again:

$$P \left\{ \begin{matrix} 0 & \infty & 1 & \\ 1-c & a & 0 & t \\ 0 & b & c-a-b & \end{matrix} \right\} = t^{1-c} P \left\{ \begin{matrix} 0 & \infty & 1 & \\ 0 & a+1-c & 0 & t \\ -1+c & b+1-c & c-a-b & \end{matrix} \right\}. \tag{14.73}$$

---

[6]Magnus, Oberhettinger and Soni, op. cit., p. 60.

Taking the P symbol on the right-hand side of (14.73) to be a hypergeometric function we immediately deduce the second solution (14.71). In a similar way, using (14.63) and (14.66) we find

$$P\left\{\begin{matrix} 1 & \infty & 0 \\ 0 & a & 0 \\ c-a-b & b & 1-c \end{matrix}\;t\right\} = P\left\{\begin{matrix} 0 & \infty & 1 \\ 0 & a & 0 & 1-t \\ c-a-b & b & 1-c \end{matrix}\right\} \qquad (14.74)$$

and

$$P\left\{\begin{matrix} \infty & 0 & 1 \\ a & 0 & 0 \\ b & 1-c & c-a-b \end{matrix}\;t\right\} = P\left\{\begin{matrix} 0 & \infty & 1 \\ a & 0 & 0 & \dfrac{1}{t} \\ b & 1-c & c-a-b \end{matrix}\right\}$$

$$= \left(\frac{1}{t}\right)^a P\left\{\begin{matrix} 0 & \infty & 1 \\ 0 & a & 0 & \dfrac{1}{t} \\ b-a & 1+a-c & c-a-b \end{matrix}\right\}.$$
$$(14.75)$$

The differential equations implied by the symbols on the left sides of (14.74) and (14.75) are the same as that implied by (14.68); interchanging the columns does not change the differential equation. Thus taking the P symbols on the right sides of (14.74) and (14.75) to be hypergeometric functions, we find two more solutions of the hypergeometric equation (14.69):

$$F(a,b;1+a+b-c;1-t); \qquad (14.76)$$

and

$$t^{-a}\,F\left(a,1+a-c;1+a-b;\frac{1}{t}\right). \qquad (14.77)$$

Proceeding in a similar way one can find altogether 24 solutions of the hypergeometric equation with various parameters and with arguments t, $1-t$, t^{-1}, $t/(t-1)$, or $1-t^{-1}$. Convenient lists of these are given by Erdélyi, et al. and by Abramowitz and Stegun.[7]

Since the hypergeometric equation has only two linearly independent solutions, there is a linear relation between any three of the 24 solutions. Some of these relations are rather trivial and can be worked out by comparing the power series term-by-term. Others are established with the aid of certain integral representations of the hypergeometric function. We shall not enter at all into the derivations here. The point we wish to make is that some of these relations can be used to effect the analytic continuation of the hypergeometric function. The power series (14.70) converges only for $|t| < 1$. Thus the solutions (14.70) and (14.71) are in the first instance given only for $|t| < 1$. On the other hand, if we employ the power series for the hypergeometric

---

[7]A. Erdélyi et al., op. cit., Vol. I, p. 105;
   M. Abramowitz and I. Stegun, op. cit., p. 563.

function in (14.77), the region of convergence is $|1/t| < 1$, i.e., $|t| > 1$. A linear relation between the solutions (14.70), (14.71) and (14.77) thus enables one to analytically continue from the region $|t| < 1$ to the region $|t| > 1$. Complete sets of relations between the 24 solutions of the hypergeometric equation are given in the references cited on the previous footnote. We list here some of the relations which are most useful for analytic continuation:

$$F(a, b, ; c; z) = (1 - z)^{c-a-b} F(c - a, c - b; c; z) \tag{14.78}$$

$$= (1 - z)^{-a} F\left(a, c - b; c; \frac{z}{z - 1}\right) \tag{14.79}$$

$$= (1 - z)^{-b} F\left(b, c - a; c; \frac{z}{z - 1}\right) \tag{14.80}$$

$$= \frac{\Gamma(c)\Gamma(c - a - b)}{\Gamma(c - a)\Gamma(c - b)} F(a, b; a + b - c + 1; 1 - z)$$

$$+ (1 - z)^{c-a-b} \frac{\Gamma(c)\Gamma(a + b - c)}{\Gamma(a)\Gamma(b)} F(c - a, c - b; c - a - b + 1; 1 - z),$$
$$(|\arg(1\text{-}z)| < \pi), \tag{14.81}$$

$$= \frac{\Gamma(c)\Gamma(b - a)}{\Gamma(b)\Gamma(c - a)} (-z)^{-a} F\left(a, 1 - c + a; 1 - b + a; \frac{1}{z}\right)$$

$$+ \frac{\Gamma(c)\Gamma(a - b)}{\Gamma(a)\Gamma(c - b)} (-z)^{-b} F\left(b, 1 - c + b; 1 - a + b; \frac{1}{z}\right), \quad (|\arg(-z)| < \pi), \tag{14.82}$$

$$= (1 - z)^{-a} \frac{\Gamma(c)\Gamma(b - a)}{\Gamma(b)\Gamma(c - a)} F\left(a, c - b; a - b + 1; \frac{1}{1 - z}\right)$$

$$+ (1 - z)^{-b} \frac{\Gamma(c)\Gamma(a - b)}{\Gamma(a)\Gamma(c - b)} F\left(b, c - a; b - a + 1; \frac{1}{1 - z}\right), \quad (|\arg(1 - z)| < \pi), \tag{14.83}$$

$$= \frac{\Gamma(c)\Gamma(c - a - b)}{\Gamma(c - a)\Gamma(c - b)} z^{-a} F\left(a, a - c + 1; a + b - c + 1; 1 - \frac{1}{z}\right)$$

$$+ \frac{\Gamma(c)\Gamma(a+b-c)}{\Gamma(a)\Gamma(b)} (1-z)^{c-a-b} \, z^{a-c} \, F\left(c-a, 1-a; c-a-b+1; 1-\frac{1}{z}\right),$$

$$(|\arg(z)| < \pi, |\arg(1-z)| < \pi). \tag{14.84}$$

Thus the hypergeometric function is an analytic function in the complex z plane, cut along the real axis from 1 to ∞, and the analytic continuation from the circle $|z| < 1$ where the power series (14.70) converges to the rest of the complex plane is given by the formulas (14.78)-(14.84). These formulas can be used to find the discontinuities across the cut; these will be due to the discontinuities across cuts in the factors $(-z)^{-a}$, $(1-z)^{-b}$, etc. Thus for $z = x + i\epsilon$, $\epsilon > 0$, $x > 1$ we have

$$(-z)^{-a} = (e^{-i\pi}x)^{-a} = e^{i\pi a}x^{-a}$$

and for $z = x - i\epsilon$

$$(-z)^{-a} = e^{-i\pi a}x^{-a},$$

so that according to (14.82) the discontinuity across the cut $x = (1, \infty)$ is given by

$$F(a, b; c; x + i\epsilon) - F(a, b; c; x - i\epsilon) =$$

$$\frac{\Gamma(c)\Gamma(b-a)}{\Gamma(b)\Gamma(c-a)} \frac{2i \sin \pi a}{x^a} F\left(a, 1-c+a; 1-b+a; \frac{1}{x}\right)$$

$$+ \frac{\Gamma(c)\Gamma(a-b)}{\Gamma(a)\Gamma(c-b)} \frac{2i \sin \pi b}{x^b} F\left(b, 1-c+b; 1-a+b; \frac{1}{x}\right). \tag{14.85}$$

These formulas can also be used to find asymptotic expansions for $F(a, b; c; z)$ for $|z| \to \infty$ or $|z| \to 1$. Using $F(a, b; c; t) \xrightarrow[t \to 0]{} 1$, as we see from the power series expansion (14.70), we find from (14.82)

$$F(a, b; c; z) \xrightarrow[z \to \infty]{} \frac{\Gamma(c)\Gamma(b-a)}{\Gamma(b)\Gamma(c-a)}(-z)^{-a} + \frac{\Gamma(c)\Gamma(a-b)}{\Gamma(a)\Gamma(c-b)}(-z)^{-b}, \tag{14.86}$$

and from (14.81)

$$F(a, b; c; z) \xrightarrow[z \to 1]{} \frac{\Gamma(c)\Gamma(c-a-b)}{\Gamma(c-a)\Gamma(c-b)} + \frac{\Gamma(c)\Gamma(a+b-c)}{\Gamma(a)\Gamma(b)}(1-z)^{c-a-b}. \tag{14.87}$$

For each of these formulas we keep only that one of the two terms on the right which is dominant in the limit indicated; this depends on the values of the parameters a, b, c.

Finally we merely mention the confluent hypergeometric function. Start with the hypergeometric equation

$$x(1-x)w'' + [c - (a+b+1)x]w' - abw = 0 \tag{14.88}$$

with three regular singular points at $0, 1, \infty$. Let $z = xb$ so that the singularities are at $z = 0, b, \infty$, and let $b \to \infty$ so that the two singularities at b and ∞ become confluent. The differential equation reduces to

$$z\frac{d^2w}{dz^2} + (c - z)\frac{dw}{dz} - aw = 0 \tag{14.89}$$

and has an essential singularity at ∞ which results from the confluence of the two singularities. The limiting process above applied to the two solutions (14.70) and (14.71) of the hypergeometric equation provides solutions for the confluent equation:

$$w_1 = {}_1F_1(a; c; z) = 1 + \frac{a}{c} \frac{z}{1!} + \frac{a(a+1)}{c(c+1)} \frac{z^2}{2!} + \cdots, \tag{14.90}$$

$$w_2 = z^{1-c} {}_1F_1(1 + a - c; 2 - c; z). \tag{14.91}$$

The confluent hypergeometric function is important in applications. The solutions of the quantum mechanical coulomb scattering problem are expressed in terms of it; this application is discussed in some detail in the book by Mott and Massey.[8]

14.5 LEGENDRE FUNCTIONS

The principal applications of Legendre functions are in real variable theory where they appear as spherical harmonics. This part of the theory has already been covered in Chapter 3. Here we wish to show the relation of Legendre functions to hypergeometric functions and how they can be analytically continued to nonintegral l and complex z outside the range $-l \le z = x \le 1$.

For integral l the nonsingular solution of Legendre's equation

$$(1 - z^2)w''(z) - 2zw'(z) + l(l+1)w(z) = 0 \tag{14.92}$$

is the Legendre polynomial, given by Rodrigues's formula,

$$P_l(z) = \frac{1}{2^l \, l!} \frac{d^l}{dz^l} \left[(z^2 - 1)^l \right] \tag{14.93}$$

–see (3.23). Using

$$(z^2 - 1)^l = \frac{1}{2\pi i} \int_C dt \, \frac{(t^2 - 1)^l}{t - z}, \tag{14.94}$$

where C is any counterclockwise oriented contour encircling z, and differentiating under the integral sign in (14.94), we find

$$P_l(z) = \frac{1}{2\pi i} \frac{1}{2^l} \int_C dt \, \frac{(t^2 - 1)^l}{(t - z)^{l+1}}, \tag{14.95}$$

which is the Schläfli integral representation for $P_l(z)$.

The point of this transformation is that (14.95) provides a solution of Legendre's equation (14.92) even when l is nonintegral, provided we choose the contour C to go around the cuts in the right way. To see this, apply the differential operator in Legendre's equation to (14.95) and collect terms, using $z = z - t + t$ and $1 - z^2 = 1 - t^2 - 2t(z - t) - (z - t)^2$:

$$(1 - z^2)P_l''(z) - 2zP_l'(z) + l(l+1)P_l(z)$$

---

[8]N. F. Mott and H. S. W. Massey, <u>The Theory of Atomic Collisions</u> (Oxford University Press, New York, 1965), 3rd edition, Chap. III, p. 53.

$$= \frac{1}{2\pi i}\frac{1}{2^l}\int_C dt\Bigg[(1-z^2)(l+1)(l+2)\frac{(t^2-1)^l}{(t-z)^{l+3}}$$

$$-2z(l+1)\frac{(t^2-1)^l}{(t-z)^{l+2}}+l(l+1)\frac{(t^2-1)^l}{(t-z)^{l+1}}\Bigg]$$

$$=\frac{1}{2\pi i}\frac{l+1}{2^l}\int_C dt\Bigg[-(l+2)\frac{(t^2-1)^{l+1}}{(t-z)^{l+3}}+2t(l+1)\frac{(t^2-1)^l}{(t-z)^{l+2}}\Bigg]$$

$$=\frac{1}{2\pi i}\frac{l+1}{2^l}\int_C dt\,\frac{d}{dt}\Bigg[\frac{(t^2-1)^{l+1}}{(t-z)^{l+2}}\Bigg]. \tag{14.96}$$

The integral of the derivative on the right side of (14.96) will vanish provided the quantity

$$\frac{(t^2-1)^{l+1}}{(t-z)^{l+2}}=\frac{(t+1)^{l+1}}{t-z}\left(\frac{t-1}{t-z}\right)^{l+1} \tag{14.97}$$

has the same value at the beginning and the end of the closed contour C. This is achieved by choosing the contour C and cuts in the t plane as indicated in Figure 14.2. The branches of the powers are chosen so that

$$(t+1)^{l+1}=r_1^{l+1}e^{i(l+1)\theta_1},$$

$$\left(\frac{t-1}{t-z}\right)^{l+1}=\left(\frac{r_2}{r_3}\right)^{l+1}e^{i(l+1)(\theta_2-\theta_3)}, \tag{14.98}$$

using the geometry as illustrated in Figure 14.3. With C a contour around $t=z$ and $t=1$ as indicated in Figure 14.2, θ_1 and $\theta_2-\theta_3$ return to their original values as t traverses C, so that the quantity (14.97) returns to its original value as t transverses C, and $P_l(z)$ as defined by (14.95) is a solution of Legendre's equation for nonintegral l. Of course for l integral the cuts disappear and (14.95) is just a Legendre polynomial.

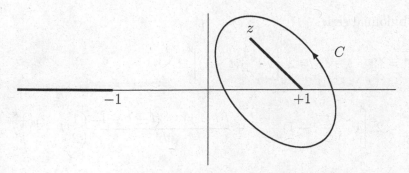

Figure 14.2: Contour choice and cuts in t plane

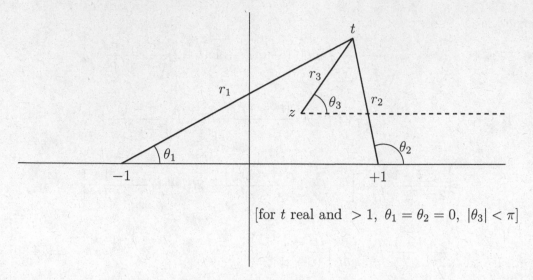

$$[\text{for } t \text{ real and } > 1, \ \theta_1 = \theta_2 = 0, \ |\theta_3| < \pi]$$

Figure 14.3: Angle analysis for equation (14.98)

Because of the cut in t along the negative real axis from $-\infty$ to -1, $P_l(z)$ as given by (14.95) will also have a branch point in z at $z = -1$ with a cut which can be taken along the negative real axis. However at $z = +1$ we have

$$\frac{(t^2 - 1)^l}{(t - z)^{l+1}} \xrightarrow[z \to 1]{} (t+1)^l \frac{1}{t - 1}, \qquad (14.99)$$

and there is no singularity at $z = +1$:

$$P_l(1) = \frac{1}{2\pi i} \frac{1}{2^l} \int_C dt \ (t+1)^l \frac{1}{t - 1}$$

$$= 1. \qquad (14.100)$$

In a similar way we can find all the derivatives of $P_l(z)$ at $z = 1$:

$$P_l^{(k)}(z)\Big|_{z=1} = \frac{1}{2\pi i} \frac{1}{2^l} (l+1)(l+2)\ldots(l+k) \int_C dt \ \frac{(t+1)^l}{(t-1)^{k+1}}. \qquad (14.101)$$

Using the binomial series

$$(t+1)^l = (2+t-1)^l = 2^l \left[1 + \frac{1}{2}(t-1)\right]^l$$

$$= 2^l \left[1 + l\frac{1}{2}(t-1) + \ldots + \frac{l(l-1)\ldots(l-k+1)}{k!}\left\{\frac{1}{2}(t-1)\right\}^k + \ldots\right] \qquad (14.102)$$

to find the residue, we obtain

$$P_l^{(k)}(z)\Big|_{z=1} = (l+1)\ldots(l+k)l(l-1)\ldots(l-k+1)\frac{1}{2^k\,k!}$$

$$= \frac{\Gamma(l+1+k)}{\Gamma(l+1)}\frac{\Gamma(-l+k)}{\Gamma(-l)}\frac{\Gamma(1)}{\Gamma(k+1)}\frac{(-1)^k}{2^k}. \tag{14.103}$$

Using these values we can form the Taylor expansion for $P_l(z)$ about the point $z=1$:

$$P_l(z) = \sum_{k=0}^{\infty} \frac{1}{k!}P_l^{(k)}(z)\Big|_{z=1}(z-1)^k$$

$$= \sum_{k=0}^{\infty} \frac{\Gamma(l+1+k)}{\Gamma(l+1)}\frac{\Gamma(-l+k)}{\Gamma(-l)}\frac{\Gamma(1)}{\Gamma(1+k)}\frac{1}{k!}\left(\frac{1-z}{2}\right)^k. \tag{14.104}$$

This has just the form of a hypergeometric series (14.70), so we finally obtain

$$\boxed{P_l(z) = F\left(l+1, -l; 1; \frac{1-z}{2}\right)} \tag{14.105}$$

as the expression for a Legendre function of arbitrary l in terms of a hypergeometric function. The analytic continuation of this function throughout the z plane can be achieved by using the formulas (14.78)-(14.84) discussed in the previous section.

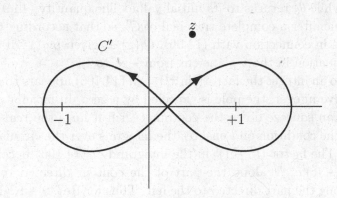

Figure 14.4: Contour used with second solution of Legendre's equation, $Q_l(z)$

To obtain a second solution of Legendre's equation we use a different contour in the integral (14.95). The function $Q_l(z)$ is defined by

$$Q_l(z) = \frac{1}{4i2^l\sin\pi l}\int_{C'}dt\,\frac{(t^2-1)^l}{(z-t)^{l+1}} \tag{14.106}$$

with the contour C' given in Figure 14.4. In (14.106), the integrand of which is not quite the same as (14.95), z is outside the contour C' and we take

$$(t^2 - 1)^l = (r_1 r_2)^l e^{il(\theta_1 + \theta_2)},$$

$$(z - t)^{l+1} = r_3^{l+1} e^{i(l+1)\theta_3}, \tag{14.107}$$

using the geometry as illustrated in Figure 14.5.

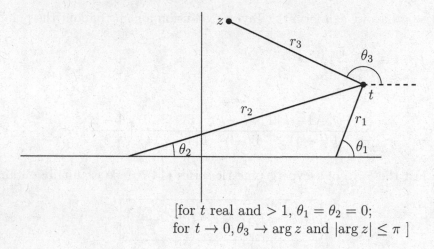

[for t real and > 1, $\theta_1 = \theta_2 = 0$;
for $t \to 0$, $\theta_3 \to \arg z$ and $|\arg z| \leq \pi$]

Figure 14.5: Angle analysis for equation 14.107

As t traverses the contour C' in the diagram above, θ_2 increases by 2π and θ_1 decreases by 2π, while θ_3 returns to its initial value. The quantity (14.97) thus returns to its initial value after a complete traversal of C', so that according to the same argument as made in connection with (14.96), $Q_l(z)$ as given by (14.106) is a solution of Legendre's equation in the z plane cut from $-\infty$ to 1.

For l equal to an integer the factor $(\sin \pi l)^{-1}$ in (14.106) appears to give a pole. For $l = 0$ or a positive integer, the pole is canceled by a zero. In fact for any l such that $\operatorname{Re} l > -1$, we can squeeze down the contour C' till it hugs the real axis, as shown in Figure 14.6; the condition on l ensures the integrals over the circular parts around ± 1 will vanish. The factor $(t^2 - 1)^l$ in the integrand of (14.106) becomes, according to (14.107), $(1 - t^2)^l e^{+i\pi l}$ along the part of the contour directed to the right and $(1 - t^2)^l e^{-i\pi l}$ along the part directed to the left. Thus for $\operatorname{Re} l > -1$, (14.106) reduces to

$$Q_l(z) = \frac{1}{2^{l+1}} \int_{-1}^{+1} dt \, \frac{(1 - t^2)^l}{(z - t)^{l+1}}. \tag{14.108}$$

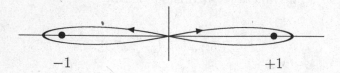

-1 $+1$

Figure 14.6: Contour squeezed along x axis while including the poles at ± 1

If, further more, l is 0 or a positive integer we can integrate by parts and use Rodrigues's formula (14.93) to obtain

$$Q_l(z) = \frac{1}{2^{l+1}} \int_{-1}^{1} dt \, (1-t^2)^l \frac{1}{l!} \frac{d^l}{dt^l} \frac{1}{z-t}$$

$$= \frac{1}{2} \int_{-1}^{1} dt \, \frac{P_l(t)}{z-t}. \tag{14.109}$$

This is the definition of $Q_l(z)$ which we used before in Section 3.4. The formula (14.106) gives a generalization for nonintegral l.

We can relate $Q_l(z)$ to a hypergeometric function by expanding (14.108) for $|z| > 1$. In this region we can expand $(z-t)^{-l-1}$ by the binomial theorem:

$$\frac{1}{(z-t)^{l+1}} = \frac{1}{z^{l+1}} \left(1 - \frac{t}{z}\right)^{-l-1}$$

$$= \frac{1}{z^{l+1}} \left[1 + (l+1)\frac{t}{z} + \frac{(l+1)(l+2)}{2!} \left(\frac{t}{z}\right)^2 + \ldots\right]. \tag{14.110}$$

Substituting this in (14.108) and using

$$\int_{-1}^{1} dt \, (1-t^2)^l \, t^{2k} = \int_{-1}^{1} d(\cos\theta) \cos^{2k}\theta \sin^{2l}\theta$$

$$= 2 \int_{0}^{\pi/2} d\theta \, \cos^{2k}\theta \sin^{2l+1}\theta$$

$$= B\left(k + \frac{1}{2}, l+1\right)$$

$$= \frac{\Gamma\left(k+\frac{1}{2}\right)\Gamma(l+1)}{\Gamma\left(l+\frac{3}{2}+k\right)}, \tag{14.111}$$

from (14.8) and (14.11), we find

$$Q_l(z) = \frac{1}{(2z)^{l+1}} \int_{-1}^{1} dt \, (1-t^2)^l \left[1 + (l+1)\frac{t}{z} + \frac{(l+1)(l+2)}{2!} \left(\frac{t}{z}\right)^2 + \ldots\right]$$

$$= \frac{1}{(2z)^{l+1}} \sum_{k=0}^{\infty} \frac{(l+1)\ldots(l+2k)}{(2k)!} \frac{1}{z^{2k}} \frac{\Gamma(k+\frac{1}{2})\Gamma(l+1)}{\Gamma(l+\frac{3}{2}+k)}$$

or

$$Q_l(z) = \frac{\Gamma\left(\frac{1}{2}\right)\Gamma(l+1)}{(2z)^{l+1}\Gamma\left(l+\frac{3}{2}\right)} \sum_{k=0}^{\infty} \frac{\Gamma\left(\frac{l+1}{2}+k\right)}{\Gamma\left(\frac{l+1}{2}\right)} \frac{\Gamma\left(\frac{l+2}{2}+k\right)}{\Gamma\left(\frac{l+2}{2}\right)} \frac{\Gamma\left(l+\frac{3}{2}\right)}{\Gamma\left(l+\frac{3}{2}+k\right)} \frac{1}{k!} \left(\frac{1}{z^2}\right)^k. \tag{14.112}$$

This is the hypergeometric series (14.70), so we find

$$\boxed{Q_l(z) = \frac{\sqrt{\pi}\,\Gamma(l+1)}{(2z)^{l+1}\Gamma\left(l+\frac{3}{2}\right)}\, F\left(\frac{l+1}{2}, \frac{l+2}{2}; l+\frac{3}{2}; \frac{1}{z^2}\right).}$$ (14.113)

This result, obtained from (14.108), which is valid for $\operatorname{Re} l > -1$, can be analytically continued in l throughout the entire l plane with the exception of the negative integers, where the poles in l are explicitly displayed in the factor $\Gamma(l+1)$.

14.6 BESSEL FUNCTIONS

The series solutions of Bessel's equation were studied in some detail in Chapter 4. Here we are interested in certain generally useful integral representations of the functions due to the physicist Sommerfeld. The method of steepest descents can be applied to these integral representations to obtain asymptotic expansions.

We shall search for a solution of Bessel's differential equation

$$z^2 u''(z) + z u'(z) + (z^2 - \lambda^2) u(z) = 0$$ (14.114)

as a contour integral of the form

$$u(z) = \int_C d\zeta\; K(z,\zeta) v(\zeta).$$ (14.115)

The functions $K(z,\zeta)$ and $v(\zeta)$ and the contour C are to be chosen so that (14.115) is a solution of (14.114). Substituting (14.115) in (14.114) we find

$$\int_C d\zeta\; \left[z^2 \frac{\partial^2 K}{\partial z^2} + z\frac{\partial K}{\partial z} + z^2 K - \lambda^2 K\right] v(\zeta) = 0.$$ (14.116)

Let $K(z,\zeta)$ satisfy the partial differential equation

$$z^2 \frac{\partial^2 K}{\partial z^2} + z\frac{\partial K}{\partial z} + z^2 K = -\frac{\partial^2 K}{\partial \zeta^2}.$$ (14.117)

Then (14.116) reduces to

$$\int_C d\zeta\; \left[\frac{\partial^2 K}{\partial \zeta^2} + \lambda^2 K\right] v(\zeta) = 0.$$ (14.118)

Integrating twice by parts this becomes

$$\int_C d\zeta\; K(z,\zeta)\left[v''(\zeta) + \lambda^2 v(\zeta)\right] = 0,$$ (14.119)

provided the end point contributions from the integration by parts vanish. To satisfy (14.119) we need only choose

$$v(\zeta) = e^{\pm i\lambda\zeta}.$$ (14.120)

Furthermore, a simple solution of the partial differential equation (14.117) for $K(z, \zeta)$ is

$$K(z, \zeta) = e^{\pm iz \sin \zeta}, \tag{14.121}$$

as can be easily checked by substitution in (14.117). Thus, if we can choose the contour C so that the integrals converge and so that the contributions from the end points on integration by parts vanish, then we have found solutions of Bessel's equation of the form

$$u(z) = \int_C d\zeta \; e^{\pm iz \sin \zeta \pm i\lambda \zeta}. \tag{14.122}$$

Try the contours L_1 and L_2 indicated in Figure 14.7. Writing ζ, z, and $\sin \zeta$ in terms of their real and imaginary parts, we have

$$\zeta = \xi + i\eta, \tag{14.123}$$
$$\sin \zeta = \cosh \eta \sin \xi + i \sinh \eta \cos \xi, \tag{14.124}$$
$$z = x + iy, \tag{14.125}$$
$$\mathrm{Re}[-iz \sin \zeta] = x \sinh \eta \cos \xi + y \cosh \eta \sin \xi. \tag{14.126}$$

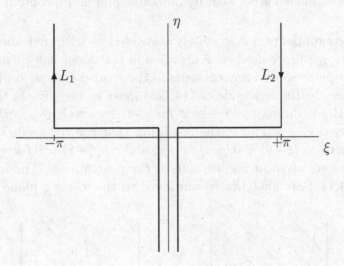

Figure 14.7: Contours for use with equation (14.122)

Along the parts of the contour going to infinity in Figure 14.7, $\sin \xi = 0$. Along the parts going to $0 - i\infty$, $\cos \xi = +1$, $\sinh \eta = \frac{1}{2}(e^\eta - e^{-\eta}) \to -\frac{1}{2}e^{-\eta} < 0$. Along the parts going to $\pm\pi + i\infty$, $\cos \xi = -1$, $\sinh \eta \to \frac{1}{2}e^\eta > 0$. Thus for $x > 0$, $\mathrm{Re}[-iz \sin \zeta]$ goes exponentially to $-\infty$ at the end points of the contours L_1 and L_2. For these contours the integral (14.122) converges very well indeed, and the contributions obtained by integration by parts in (14.118), (14.119) vanish. Thus for $\mathrm{Re}\, z > 0$ we have found two solutions of Bessel's equation, called Hankel functions:

$$H_\lambda^{(1)}(z) = -\frac{1}{\pi} \int_{L_1} d\zeta \; e^{-iz \sin \zeta + i\lambda \zeta}, \tag{14.127}$$

$$H_\lambda^{(2)}(z) = -\frac{1}{\pi} \int_{L_2} d\zeta \; e^{-iz \sin \zeta + i\lambda \zeta}. \tag{14.128}$$

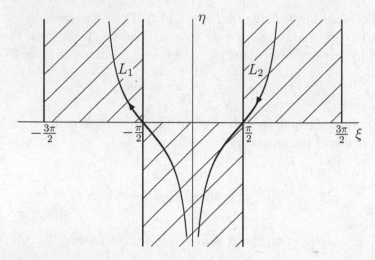

Figure 14.8: Shaded areas identify allowable and useful contour regions

In fact the contours are not as closely restricted as indicated above. For z real and > 0, so that $y = 0$, we need $\cos \xi$ negative in the upper half-plane and positive in the lower half-plane so that any contours in the shaded regions in Figure 14.8 will work. More generally, the right side of (14.126) must be negative at the extremities of the contour. Replacing $\sinh \eta \to \pm\frac{1}{2}e^{\pm\eta}$ for $\eta \to \pm\infty$, $\cosh \eta \to \frac{1}{2}e^{\pm\eta}$, and writing $z = re^{i\theta}$, $x = r\cos\theta$, $y = r\sin\theta$, the condition that the right side of (14.126) be negative reduces to $\cos(\xi - \theta) < 0$ for $\eta \to +\infty$ and $\cos(\xi + \theta) > 0$ for $\eta \to -\infty$. Thus suitable contours are given in Figure 14.9 for the general case. The functions given by (14.127), (14.128) are analytically continued to the whole z plane by moving, if

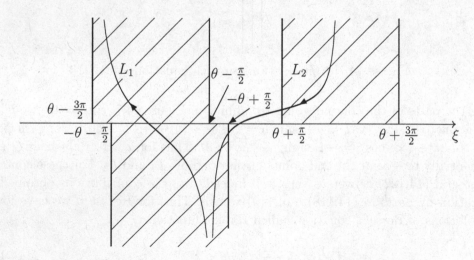

Figure 14.9: Shaded regions for generally allowed contours

necessary, the contours within the allowed regions, the boundaries of which also move as $\theta = \arg(z)$ changes.

In terms of the Hankel functions, we can define Bessel and Neumann functions by

$$H_\lambda^{(1)}(z) = J_\lambda(z) + iN_\lambda(z), \qquad (14.129)$$

$$H_\lambda^{(2)}(z) = J_\lambda(z) - iN_\lambda(z) \qquad (14.130)$$

or

$$J_\lambda(z) = \frac{1}{2}\left[H_\lambda^{(1)}(z) + H_\lambda^{(2)}(z)\right], \qquad (14.131)$$

$$N_\lambda(z) = \frac{1}{2i}\left[H_\lambda^{(1)}(z) - H_\lambda^{(2)}(z)\right]. \qquad (14.132)$$

We shall shortly show that all these definitions are equivalent to our earlier series definitions of Bessel, Neumann, and Hankel functions in Chapter 4.

For the case of the Bessel function $J_\lambda(z)$ the lower parts of the contours cancel when we add $H_\lambda^{(1)}(z)$ and $H_\lambda^{(2)}(z)$ so that we obtain

$$J_\lambda(z) = -\frac{1}{2\pi}\int_L d\zeta \; e^{-iz\sin\zeta + i\lambda\zeta}, \qquad (14.133)$$

where the contour L is appropriate for $\operatorname{Re} z > 0$ as shown in Figure 14.10.

For the special case $\lambda = n =$ integer, the two vertical parts of the contour cancel each other because of the periodicity of the integrand, and we find

$$J_n(z) = \frac{1}{2\pi}\int_{-\pi}^{\pi} d\zeta \; e^{-iz\sin\zeta + in\zeta}$$

$$= \frac{1}{2\pi}\int_{-\pi}^{\pi} d\zeta \; \cos[z\sin\zeta - n\zeta]. \qquad (14.134)$$

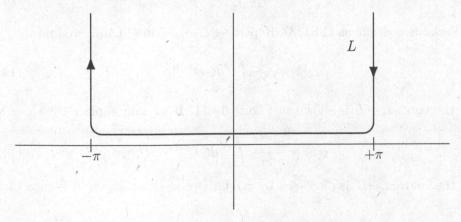

Figure 14.10: Contour choice for the Bessel function $J_\lambda(z)$

The first of these forms implies that $J_n(z)$ is the Fourier transform of $e^{-iz\sin\zeta}$, so that the latter has the Fourier expansion

$$e^{-iz\sin\zeta} = \sum_{n=-\infty}^{\infty} J_n(z)e^{-in\zeta}. \qquad (14.135)$$

This is then a generating function for Bessel functions of integral index. It also provides the expansion of a plane wave in cylindrical waves:

$$e^{i\mathbf{k}\cdot\mathbf{r}} = e^{ikr\cos\theta} = e^{-ikr\sin(\theta-\frac{\pi}{2})}$$

$$= \sum_{n=-\infty}^{\infty} i^n J_n(kr)e^{-in\theta}. \qquad (14.136)$$

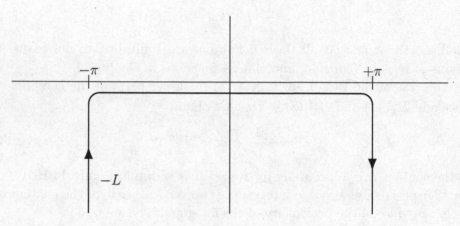

Figure 14.11: Contour choice for the Bessel function $J_{-\lambda}(z)$

We now establish some relations between $J_\lambda(z)$, $N_\lambda(z)$, $H_\lambda^{(1)}(z)$ and $H_\lambda^{(2)}(z)$. $J_\lambda(z)$ is given by (14.133). Reversing the sign of λ we also have

$$J_{-\lambda}(z) = -\frac{1}{2\pi} \int_L d\zeta \, e^{-iz\sin\zeta - i\lambda\zeta}, \qquad (14.137)$$

with the same contour as (14.133). Replacing $\zeta \to -\zeta$ in (14.137) we find

$$J_{-\lambda}(z) = \frac{1}{2\pi} \int_{-L} d\zeta \, e^{iz\sin\zeta + i\lambda\zeta}, \qquad (14.138)$$

where the contour $-L$ is shown in Figure 14.11. If we now replace $\zeta \to \zeta - \pi$, we obtain

$$J_{-\lambda}(z) = \frac{e^{-i\pi\lambda}}{2\pi} \int_{-L'} d\zeta \, e^{-iz\sin\zeta + i\lambda\zeta}. \qquad (14.139)$$

where the contour $-L'$ is displaced by π relative to $-L$ as shown in Figure 14.12.

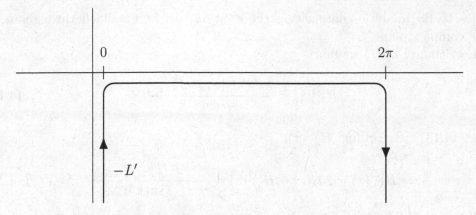

Figure 14.12: Shifting of contour $-L$ by π to obtain $-L'$

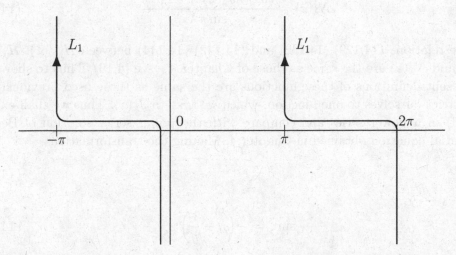

Figure 14.13: New contours, L_1 and L_1', from combining contours L and $-L'$

Combining (14.133) and (14.139) we find

$$J_\lambda(z) - e^{i\pi\lambda} J_{-\lambda}(z) = -\frac{1}{2\pi}\left[\int_{L_1} d\zeta - \int_{L_1'} d\zeta\right] e^{-iz\sin\zeta + i\lambda\zeta}$$

$$= -\frac{1}{2\pi}(1 - e^{i2\pi\lambda})\int_{L_1} d\zeta\, e^{-iz\sin\zeta + i\lambda\zeta}, \qquad (14.140)$$

where the contours L_1 and L_1' which arise from combining L and $-L'$ are shown in Figure 14.13. The circumstance that L_1 and L_1' are displaced from each other by 2π leads to the second form of (14.140). Now L_1 is the original contour used in (14.127) to define the Hankel function $H_\lambda^{(1)}(z)$ so that (14.140) becomes

$$J_\lambda(z) - e^{i\pi\lambda} J_{-\lambda}(z) = \frac{1}{2}\left(1 - e^{i2\pi\lambda}\right) H_\lambda^{(1)}(z). \qquad (14.141)$$

With the contours chosen as above, our derivation of this relation holds only for

Re $z > 0$. By analytic continuation the relation can be extended throughout the whole complex plane.

Rewriting (14.141) we find

$$H_\lambda^{(1)}(z) = \frac{e^{-i\pi\lambda}J_\lambda(z) - J_{-\lambda}(z)}{-i\sin\pi\lambda} \qquad (14.142)$$

Using (14.131) we can find $H_\lambda^{(2)}(z)$:

$$H_\lambda^{(2)}(z) = 2J_\lambda(z) - H_\lambda^{(1)}(z) = \frac{e^{i\pi\lambda}J_\lambda(z) - J_{-\lambda}(z)}{i\sin\pi\lambda}. \qquad (14.143)$$

Finally, from (14.132) we obtain $N_\lambda(z)$:

$$N_\lambda(z) = \frac{\cos\pi\lambda J_\lambda(z) - J_{-\lambda}(z)}{\sin\pi\lambda}. \qquad (14.144)$$

The relations (14.129)-(14.132) and (14.142)-(14.144) between $H_\lambda^{(1)}(z)$, $H_\lambda^{(2)}(z)$, $J_\lambda(z)$, and $N_\lambda(z)$ are the same as those of Chapter 4 – see (4.19). Thus to show that the present definitions of these functions are the same as those used previously, we can restrict ourselves to one function, which we take to $J_\lambda(z)$. Thus we shall expand (14.133) in a power series and compare with the power series solution of Bessel's differential equation obtained in Chapter 4. Making the transformation

$$\eta = e^{-i\zeta} \qquad (14.145)$$

$$\sin\zeta = \frac{i}{2}\left(\eta - \frac{1}{\eta}\right), \qquad (14.146)$$

the integral representation (14.133) is transformed into a different integral representation,

$$J_\lambda(z) = \frac{1}{2\pi i}\int_C d\eta\ \eta^{-\lambda-1}\exp\left[\frac{z}{2}\left(\eta - \frac{1}{\eta}\right)\right], \qquad (14.147)$$

with the contour C indicated in Figure 14.14. Making a further transformation

$$\eta = \frac{2v}{z}, \qquad (14.148)$$

(14.147) becomes

$$J_\lambda(z) = \frac{1}{2\pi i}\left(\frac{z}{2}\right)^\lambda\int_C dv\ v^{-\lambda-1}\exp\left[v - \frac{z^2}{4v}\right], \qquad (14.149)$$

where the contour has been moved back to C by using Cauchy's theorem. Using the expansion

$$e^{-z^2/4v} = \sum_{r=0}^\infty \frac{(-1)^r}{r!}\frac{z^{2r}}{2^{2r}v^r}, \qquad (14.150)$$

we find

$$J_\lambda(z) = \left(\frac{z}{2}\right)^\lambda \sum_{r=0}^\infty \frac{(-1)^r}{r!} \left(\frac{z}{2}\right)^{2r} \frac{1}{2\pi i} \int_C dv \, e^v v^{-\lambda-r-1}. \tag{14.151}$$

Now the integral in this expression is just one of the integral representations of the gamma function. In fact, combining (14.3) and (14.12) we find

$$\frac{1}{\Gamma(z)} = \frac{1}{2\pi i} \int_C dt \, e^t t^{-z}. \tag{14.152}$$

Using this in (14.151), we finally obtain

$$J_\lambda(z) = \left(\frac{z}{2}\right)^\lambda \sum_{r=0}^\infty \frac{(-1)^r}{r! \, \Gamma(\lambda+r+1)} \left(\frac{z}{2}\right)^{2r}, \tag{14.153}$$

which is the series for the Bessel function – see (4.13). This completes our demonstration of the equivalence of Sommerfeld's integral representations and the power series solutions of Bessel's equation.

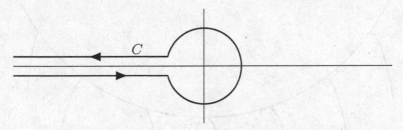

Figure 14.14: Contour for use with equation (14.147)

14.7 ASYMPTOTIC EXPANSIONS FOR BESSEL FUNCTIONS

To obtain these asymptotic expansions, we employ the method of steepest descents, introduced in Section 14.3 in connection with the gamma function, to Sommerfeld's integral representations (14.127) and (14.128) for $H_\lambda^{(1)}(z)$ and $H_\lambda^{(2)}(z)$.

In order to cope with complex variables z and ζ and the contour integral, let us first consider a general case,

$$\int_C d\zeta \, e^{f(z,\zeta)}. \tag{14.154}$$

One might first think that we should find the maximum of $\operatorname{Re} f(z,\zeta)$ and replace $f(z,\zeta)$ by a good approximation in the neighborhood of this point. However, we shall assume $f(z,\zeta)$ is an analytic function of ζ, and we must remember that the real and imaginary parts, u and v, of an analytic function,

$$f(z,\zeta) = u(z,\xi,\eta) + iv(z,\xi,\eta), \quad \zeta = \xi + i\eta, \tag{14.155}$$

satisfy Laplace's equation

$$\frac{\partial^2 u}{\partial \xi^2} = -\frac{\partial^2 u}{\partial \eta^2}, \tag{14.156}$$

so that $u(z, \xi, \eta)$ does not have a true maximum. In fact there can only be saddle points. Thus the method of steepest descents consists in using Cauchy's theorem to deform the contour C so that it goes through the saddle point in the direction such that $u(z, \xi, \eta)$ drops off as rapidly as possible on both sides as illustrated in Figure 14.15 (hence the name steepest descents). At the saddle point we have

$$\frac{\partial u}{\partial \xi} = \frac{\partial u}{\partial \eta} = 0. \tag{14.157}$$

Using the Cauchy-Riemann conditions we then find

$$\frac{\partial v}{\partial \xi} = \frac{\partial v}{\partial \eta} = 0, \tag{14.158}$$

so that at the saddle point

$$\frac{df(z, \zeta)}{d\zeta} = 0. \tag{14.159}$$

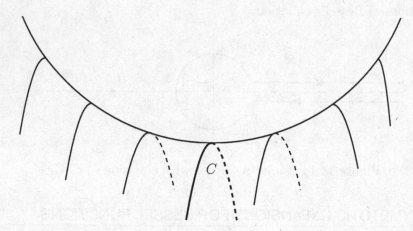

Figure 14.15: Deforming the contour until a saddle point is determined

If ζ_0 is the location of the saddle point, the Taylor expansion of f in the vicinity of this point assumes the form

$$f(z, \zeta) = f(z, \zeta_0) + \frac{1}{2}\frac{d^2 f(z, \zeta_0)}{d\zeta_0^2}(\zeta - \zeta_0)^2 + \dots . \tag{14.160}$$

If the dependence on z is appropriate, and if C goes through the saddle point in the right direction, only the near vicinity of $\zeta = \zeta_0$ will contribute for large $|z|$ and we need only keep the two terms given in (14.160) of the Taylor expansion. If the dependence on z is appropriate, higher terms in the series (14.160) will lead to higher terms (in the form of series of increasing powers of $1/z$) in the asymptotic expansion, as for the gamma function – see Section 14.3.

Of course one has to check the z dependence for any particular case, but supposing that it is appropriate, let us find a general result for the asymptotic expansion, keeping only the quadratic term in (14.160). We can choose the proper direction for the integration as follows. Suppose the phase of $f''(\zeta_0)$ is φ, so that

$$\frac{d^2 f(\zeta_0)}{d\zeta_0^2} = |f''(\zeta_0)| e^{i\varphi}. \tag{14.161}$$

Then integrate ζ in one or the other of the two opposite directions

$$\zeta - \zeta_0 = \rho e^{i\left(\frac{\pi}{2} - \frac{\varphi}{2}\right)} \quad \text{or} \quad \rho e^{i\left(-\frac{\pi}{2} - \frac{\varphi}{2}\right)}, \tag{14.162}$$

where ρ is a real number. Then the second term in (14.160) becomes

$$\frac{1}{2} f''(\zeta_0)(\zeta - \zeta_0)^2 = -\frac{1}{2} |f''(\zeta_0)| \rho^2, \tag{14.163}$$

so that the approximation to the $e^{f(z,\zeta)}$ becomes

$$e^{f(z,\zeta)} = e^{f(z,\zeta_0)} e^{-\frac{1}{2}|f''(\zeta_0)|\rho^2}. \tag{14.164}$$

This falls off with increasing ρ^2 as rapidly as possible. Choice of a different path in (14.162) leads to a less rapid fall off with ρ^2: including an additional phase $e^{i\psi/2}$ in (14.162) would lead to the replacement

$$\exp\left\{-\frac{1}{2} |f''(\zeta_0)| \rho^2\right\} \rightarrow \exp\left\{-\frac{1}{2} |f''(\zeta_0)| \rho^2 [\cos\psi + i\sin\psi]\right\} \tag{14.165}$$

in (14.164). Which of the two integration directions (14.162) is to be chosen depends on where the ends of the contour C are tied down. Thus choice of the wrong direction (14.162) will imply crossing three times the "mountain range" of Figure 14.15 rather than just going once through the "pass". With the approximation (14.164) the integral (14.154) is easy to calculate and we find

$$\int_C d\zeta \, e^{f(z,\zeta)} \sim e^{f(z,\zeta_0)} e^{i\left(\pm\frac{\pi}{2} - \frac{\varphi}{2}\right)} \int_{-\infty}^{\infty} d\rho \, e^{-\frac{1}{2}|f''(\zeta_0)|\rho^2}$$

$$= e^{f(z,\zeta_0)} e^{i\left(\pm\frac{\pi}{2} - \frac{\varphi}{2}\right)} \sqrt{\frac{2\pi}{|f''(\zeta_0)|}}. \tag{14.166}$$

Here we took the limits on the ρ integral on the right of (14.166) to be $\pm\infty$ since only the region near the saddle point makes an appreciable contribution to the integral. Just as for the gamma function – see (14.36)ff. - if we keep a few more terms in the Taylor expansion (14.160) we can (a) check that the approximations become better and better for larger $|z|$ [perhaps only in a certain range of $\arg(z)$] and (b) calculate, if desired, higher terms in the asymptotic expansion.

We can now apply these ideas to the Hankel function

$$H_\lambda^{(1)}(z) = -\frac{1}{\pi} \int_{L_1} d\zeta \, e^{-iz\sin\zeta + i\lambda\zeta}. \tag{14.167}$$

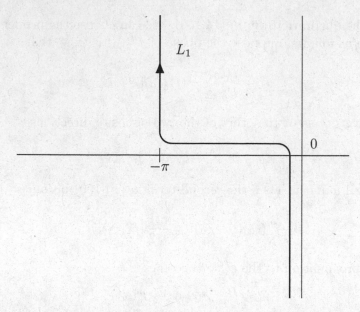

Figure 14.16: Contour used in equation (14.167) for $H_\lambda^{(1)}(z)$

For this case we have

$$f(z,\zeta) = -iz\sin\zeta + i\lambda\zeta,$$

$$\frac{df(z,\zeta)}{d\zeta} = -iz\cos\zeta + i\lambda,$$

$$\frac{d^2 f(z,\zeta)}{d\zeta^2} = iz\sin\zeta,$$

$$\frac{d^3 f(z,\zeta)}{d\zeta^3} = iz\cos\zeta$$

$$\vdots$$

(14.168)

where the contour for L_1 as shown in Figure 14.16 has been used. The location of the saddle point is given by (14.159), i.e.,

$$\cos\zeta_0 = \frac{\lambda}{z}.$$

(14.169)

For $|z| \to \infty$ this becomes

$$\cos\zeta_0 \to 0, \quad \zeta_0 \to -\frac{\pi}{2},$$

(14.170)

where we have chosen the zero of $\cos\zeta_0$ on the contour L_1. Evaluating the function

$f(z, \zeta)$ and its derivatives (14.168) at the saddle point ζ_0, we find for the Taylor series (14.160)

$$f(z, \zeta) = -iz \sin \zeta_0 + i\lambda\zeta_0 + \frac{1}{2!}iz \sin \zeta_0 \, (\zeta - \zeta_0)^2$$

$$+ \frac{1}{3!}iz \cos \zeta_0 \, (\zeta - \zeta_0)^3 - \frac{1}{4!}iz \sin \zeta_0 \, (\zeta - \zeta_0)^4 + \dots . \qquad (14.171)$$

If we make a change of variable $\zeta' = \sqrt{z}(\zeta - \zeta_0)$ this becomes

$$f(z, \zeta) = -iz \sin \zeta_0 + i\lambda\zeta_0 + \frac{1}{2!}i \sin \zeta_0 \zeta'^2$$

$$+ \frac{1}{3!}\frac{i \cos \zeta_0}{\sqrt{z}}\zeta'^3 - \frac{1}{4!}\frac{i \sin \zeta_0}{z}\zeta'^4 + \dots , \qquad (14.172)$$

and it is clear that the higher terms provide an expansion in powers of $z^{-\frac{1}{2}}$ (or z^{-1} if the odd powers cancel). We shall not pursue these higher terms further; if it is desired to keep them, one would also have to expand ζ_0 in inverse powers of z.

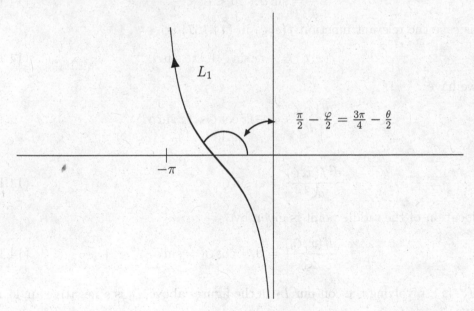

Figure 14.17: Contour direction and associated phase arguments

Keeping only the first two terms in the Taylor expansion (14.171), we can use the general result (14.166). In this formula φ is the phase of $\frac{d^2 f(z,\zeta_0)}{d\zeta_0^2}$. According to (14.168) and (14.170) this is given by

$$\varphi \xrightarrow[|z|\to\infty]{} -\frac{\pi}{2} + \theta, \qquad (14.173)$$

where $\theta = \arg(z)$. As we see from Figure 14.17, for $\mathrm{Re}\, z > 0$, i.e., $-\pi/2 < \theta < \pi/2$, the correct direction for the contour (14.162) is obtained by choosing $+\pi/2$ in the phase factor of (14.162). Thus using (14.166), (14.168), (14.170), we find for the leading term in the asymptotic expansion of (14.167) the expression

$$H_\lambda^{(1)}(z) \sim -\frac{1}{\pi} \exp\left[iz - i\lambda\frac{\pi}{2} + i\left(\frac{\pi}{2} + \frac{\pi}{4} - \frac{\theta}{2}\right)\right] \sqrt{\frac{2\pi}{|z|}}$$

$$= \sqrt{\frac{2}{\pi z}} \exp\left[i\left(z - \frac{\lambda\pi}{2} - \frac{\pi}{4}\right)\right]. \tag{14.174}$$

A similar argument for $H_\lambda^{(2)}(z)$ leads to the expression

$$H_\lambda^{(2)}(z) \sim \sqrt{\frac{2}{\pi z}} \exp\left[-i\left(z - \frac{\lambda\pi}{2} - \frac{\pi}{4}\right)\right]. \tag{14.175}$$

The asymptotic expansions above hold for $|z| \to \infty$, λ fixed; thus they are good for $|z| \gg |\lambda|$. To obtain formulas good for $|\lambda| \lesssim |z|$ we make a slightly different calculation. Suppose λ and $z = x$ are both large and positive and $\lambda < x$ so that we can write

$$\frac{\lambda}{x} = \sin\alpha, \quad 0 < \alpha < \frac{\pi}{2}. \tag{14.176}$$

In this case the relevant function $f(z,\zeta)$ in (14.167) is

$$f(x,\zeta) = -ix\sin\zeta + ix\zeta\sin\alpha, \tag{14.177}$$

and we have

$$\frac{df(x,\zeta)}{d\zeta} = -ix\cos\zeta + ix\sin\alpha,$$

$$\frac{d^2 f(x,\zeta)}{d\zeta^2} = ix\sin\zeta. \tag{14.178}$$

The location of the saddle point is given by

$$\frac{df(x,\zeta_0)}{d\zeta_0} = 0, \quad \cos\zeta_0 = \sin\alpha. \tag{14.179}$$

For $H_\lambda^{(1)}(x)$, involving the contour L_1 in the figures above, ζ_0 is a negative angle, and we have

$$\sin\zeta_0 = -\cos\alpha, \quad \zeta_0 = \alpha - \frac{\pi}{2}. \tag{14.180}$$

From (14.166), (14.167), (14.177) and (14.178) we then find

$$H_\lambda^{(1)} \sim -\frac{1}{\pi} \exp\left[+ix\cos\alpha + ix\left(\alpha - \frac{\pi}{2}\right)\sin\alpha + i\left(\frac{\pi}{2} + \frac{\pi}{4}\right)\right] \sqrt{\frac{2\pi}{x\cos\alpha}}$$

$$= \sqrt{\frac{2}{\pi x\cos\alpha}} \, e^{-i\pi/4} \exp\left\{ix\left[\cos\alpha + \left(\alpha - \frac{\pi}{2}\right)\sin\alpha\right]\right\}. \tag{14.181}$$

For x and λ real we also have

$$H_\lambda^{(2)}(x) = [H_\lambda^{(1)}(x)]^*$$

$$\sim \sqrt{\frac{2}{\pi x \cos \alpha}} \, e^{+i\pi/4} \, \exp\left\{-ix\left[\cos\alpha + \left(\alpha - \frac{\pi}{2}\right)\sin\alpha\right]\right\}. \qquad (14.182)$$

These formulas are due to the physicist Debye. For $\lambda = x\sin\alpha$ finite as $x \to \infty$, we have $\alpha \to 0$. The formulas (14.181) and (14.182) then reduce to (14.174) and (14.175), respectively.

Other asymptotic formulas, valid for other ranges of the variable z and index λ, are given in the mathematical handbooks referred to in Section 14.1. The higher order terms in the asymptotic expansions are also given in these references; usually these higher order terms are not needed in applications.

PROBLEMS

14.1 Find the solutions of the differential equation

$$z^2(z^2 - 1)u''(z) + z\left(2z^2 + \frac{1}{3}z - 1\right)u'(z) + \frac{1}{4}u(z) = 0$$

a) in terms of hypergeometric functions of argument $\dfrac{2z}{z-1}$.

b) in terms of hypergeometric functions of argument $\dfrac{z-1}{2z}$.

c) in terms of hypergeometric functions of argument $\dfrac{z+1}{2z}$.

14.2 Legendre's equation is

$$(1 - z^2)w''(z) - 2zw'(z) + l(l+1)w(z) = 0.$$

a) Describe the location and nature of its singularities.

b) Write the equation in P notation form and check

$$\alpha + \alpha' + \beta + \beta' + \gamma + \gamma' = 1.$$

c) Use the transformation properties of the P symbol to derive the solutions

$$w_1(z) = F\left(l+1, -l; 1; \frac{1-z}{2}\right),$$

$$w_2(z) = \frac{1}{z^{l+1}}F\left(\frac{l+1}{2}, \frac{l+2}{2}; l+\frac{3}{2}; \frac{1}{z^2}\right).$$

For $w_2(z)$ first change the independent variable to $s = z^2$.

14.3 a) The contours C and C', as shown in Figure 14.18, were used to define $P_l(z)$ and $Q_l(z)$. The contour $C + C'$ encloses -1 and z but excludes +1. Use this contour in the integral (14.95), change variables to $t' = -t$, and derive

$$P_l(-z) = e^{\mp l\pi i} P_l(z) - \frac{2}{\pi} \sin \pi l\, Q_l(z),$$

where the $\mp$ sign corresponds to Im $z \gtrless 0$.

b) Show that $P_l(z) = P_{-l-1}(z)$ and hence derive

$$Q_l(z) - Q_{-l-1}(z) = \pi(\cot \pi l)P_l(z).$$

This formula is necessary in Regge pole theory.

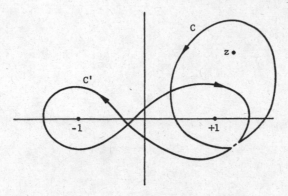

Figure 14.18: Contours to use for Problem 14.3

14.4 A charge particle moving in a circle at constant relativistic speed emits radiation in harmonics of the fundamental frequency of revolution. The power radiated in the nth harmonic can be expressed in terms of the integrals

$$I_1 = \int_0^{2\pi} du \ \cos u \ \exp\left[-inu + i\alpha \cos u\right]$$

and

$$I_2 = \int_0^{2\pi} du \ \sin u \ \exp\left[-inu + i\alpha \cos u\right],$$

where α is a real parameter related to the velocity of the particle and the angle at which the radiation is observed. Express I_1 and I_2 in terms of a well-known special function and its derivatives.

CHAPTER 15

Integral Transforms in the Complex Plane

15.1 INTRODUCTION

Various integral transforms have already been introduced and studied for the real variable case– Fourier transforms in Section 2.9 and Fourier Bessel transforms in Section 4.13 and at the end of Section 6.3. We know that these transforms appear naturally in the solution by separation of variables of partial differential equations involving the Laplacian, and that the integral transforms, rather than a summation of discrete eigenfunctions, appear when we deal with an infinite rather than a finite region.

Complex variables enter at various levels in the study of Fourier transforms. The simplest applications involve merely the evaluation of the relevant integrals by contour integration techniques. More sophisticated applications involve the analytic continuation of the Fourier transform into the complex plane of the transform variable. In such applications dispersion relation methods and the so-called Wiener-Hopf method are powerful techniques for the solution of certain types or problems.

We recall the basic formulas for Fourier transforms here. A more or less arbitrary function $f(x)$ can be expanded in the form

$$f(x) = \frac{1}{\sqrt{2\pi}} \int_{-\infty}^{\infty} dk \, \widetilde{f}(k) \, e^{ikx}, \tag{15.1}$$

where the Fourier transform $\widetilde{f}(k)$ is given by

$$\widetilde{f}(k) = \frac{1}{\sqrt{2\pi}} \int_{-\infty}^{\infty} dx \, f(x) \, e^{-ikx}. \tag{15.2}$$

The completeness relation for the complete set e^{ikx} is

$$\frac{1}{2\pi} \int_{-\infty}^{\infty} dk \, e^{ik(x-x')} = \delta(x - x'). \tag{15.3}$$

The Fourier transform of a derivative is algebraic, after using integration by parts:

$$\frac{1}{\sqrt{2\pi}} \int_{-\infty}^{\infty} dx \; e^{-ikx} \frac{df(x)}{dx} = \frac{1}{\sqrt{2\pi}} e^{-ikx} f(x) \Big|_{-\infty}^{\infty} - \frac{1}{\sqrt{2\pi}} \int_{-\infty}^{\infty} dx \; \frac{d}{dx} \left(e^{-ikx} \right) f(x)$$

$$= ik\tilde{f}(k) \tag{15.4}$$

provided $f(\infty) = f(-\infty) = 0$. Finally, the Fourier transform of a convolution is a product: If

$$h(x) = \int_{-\infty}^{\infty} dx' \; f(x - x')g(x'), \tag{15.5}$$

then we have

$$\tilde{h}(k) = \frac{1}{\sqrt{2\pi}} \int_{-\infty}^{\infty} dx \; h(x) \; e^{-ikx}$$

$$= \frac{1}{\sqrt{2\pi}} \int_{-\infty}^{\infty} dx \; e^{-ikx} \int_{-\infty}^{\infty} dx' \; f(x - x')g(x')$$

$$= \frac{1}{\sqrt{2\pi}} \int_{-\infty}^{\infty} dx' \; e^{-ikx'} g(x') \int_{-\infty}^{\infty} dx \; e^{-ik(x-x')} f(x - x')$$

$$= \sqrt{2\pi} \; \tilde{g}(k)\tilde{f}(k). \tag{15.6}$$

The multidimensional generalizations of all these results are straightforward:

$$f(\mathbf{r}) = \frac{1}{(2\pi)^{\frac{3}{2}}} \int d^3k \; \tilde{f}(\mathbf{k}) \; e^{i\mathbf{k}\cdot\mathbf{r}} \tag{15.7}$$

$$\tilde{f}(\mathbf{k}) = \frac{1}{(2\pi)^{\frac{3}{2}}} \int d^3x \; f(\mathbf{r}) \; e^{-i\mathbf{k}\cdot\mathbf{r}}, \tag{15.8}$$

etc.

15.2 THE CALCULATION OF GREEN'S FUNCTIONS BY FOURIER TRANSFORM METHODS

15.2.1 The Helmholtz Equation

Consider first the Green's function for the Helmholtz equation. For the case of all space, i.e., no boundary conditions in finite regions, $G(\mathbf{r}, \mathbf{r}')$ is a function of the relative coordinate $\mathbf{r} - \mathbf{r}'$ and satisfies

$$\boldsymbol{\nabla}^2 G(\mathbf{r} - \mathbf{r}') + \lambda^2 G(\mathbf{r} - \mathbf{r}') = \delta^3(\mathbf{r} - \mathbf{r}'). \tag{15.9}$$

Fourier transforming (15.9),

$$G(\mathbf{r} - \mathbf{r}') = \frac{1}{(2\pi)^{\frac{3}{2}}} \int d^3k \; \widetilde{G}(\mathbf{k}) e^{i\mathbf{k} \cdot (\mathbf{r} - \mathbf{r}')}, \tag{15.10}$$

$$\widetilde{G}(\mathbf{k}) = \frac{1}{(2\pi)^{\frac{3}{2}}} \int d^3x \; G(\mathbf{r} - \mathbf{r}') e^{-i\mathbf{k} \cdot (\mathbf{r} - \mathbf{r}')}, \tag{15.11}$$

and using the Fourier representation for the δ function,

$$\delta(\mathbf{r} - \mathbf{r}') = \frac{1}{(2\pi)^3} \int d^3k \; e^{i\mathbf{k} \cdot (\mathbf{r} - \mathbf{r}')}, \tag{15.12}$$

we find

$$(\lambda^2 - k^2)\widetilde{G}(\mathbf{k}) = \frac{1}{(2\pi)^{\frac{3}{2}}}. \tag{15.13}$$

The general solution of this equation is

$$\widetilde{G}(\mathbf{k}) = \frac{1}{(2\pi)^{\frac{3}{2}}} \left[\frac{1}{\lambda^2 - k^2} + C(\mathbf{k})\delta(\lambda^2 - k^2) \right], \tag{15.14}$$

where $C(\mathbf{k})$ is arbitrary, as we see from the general result $x\delta(x) = 0$. The undetermined multiple of $\delta(\lambda^2 - k^2)$ is equivalent to the lack of specification of the contour around the poles at $k = \pm\lambda$. In fact, substituting (15.14) in (15.10) with $\mathbf{r}' = 0$ we find

$$G(\mathbf{r}) = \frac{1}{(2\pi)^3} \int d^3k \; \frac{e^{i\mathbf{k} \cdot \mathbf{r}}}{\lambda^2 - k^2}, \tag{15.15}$$

and different values of C in (15.14) correspond to different contours around the poles in (15.15).

Before deciding on the contour, we can do the angular integrals in (15.15). Choosing the z axis in $\mathbf{k}$ space along $\mathbf{r}$ we find

$$G(\mathbf{r}) = \frac{1}{(2\pi)^3} 2\pi \int_0^\infty k^2 dk \int_{-1}^1 d(\cos\theta) \frac{e^{ikr\cos\theta}}{\lambda^2 - k^2}$$

$$= \frac{1}{(2\pi)^2 ir} \int_0^\infty k dk \frac{e^{ikr} - e^{-ikr}}{\lambda^2 - k^2}$$

$$= \frac{1}{(2\pi)^2 ir} \int_{-\infty}^\infty k dk \frac{e^{ikr}}{\lambda^2 - k^2}$$

$$= \frac{1}{(2\pi)^2} \frac{1}{r} \frac{\partial}{\partial r} \int_{-\infty}^\infty dk \frac{e^{ikr}}{k^2 - \lambda^2}. \tag{15.16}$$

Before performing the last integration of k in (15.16) we must choose the contour

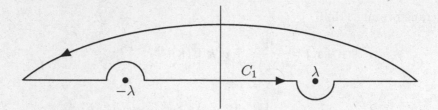

Figure 15.1: Consideration of contour, C_1, around the poles, $\pm\lambda$

around the poles. Choosing the contour C_1, as shown in Figure 15.1, and remembering that $r > 0$ so that we can complete the contour in the upper half-plane, we find the outgoing wave Green's function

$$G_{\text{out}}(\mathbf{r}) = \frac{1}{(2\pi)^2}\frac{1}{r}\frac{\partial}{\partial r}\int_{C_1}dk\,\frac{e^{ikr}}{k^2 - \lambda^2}$$

$$= \frac{1}{(2\pi)^2}\frac{1}{r}\frac{\partial}{\partial r}\left[2\pi i\frac{e^{i\lambda r}}{2\lambda}\right]$$

$$= -\frac{1}{4\pi}\frac{e^{i\lambda r}}{r}, \tag{15.17}$$

in agreement with (9.12).

As an equivalent procedure to choosing the contour C_1 above, we can leave the contour along the real k axis and displace the poles slightly, as shown in Figure 15.2:

$$\lambda \to \lambda + i\epsilon'$$
$$-\lambda \to -\lambda - i\epsilon'$$
$$\lambda^2 \to (\lambda + i\epsilon')^2 = \lambda^2 + i\epsilon, \tag{15.18}$$

where ϵ and ϵ' are real, positive infinitesimals. Thus the Fourier transform of the outgoing wave Green's function is

$$G_{\text{out}}(\mathbf{k}) = \frac{1}{(2\pi)^{\frac{3}{2}}}\frac{1}{\lambda^2 + i\epsilon - k^2}$$

$$= \frac{1}{(2\pi)^{\frac{3}{2}}}\left[\frac{\mathcal{P}}{\lambda^2 - k^2} - i\pi\delta(\lambda^2 - k^2)\right]. \tag{15.19}$$

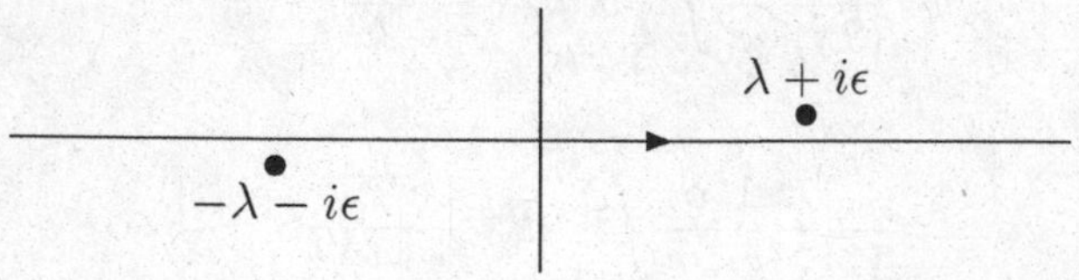

Figure 15.2: Poles displaced with contour along real k axis

This is of the form (15.14) with $C = -i\pi$.

Other choices are possible. With the contour C_2, as shown in Figure 15.3, we obtain the ingoing wave Green's function

$$G_{\text{in}}(\mathbf{r}) = \frac{1}{(2\pi)^2} \frac{1}{r} \frac{\partial}{\partial r} \int_{C_2} dk \, \frac{e^{ikr}}{k^2 - \lambda^2}$$

$$= -\frac{1}{4\pi} \frac{e^{-i\lambda r}}{r}. \tag{15.20}$$

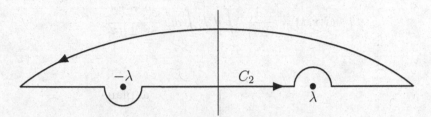

Figure 15.3: Consideration of contour, C_2, around the poles, $\pm\lambda$

The Fourier transform of this ingoing wave Green's function is

$$\widetilde{G}_{\text{in}}(\mathbf{k}) = \frac{1}{(2\pi)^{\frac{3}{2}}} \frac{1}{\lambda^2 - i\epsilon - k^2}. \tag{15.21}$$

Many other choices are possible. The physics problem determines via the boundary conditions which Green's function is to be chosen. Usually this will be the outgoing wave Green's function. We note that the Green's functions (15.17) and (15.20) found here are identical to those found in Section 9.2 without the aid of Fourier transforms. We have seen here that the outgoing (ingoing) wave boundary conditions are equivalent to the prescriptions for going around the pole $\lambda^2 \to \lambda^2 \pm i\epsilon$ in the Fourier transform Green's functions (15.19) and (15.21).

15.2.2 The Wave Equation

The four-dimensional Green's function for the wave equation can be found in a similar way. The Green's function for all space is a function of coordinate differences $\mathbf{r} - \mathbf{r}'$, $t - t'$ and satisfies

$$\left[\boldsymbol{\nabla}^2 - \frac{1}{c^2} \frac{\partial^2}{\partial t^2} \right] G(\mathbf{r} - \mathbf{r}', t - t') = \delta^3(\mathbf{r} - \mathbf{r}')\delta(t - t'). \tag{15.22}$$

Fourier transforming (15.22),

$$G(\mathbf{r} - \mathbf{r}', t - t') = \frac{1}{(2\pi)^2} \int d^3k \int d\omega \, \widetilde{G}(\mathbf{k}, \omega) e^{i\mathbf{k} \cdot (\mathbf{r} - \mathbf{r}') - i\omega(t - t')}, \tag{15.23}$$

$$\delta^3(\mathbf{r} - \mathbf{r}')\delta(t - t') = \frac{1}{(2\pi)^4} \int d^3k \int d\omega \, e^{i\mathbf{k} \cdot (\mathbf{r} - \mathbf{r}') - i\omega(t - t')}, \tag{15.24}$$

we find

$$\left[-k^2 + \frac{\omega^2}{c^2}\right] \widetilde{G}(\mathbf{k}, \omega) = \frac{1}{(2\pi)^2}, \tag{15.25}$$

so that

$$\widetilde{G}(\mathbf{k}, \omega) = \frac{1}{(2\pi)^2}\left[\frac{1}{\frac{\omega^2}{c^2} - k^2} + D\delta\left(\frac{\omega^2}{c^2} - k^2\right)\right]. \tag{15.26}$$

The undetermined multiple of $\delta\left(\frac{\omega^2}{c^2} - k^2\right)$ is equivalent to different prescriptions for going around the poles in Fourier transform integral

$$G(\mathbf{r}, t) = \frac{1}{(2\pi)^4}\int d^3k \int d\omega \, \frac{e^{i\mathbf{k}\cdot\mathbf{r} - i\omega t}}{\frac{\omega^2}{c^2} - k^2}. \tag{15.27}$$

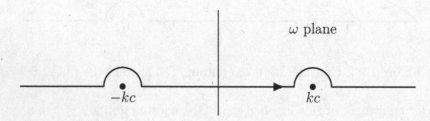

Figure 15.4: Contour choice above the poles for ω plane

The so-called retarded Green's function $G_{\text{ret}}(\mathbf{r}, t)$ satisfies the condition

$$G_{\text{ret}}(\mathbf{r}, t) = 0, \quad t < 0. \tag{15.28}$$

In order to achieve this, we choose a contour in the ω plane above the poles as shown in Figure 15.4. For $t < 0$ we must complete the ω contour in the upper half-plane. Since this contour encloses no poles we obtain (15.28). For $t > 0$ we complete the contour in the lower half-plane and obtain contributions from both poles $\omega = \pm kc$:

$$G_{\text{ret}}(\mathbf{r}, t) = \frac{-2\pi i}{(2\pi)^4}\int d^3k \, e^{i\mathbf{k}\cdot\mathbf{r}}\left[\frac{e^{-ikct}}{2k/c} + \frac{e^{ikct}}{-2k/c}\right]$$

$$= -\frac{1}{4\pi r}\delta\left(\frac{r}{c} - t\right), \quad t > 0, \tag{15.29}$$

where the second line follows from a computation identical to that in (9.71). In a similar way, choosing the contour in the ω plane below both poles, as shown in Figure 15.5, leads to the advanced Green's function

$$G_{\text{adv}}(\mathbf{r}, t) = \begin{cases} -\dfrac{1}{4\pi r}\delta\left(\dfrac{r}{c} + t\right), & t < 0, \\ \\ 0, & t > 0 \end{cases} \tag{15.30}$$

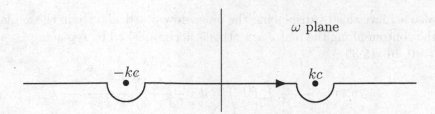

Figure 15.5: Contour choice below the poles for ω-plane

in agreement with (9.80).

15.2.3 The Klein-Gordon Equation

The Klein-Gordon equation

$$\left[\nabla^2 - \frac{1}{c^2}\frac{\partial^2}{\partial t^2} - \frac{m^2 c^2}{\hbar^2}\right]\psi(\mathbf{r}, t) = 0 \qquad (15.31)$$

is a generalization of the wave equation important in quantum field theory. We can find the Green's function for this equation by Fourier transformation, as above for the wave equation. Setting $\hbar = c = 1$ for notational simplicity, the differential equation for the Green's function is

$$\left[\nabla^2 - \frac{\partial^2}{\partial t^2} - m^2\right]G(\mathbf{r} - \mathbf{r}', t - t') = \delta^3(\mathbf{r} - \mathbf{r}')\delta(t - t'). \qquad (15.32)$$

Fourier transforming,

$$G(\mathbf{r} - \mathbf{r}', t - t') = \frac{1}{(2\pi)^2}\int d^3k \int d\omega\, \widetilde{G}(\mathbf{k}, \omega)e^{i\mathbf{k}\cdot(\mathbf{r}-\mathbf{r}')-i\omega(t-t')}, \qquad (15.33)$$

(15.32) becomes

$$(\omega^2 - k^2 - m^2)\widetilde{G}(\mathbf{k}, \omega) = \frac{1}{(2\pi)^2}. \qquad (15.34)$$

Thus as generalization of (15.27) we find

$$G(\mathbf{r}, t) = \frac{1}{(2\pi)^4}\int d^3k \int d\omega\, \frac{e^{i\mathbf{k}\cdot\mathbf{r}-i\omega t}}{\omega^2 - k^2 - m^2}. \qquad (15.35)$$

If we choose the contour above the poles in the ω plane, as shown in Figure 15.6, we obtain the retarded Green's function, which satisfies the condition

$$G_{\text{ret}}(\mathbf{r}, t) = 0, \quad t < 0. \qquad (15.36)$$

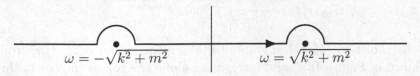

Figure 15.6: Contour above the poles in the ω plane for equation (15.35)

We can also achieve this by displacing the poles downward slightly in the ω plane and leaving the contour along the real ω axis; this is accomplished by replacing $\omega \to \omega + i\epsilon$, $\epsilon > 0$, $\epsilon \to 0$, in (15.35):

$$G_{\text{ret}}(\mathbf{r}, t) = \frac{1}{(2\pi)^4} \int d^3k \int_{-\infty}^{\infty} d\omega \, \frac{e^{i\mathbf{k}\cdot\mathbf{r} - i\omega t}}{(\omega + i\epsilon)^2 - k^2 - m^2}. \tag{15.37}$$

In order to actually evaluate this integral, it is convenient to first integrate over $\mathbf{k}$. Integrating over the angles of $\mathbf{k}$ we find from (15.37)

$$G_{\text{ret}}(\mathbf{r}, t) = \frac{i}{(2\pi)^3 r} \int_{-\infty}^{\infty} d\omega \int_0^{\infty} dk \, k(e^{ikr} - e^{-ikr}) \frac{e^{-i\omega t}}{k^2 + m^2 - (\omega + i\epsilon)^2}$$

$$= \frac{i}{(2\pi)^3 r} \int_{-\infty}^{\infty} d\omega \int_{-\infty}^{\infty} dk \, k \frac{e^{ikr - i\omega t}}{k^2 + m^2 - (\omega + i\epsilon)^2}. \tag{15.38}$$

Since $r > 0$, we can complete the contour for the k integration by an infinite semicircle in the upper half-plane. The denominator has poles for

$$k^2 = (\omega + i\epsilon)^2 - m^2 = \omega^2 - m^2 + 2i\omega\epsilon, \quad \epsilon > 0, \quad \epsilon \to 0, \tag{15.39}$$

keeping the linear term in *epsilon* as the higher orders go to zero faster. Taking the square root of this expression and picking out the poles which lie in the upper half-plane, we find

$$\omega > m: \quad k = \pm\left(\sqrt{\omega^2 - m^2} + i\epsilon\right) \qquad k = \sqrt{\omega^2 - m^2} + i\epsilon$$

$$-m < \omega < m: \quad k = \pm i\sqrt{m^2 - \omega^2} \qquad k = i\sqrt{m^2 - \omega^2}$$

$$\omega < -m: \quad k = \pm\left(\sqrt{\omega^2 - m^2} - i\epsilon\right) \qquad k = -\sqrt{\omega^2 - m^2} + i\epsilon \tag{15.40}$$

In these expressions ϵ stands for different things, but it is always a positive infinitesimal. If $\sqrt{\omega^2 - m^2}$ is defined by the expressions in the rightmost column of (15.40),

$$\sqrt{\omega^2 - m^2} = \begin{cases} +\sqrt{\omega^2 - m^2}, & \omega > m \\ i\sqrt{m^2 - \omega^2}, & -m < \omega < m \, , \\ -\sqrt{\omega^2 - m^2}, & \omega < -m \end{cases} \tag{15.41}$$

the evaluation of the k integral in (15.38) by the residue theorem gives

$$G_{\text{ret}}(\mathbf{r}, t) = -\frac{1}{8\pi^2 r} \int_{-\infty}^{\infty} d\omega \, e^{i\sqrt{\omega^2 - m^2} \, r - i\omega t}. \tag{15.42}$$

Actually the choice in (15.41) amounts to taking the cut in $\sqrt{\omega^2 - m^2}$ from $-m$ to $+m$, as shown in Figure 15.7, and integrating along the <u>top</u> of the cut in this region:

$$\sqrt{\omega^2 - m^2} = (r_1 r_2)^{\frac{1}{2}} e^{i(\theta_1 + \theta_2)/2}. \tag{15.43}$$

To proceed to the next step, it is convenient to introduce the quantity

$$H(r,t) = \frac{1}{2\pi i} \int_{-\infty}^{\infty} d\omega \, \frac{e^{i\sqrt{\omega^2 - m^2}\, r - i\omega t}}{\sqrt{\omega^2 - m^2}}. \tag{15.44}$$

The Green's function $G(\mathbf{r}, t)$ is then given by

$$G(\mathbf{r}, t) = -\frac{1}{4\pi} \frac{1}{r} \frac{\partial}{\partial r} H(r,t). \tag{15.45}$$

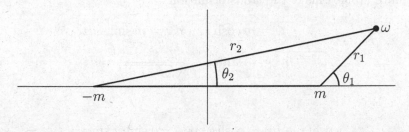

Figure 15.7: Geometry of cut from $-m$ to $+m$ in reference to equation (15.43)

For $\omega \to \infty$ the m in the exponential of (15.42) or (15.44) can be neglected and the exponential becomes

$$e^{i\omega(r-t)}. \tag{15.46}$$

For $t < r$, i.e., $r - t > 0$, the contour for the ω integral in (15.42) or (15.44) can thus be completed by an infinite semicircle in the upper half-plane, and since the contour goes above the cut $(-m, m)$, and since there are no other singularities in the upper half-plane, we find

$$G(\mathbf{r}, t) = H(r, t) = 0, \quad t < r. \tag{15.47}$$

This is consistent with, but more than, the condition (15.36). For $t > r$ we can complete the contour by an infinite semicircle in the lower half-plane and then use Cauchy's theorem to pull the contour in until it is just wrapped around the cut $(-m, m)$, as shown in Figure 15.8:

$$H(r, t) = \frac{1}{2\pi i} \int_C d\omega \, \frac{e^{i\sqrt{\omega^2 - m^2}\, r - i\omega t}}{\sqrt{\omega^2 - m^2}}. \tag{15.48}$$

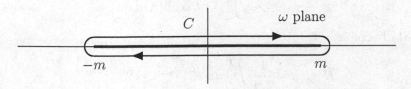

Figure 15.8: Contour shifted to wrap just around the cut line from $-m$ to $+m$

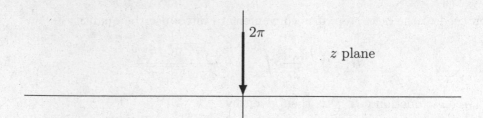

Figure 15.9: Cut from change of variables from ω plane to z plane

To evaluate (15.48) make the transformation

$$\omega = m\cosh z, \quad d\omega = m\sinh z\, dz,$$

$$\sqrt{\omega^2 - m^2} = m\sinh z, \quad \frac{d\omega}{\sqrt{\omega^2 - m^2}} = dz. \tag{15.49}$$

Using

$$\cosh(x+iy) = \cosh x\cos y = i\sinh x\sin y,$$

$$\sinh(x+iy) = \sinh x\cos y + i\cosh x\sin y, \tag{15.50}$$

it is easy to check that the contour from $2\pi i$ to 0 in the z plane maps into the contour C in the ω plane, as shown in Figure 15.9, with the correct values for $\sqrt{\omega^2 - m^2}$ on the top and bottom of the cut. Thus we find

$$H(r,t) = \frac{1}{2\pi i}\int_{2\pi i}^{0} dz\; e^{im[r\sinh z - t\cosh z]}. \tag{15.51}$$

Now we can write

$$r\sinh z - t\cosh z = \sqrt{t^2 - r^2}\left[\frac{r}{\sqrt{t^2 - r^2}}\sinh z - \frac{t}{\sqrt{t^2 - r^2}}\cosh z\right], \tag{15.52}$$

and if we define θ by

$$\cosh\theta = \frac{t}{\sqrt{t^2 - r^2}}, \quad \sinh\theta = \frac{r}{\sqrt{t^2 - r^2}} \tag{15.53}$$

and use the identities for hyperbolic functions, we can rewrite (15.52) in the form

$$r\sinh z - t\cosh z = \sqrt{t^2 - r^2}\left[\sinh z\sinh\theta - \cosh z\cosh\theta\right]$$

$$= -\sqrt{t^2 - r^2}\cosh(z - \theta). \tag{15.54}$$

Thus (15.51) becomes

$$H(r,t) = \frac{1}{2\pi i}\int_{2\pi i}^{0} dz\; e^{-im\sqrt{t^2 - r^2}\cosh(z-\theta)}$$

$$= \frac{1}{2\pi i}\int_{2\pi i - \theta}^{-\theta} dz'\; e^{-im\sqrt{t^2 - r^2}\cosh z'}, \tag{15.55}$$

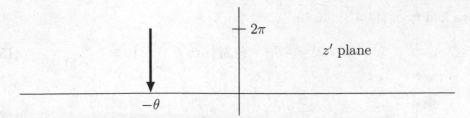

Figure 15.10: Change of variables for the integration of equation (15.35)

with change of variables to $z' = z - \theta$ as shown in Figure 15.10. Now since there are no singularities inside the contour C' of Figure 15.11, we have

$$\int_{C'} dz' \, e^{-im\sqrt{t^2-r^2} \, \cosh z'} = 0, \tag{15.56}$$

and since $\cosh(z + 2\pi i) = \cosh z$, the contributions from the top and bottom of the contour C' cancel. Thus combining (15.55), (15.56) we find for $H(r,t)$

$$H(r,t) = \frac{1}{2\pi i} \int_{2\pi i}^{0} dz' \, e^{-im\sqrt{t^2-r^2} \, \cosh z'}. \tag{15.57}$$

It is interesting that only after this long series of transformations does the dependence of $G(\mathbf{r},t)$ and $H(r,t)$ on the single variable combination $\sqrt{t^2 - r^2}$ (as required by special relativity) actually appear.

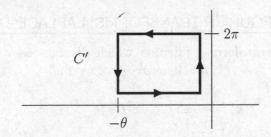

Figure 15.11: Contour for use with equation (15.56)

The simple transformation $z' = iy$, $\cosh z' = \cos y$ now puts (15.57) in the form

$$H(r,t) = -\frac{1}{2\pi} \int_{0}^{2\pi} dy \, e^{-im\sqrt{t^2-r^2} \, \cos y}, \tag{15.58}$$

and this has the form of the integral representation (14.134) for the Bessel function J_0:

$$H(r,t) = -J_0\left(m\sqrt{t^2 - r^2}\right), \quad t > r. \tag{15.59}$$

Introducing the step function

$$\theta(t - r) = \begin{cases} 0, & t < r \\ 1, & t > r \end{cases}, \tag{15.60}$$

we thus find from (15.47), (15.59)

$$H(r,t) = -\theta(t-r)J_0\left(m\sqrt{t^2-r^2}\right).$$

(15.61)

Using

$$\frac{\partial}{\partial r}\theta(t-r) = -\delta(t-r)$$

(15.62)

and

$$\frac{\partial}{\partial r}J_0\left(m\sqrt{t^2-r^2}\right) = \frac{-mr}{\sqrt{t^2-r^2}}J_0'\left(m\sqrt{t^2-r^2}\right)$$

(15.63)

and also

$$J_1(x) = -J_0'(x)$$

(15.64)

and

$$J_0(0) = 1,$$

(15.65)

we then find, according to (15.45),

$$G_{\text{ret}}(\mathbf{r},t) = -\frac{\delta(t-r)}{4\pi r} + \theta(t-r)\frac{m}{4\pi\sqrt{t^2-r^2}}J_1\left(m\sqrt{t^2-r^2}\right).$$

(15.66)

This reduces to the retarded Green's function (15.29) for the wave equation when $m = 0$. For $m \neq 0$ we have in addition to the sharp wavefront at $r = t$ (or $r = ct$ for $c \neq 1$) a contribution at $r < t$ (or $r < ct$ for $\neq 1$) behind the wavefront.

15.3   ONE-SIDED FOURIER TRANSFORMS; LAPLACE TRANSFORMS

If we try to Fourier transform a function which diverges as $x \to \infty$, we get into trouble. Thus with $f(x) \sim e^{ax}$, $a > 0$, as $x \to \infty$, we find

$$\widetilde{f}(k) = \frac{1}{\sqrt{2\pi}}\int_{-\infty}^{\infty} dx\, f(x)e^{-ikx} \to \infty.$$

(15.67)

In order to cope with such functions we can proceed as follows: Suppose $f(x)$ diverges less badly than e^{cx} so that

$$f(x)e^{-cx} \sim e^{-\epsilon x}, \quad x \to \infty, \epsilon > 0.$$

(15.68)

We can then try to Fourier transform the function $f(x)e^{-cx}$, which differs in a known way from the function $f(x)$ of interest:

$$\widetilde{F}(k) = \frac{1}{\sqrt{2\pi}}\int_{-\infty}^{\infty} dx\, f(x)e^{-cx}e^{-ikx}.$$

(15.69)

However, the exponential e^{-cx}, which ensures convergence of the integral (15.69) at $x \to +\infty$, will make the convergence at $x \to -\infty$ worse; in fact unless $f(x)$ falls off more rapidly than e^{cx} for $x \to -\infty$, the integral (15.69) will now diverge at $x \to -\infty$.

To take care of this, we discard half of the function $f(x)$, defining a new function $F(x)$ by

$$F(x) = \begin{cases} f(x)e^{-cx}, & x > 0 \\ \\ 0, & x < 0. \end{cases} \qquad (15.70)$$

The Fourier transform of this function

$$\widetilde{F}(k) = \frac{1}{\sqrt{2\pi}} \int_0^\infty dx \; f(x)e^{-cx}e^{-ikx} \qquad (15.71)$$

now exists provided (15.68) is satisfied.

Since this method is employed mainly for problems involving the time, we write t instead of x and reverse the sign of the Fourier transform variable $k \to -\omega$, in conformity with the conventions usually employed by physicists. For a function $f(t)$, taken to vanish for $t < 0$ and to diverge less rapidly than e^{ct} as $t \to \infty$,

$$f(t) = 0, \qquad\qquad t < 0 \qquad (15.72)$$

$$f(t)e^{-ct} \to e^{-\epsilon t}, \qquad\qquad t \to \infty, \quad \epsilon > 0, \qquad (15.73)$$

we find the pair of Fourier transform equations

$$f(t)e^{-ct} = \frac{1}{\sqrt{2\pi}} \int_{-\infty}^\infty d\omega \; e^{-i\omega t}\widetilde{F}(\omega) \qquad (15.74)$$

$$\widetilde{F}(\omega) = \frac{1}{\sqrt{2\pi}} \int_0^\infty dt \; e^{i\omega t} f(t)e^{-ct}, \qquad (15.75)$$

or equivalently

$$f(t) = \frac{1}{\sqrt{2\pi}} \int_{-\infty}^\infty d\omega \; e^{-i(\omega+ic)t}\widetilde{F}(\omega), \qquad (15.76)$$

$$\widetilde{F}(\omega) = \frac{1}{\sqrt{2\pi}} \int_0^\infty dt \; e^{i(\omega+ic)t} f(t). \qquad (15.77)$$

If the integral (15.75) converges for a given value of c, it will converge better if c is replaced by a larger value. Thus we can use the integral representation (15.77) to effect an analytic continuation into the complex ω plane: Just define

$$\widetilde{f}(\omega) = \frac{1}{\sqrt{2\pi}} \int_0^\infty dt \; e^{i\omega t} f(t). \qquad (15.78)$$

This integral converges in the region

$$\text{Im}\,\omega \geq c \qquad (15.79)$$

according to the (15.73). That $\widetilde{f}(\omega)$ is an analytic function of ω follows from the uniform convergence[1] of the integral in this region:

$$\left| \widetilde{f}(\omega) - \frac{1}{\sqrt{2\pi}} \int_0^T dt \; e^{i\omega t} f(t) \right| = \left| \frac{1}{\sqrt{2\pi}} \int_T^\infty dt \; e^{i\omega t} f(t) \right|$$

$$\leq \frac{1}{\sqrt{2\pi}} \int_T^\infty dt \; \left| e^{i\omega t} f(t) \right|$$

$$= \frac{1}{\sqrt{2\pi}} \int_T^\infty dt \; e^{-(\text{Im}\,\omega)t} \left| f(t) \right|$$

$$\leq \frac{1}{\sqrt{2\pi}} \int_T^\infty dt \; e^{-ct} \left| f(t) \right|. \qquad (15.80)$$

We assume that c is large enough so that this last integral converges. It can then be made arbitrarily small by making T sufficiently large, the same value of T serving for all ω in the region (15.79). In general we expect to be able to analytically continue $\widetilde{f}(\omega)$ to the region $\text{Im}\,\omega < c$; in this region it may have singularities, e.g., poles or cuts.

We can now rewrite the Fourier transform pair (15.76), (15.77) in terms of the analytically continued Fourier transform $\widetilde{f}(\omega)$, (15.78):

$$f(t) = \frac{1}{\sqrt{2\pi}} \int_C d\omega \cdot e^{-i\omega t} \widetilde{f}(\omega), \qquad (15.81)$$

$$\widetilde{f}(\omega) = \frac{1}{\sqrt{2\pi}} \int_0^\infty dt \; e^{i\omega t} f(t). \qquad (15.82)$$

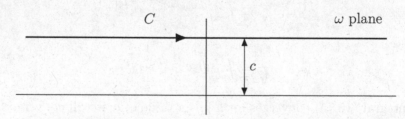

Figure 15.12: Contour used with equation (15.81)

The contour C in (15.81) is a straight line a distance c above the axis in the complex ω plane as shown in Figure 15.12. The formulas (15.81), (15.82) are just the ordinary Fourier transform formulas except that ω is complex with $\text{Im}\,\omega > 0$ sufficiently to

---

[1] The theorem that a uniformly convergent integral $\int_{-\infty}^\infty dt \; f(t,\omega)$ of a function $f(t,\omega)$, analytic in ω for each t, is itself analytic, is proved, for example, by Titchmarsh, op. cit., p. 100.

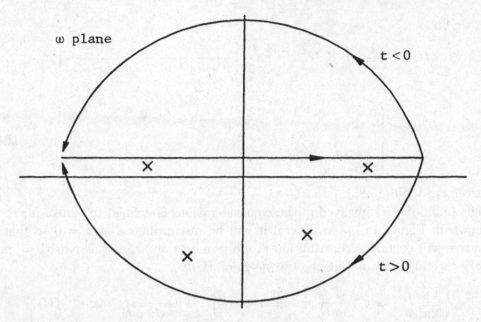

Figure 15.13: Expansion of contour to the upper and lower half-planes

make the integral (15.82) converge and $f(t) = 0$, $t < 0$. Since $\widetilde{f}(\omega)$ is an analytic function of ω, we can use Cauchy's theorem to move the contour C in (15.81); in particular we can move the contour parallel to the path C in Figure 15.12, as long as we do not cross over any singularities of $\widetilde{f}(\omega)$.

The integral (15.81) for $f(t)$ is usually done as a contour integral. For $t > 0$ we can close the contour in the lower half-plane and pick up contributions from poles of $\widetilde{f}(\omega)$ below C. For $t < 0$ we close the contour in the upper half-plane. We are supposed to obtain 0 since $f(t) = 0$ for $t < 0$. Thus C must be high enough so that there are no singularities of $\widetilde{f}(\omega)$ above C. The contours for the two cases, $t > 0$ and $t < 0$, are illustrated in Figure 15.13.

As a simple example, consider the function

$$f(t) = \begin{cases} 0, & t < 0 \\ \\ e^{at}\cos\omega_0 t, & t > 0 \end{cases}. \tag{15.83}$$

Then we have

$$\widetilde{f}(\omega) = \frac{1}{\sqrt{2\pi}} \int_0^\infty dt\, e^{i\omega t} e^{at} \frac{1}{2} \left[e^{i\omega_0 t} + e^{-i\omega_0 t} \right]. \tag{15.84}$$

For $\mathrm{Im}\,\omega > a$ this integral converges and we find

$$\widetilde{f}(\omega) = \frac{1}{\sqrt{2\pi}} \frac{i}{2} \left[\frac{1}{\omega + \omega_0 - ia} + \frac{1}{\omega - \omega_0 - ia} \right]. \tag{15.85}$$

This function has poles at $\omega = \pm\omega_0 + ia$. Using (15.81) with a contour C above these poles we find

$$f(t) = \frac{1}{\sqrt{2\pi}} \int_C d\omega \ e^{-i\omega t} \frac{1}{\sqrt{2\pi}} \frac{i}{2} \left[\frac{1}{\omega + \omega_0 - ia} + \frac{1}{\omega - \omega_0 - ia} \right]$$

$$= \begin{cases} 0, & t < 0 \\ -\dfrac{2\pi i}{2\pi} \dfrac{i}{2} \left[e^{+i\omega_0 t + at} + e^{-i\omega_0 t + at} \right], & t > 0. \end{cases} \qquad (15.86)$$

in agreement with (15.83).

The function $f(t)$ involved in these considerations is defined to vanish for $t < 0$, as shown in Figure 15.14. In general it will be discontinuous at $t = 0$ so that its derivative will contain a contribution proportional to $\delta(t)$. For this reason we must be very careful in calculating the transform of df/dt:

$$\frac{1}{\sqrt{2\pi}} \int_0^\infty d\omega \ e^{i\omega t} \frac{df(t)}{dt} = \frac{1}{\sqrt{2\pi}} e^{i\omega t} f(t) \Big|_0^\infty - i\omega \frac{1}{\sqrt{2\pi}} \int_0^\infty d\omega \ e^{i\omega t} f(t)$$

$$= -\frac{1}{\sqrt{2\pi}} f(0) - i\omega \widetilde{f}(\omega). \qquad (15.87)$$

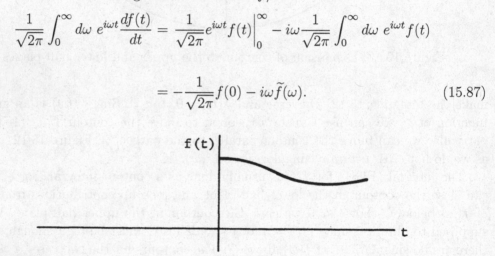

Figure 15.14: Function $f(t)$ from equation (15.86) only defined for $t > 0$

The formalism developed above as an extension of Fourier transforms, essentially the analytic continuation of $\widetilde{f}(\omega)$ into the complex ω plane for functions $f(t)$ which vanish for $t < 0$, is an important tool for the solution of certain types of initial value problems. Unfortunately perhaps, a further unnecessary transformation is usually made to obtain the so-called Laplace transform. A rotation through $90°$ is made in the complex ω plane.

$$s = -i\omega, \qquad (15.88)$$

and in terms of the new variable s and $F(s) = \sqrt{2\pi} \widetilde{f}(\omega)$, (15.81), (15.82) become

$$f(t) = \frac{1}{2\pi i} \int_{c-i\infty}^{c+i\infty} ds \ e^{st} F(s), \qquad (15.89)$$

$$F(s) = \int_0^\infty dt \ e^{-st} f(t). \qquad (15.90)$$

The contour $(c-i\infty, c+i\infty)$ in (15.89) must be chosen to the right of any singularities of $F(s)$; to the right of this line, $\mathrm{Re}\, s > c$, $F(s)$ is an analytic function. For $t < 0$ the contour in (15.89) can be completed by an infinite semicircle to the right and we obtain $f(t) = 0$, $t < 0$.

The possible discontinuity of $f(t)$ at $t = 0$ leads to contributions to the Laplace transforms of $df(t)/dt$, $d^2 f(t)/dt^2$, etc. For $\mathrm{Re}\, s > c$

$$\int_0^\infty dt\, e^{-st} \frac{df(t)}{dt} = -f(0) + sF(s),$$

$$\int_0^\infty dt\, e^{-st} \frac{d^2 f(t)}{dt^2} = -f'(0) - sf(0) + s^2 F(s), \tag{15.91}$$

$$\vdots$$

The Laplace transform of a convolution of the form

$$h(t) = \int_0^t d\tau\, f(\tau) g(t - \tau) \tag{15.92}$$

is a product:

$$H(s) = \int_0^\infty dt\, e^{-st} h(t)$$

$$= \int_0^\infty dt\, e^{-st} \int_0^t d\tau\, f(\tau) g(t - \tau)$$

$$= \int_0^\infty d\tau\, e^{-s\tau} f(\tau) \int_\tau^\infty dt\, e^{-s(t-\tau)} g(t - \tau)$$

$$= \int_0^\infty d\tau\, e^{-s\tau} f(\tau) \int_0^\infty dt'\, e^{-st'} g(t')$$

$$H(s) = F(s)G(s). \tag{15.93}$$

15.4 LINEAR DIFFERENTIAL EQUATIONS WITH CONSTANT COEFFICIENTS

Linear differential equations with constant coefficients are readily solved using Laplace transforms. Suppose we wish to solve

$$\frac{d^2 y(t)}{dt^2} + a \frac{dy(t)}{dt} + by(t) = f(t) \tag{15.94}$$

in terms of initial values $y(0)$, $y'(0)$. If the Laplace transforms of $y(t)$ and $f(t)$ are $Y(s)$ and $F(s)$, respectively, we find, using (15.91), for the Laplace transform of (15.94)

$$(s^2 + as + b)Y(s) = y(0)(s + a) + y'(0) + F(s). \tag{15.95}$$

The differential equation is converted to an algebraic equation, which is readily solved:

$$Y(s) = \frac{y(0)(s + a) + y'(0) + F(s)}{s^2 + as + b}. \tag{15.96}$$

When this expression is analytically continued to the left of the line $\operatorname{Re} s = c$ in the transform (15.89), singularities appear, due to the poles of $(s^2 + as + b)^{-1}$ and the singularities of $F(s)$. Suppose the zeros of $s^2 + as + b$ are s_1 and s_2 with $s_1 \neq s_2$ so that $s^2 + as + b = (s - s_1)(s - s_2)$. The residues of $(s^2 + as + b)^{-1}$ at the two poles are then $(2s_1 + a)^{-1}$ and $(2s_2 + a)^{-1}$. Substituting (15.96) in the inverse transform (15.89),

$$y(t) = \frac{1}{2\pi i} \int_{c-i\infty}^{c+i\infty} ds \, e^{st} Y(s), \tag{15.97}$$

and completing the contour with an infinite semicircle to the left, we find, for $t > 0$,

$$y(t) = \frac{y(0)(s_1 + a) + y'(0) + F(s_1)}{2s_1 + a} e^{s_1 t}$$

$$+ \frac{y(0)(s_2 + a) + y'(0) + F(s_2)}{2s_2 + a} e^{s_2 t}$$

$$+ \sum_{j=2}^{n} \frac{\operatorname{Res}(F, s_j)}{s_j^2 + as_j + b} e^{s_j t}, \tag{15.98}$$

where it has been assumed as an example that $F(s)$ has simple poles with residues $\operatorname{Res}(F, s_j)$ at the locations s_j, $j = 2, \dots, n$.

15.5   INTEGRAL EQUATIONS OF CONVOLUTION TYPE

Integral equations of the type

$$f(t) = g(t) + \int_0^t d\tau \, k(t - \tau) f(\tau) \tag{15.99}$$

can be easily solved with Laplace transforms. Taking the Laplace transform of (15.99) and using the formula (15.93) for the Laplace transform of a convolution, we find

$$F(s) = G(s) + K(s)F(s), \tag{15.100}$$

which is easily solved:

$$F(s) = \frac{G(s)}{1 - K(s)}. \tag{15.101}$$

The inverse Laplace transform then gives

$$f(t) = \frac{1}{2\pi i} \int_{c-i\infty}^{c+i\infty} ds\, e^{st} \frac{G(s)}{1 - K(s)}. \tag{15.102}$$

Various tables of Laplace transforms[2] for many functions greatly facilitate the use of this tool in practice.

15.6   MELLIN TRANSFORMS

A further change of variable in the Laplace transform (15.89), (15.90) leads to the so-called Mellin transform. In order to end up with the variable names as usually used in Regge theory in high energy physics, first replace s by j and f by $\tilde{f}$ in (15.89), (15.90):

$$\tilde{f}(t) = \frac{1}{2\pi i} \int_{c-i\infty}^{c+i\infty} dj\, e^{jt} F(j), \tag{15.103}$$

$$F(j) = \int_0^\infty dt\, \frac{1}{e^{jt}} \tilde{f}(t). \tag{15.104}$$

Now change variables,

$$t = \ln\left(\frac{s}{s_0}\right), \quad e^t = \frac{s}{s_0},$$

$$\tilde{f}(t) = f(s), \tag{15.105}$$

with s_0 a constant, to obtain

$$f(s) = \frac{1}{2\pi i} \int_{c-i\infty}^{c+i\infty} dj\, \left(\frac{s}{s_0}\right)^j F(j), \tag{15.106}$$

$$F(j) = \int_{s_0}^\infty \frac{ds}{s_0} \left(\frac{s_0}{s}\right)^{j+1} f(s). \tag{15.107}$$

Evidently the Mellin transform is essentially a Laplace transform, which in turn is essentially a Fourier transform. For the application in high energy physics s is the (energy)^2 and j is the angular momentum in the "crossed channel".

The transforms in the form (15.106), (15.107) are convenient for solving an integral equation of the form

$$f(s) = g(s) + \int_{s_0}^s \frac{ds'}{s'} k\left(\frac{s}{s'}\right) f(s'), \tag{15.108}$$

---

[2]See, for example, Erdélyi, et al., op. cit.

with a kernel which is a function for the ratio s/s'. Applying the transform (15.107) to a convolution integral of the sort that appears on the right of (15.108) we find

$$\int_{s_0}^{\infty} \frac{ds}{s_0} \left(\frac{s_0}{s}\right)^{j+1} \int_{s_0}^{s} \frac{ds'}{s'} k\left(\frac{s}{s'}\right) f(s')$$

$$= \int_{s_0}^{\infty} \frac{ds'}{s_0} \left(\frac{s_0}{s'}\right)^{j+1} f(s') \int_{1}^{\infty} d\left(\frac{s}{s'}\right) \left(\frac{s'}{s}\right)^{j+1} k\left(\frac{s}{s'}\right)$$

$$= F(j)K(j), \tag{15.109}$$

where $K(j)$ is calculated with the scale factor $s_0 = 1$. Applying (15.107) to the integral equation (15.108) thus gives

$$F(j) = G(j) + F(j)K(j), \tag{15.110}$$

$$F(j) = \frac{G(j)}{1 - K(j)}, \tag{15.111}$$

and the inverse transform (15.106) yields the solution of the integral equation (15.108):

$$f(s) = \frac{1}{2\pi i} \int_{c-i\infty}^{c+i\infty} dj \left(\frac{s}{s_0}\right)^j \frac{G(j)}{1 - K(j)}. \tag{15.112}$$

15.7 PARTIAL DIFFERENTIAL EQUATIONS

The Laplace transform method can be profitably applied to initial value problems in heat conduction. Many examples of this sort are discussed in the book by Carslaw and Jaeger.[3] We discuss here only a relatively simple example which illustrates how Laplace transforms can be used in the solution of partial differential equations. We want to solve the one-dimensional heat conduction equation

$$\frac{\partial^2 u(x,t)}{\partial x^2} = \frac{\partial u(x,t)}{\partial t}, \tag{15.113}$$

with the boundary conditions

$$u(0,t) \quad \text{given},$$
$$u(l,t) \quad \text{given}, \tag{15.114}$$

and the initial condition

$$u(x,0) \quad \text{given}. \tag{15.115}$$

---

[3] H. S. Carslaw and J. C. Jaeger, <u>Conduction of Heat in Solids</u>, Second Edition (Clarendon Press, Oxford, 1959).

We carry out the Laplace transform in the time variable:

$$u(x,t) = \frac{1}{2\pi i} \int_{c-i\infty}^{c+i\infty} ds \; e^{st} \, \widetilde{u}(x,s), \tag{15.116}$$

$$\widetilde{u}(x,s) = \int_0^\infty dt \; e^{-st} \, u(x,t). \tag{15.117}$$

From (15.91) we find for the Laplace transform of the time derivative

$$\int_0^\infty dt \; e^{-st} \frac{\partial u(x,t)}{\partial t} = -u(x,0) + s\widetilde{u}(x,s). \tag{15.118}$$

The Laplace transform of the partial differential equation (15.113) is thus

$$\frac{\partial^2 \widetilde{u}(x,s)}{\partial x^2} - s\widetilde{u}(x,s) = -u(x,0). \tag{15.119}$$

The function on the right-hand side of (15.119) is a known function, given by the initial condition (15.115). Thus (15.119) together with the boundary conditions (15.114) is the type of problem which can be solved by Green's function techniques. According to the discussion in Section 8.3 the appropriate Green's function $G(x,x')$ satisfies the differential equation

$$\frac{d^2 G(x,x')}{dx^2} - sG(x,x') = \delta(x-x'), \tag{15.120}$$

the boundary conditions

$$G(0,x') = 0,$$
$$G(l,x') = 0, \tag{15.121}$$

and the joining conditions

$$G(x,x') \quad \text{continuous in } x \text{ at } x = x',$$

$$\left.\frac{dG(x,x')}{dx}\right|_{x=x'+\epsilon} - \left.\frac{dG(x,x')}{dx}\right|_{x=x'-\epsilon} = 1. \tag{15.122}$$

The solutions of (15.120) for $x \neq x'$ which satisfy the boundary conditions (15.121) are

$$G(x,x') = \begin{cases} A \sinh \sqrt{s}x, & x < x' \\ B \sinh \sqrt{s}(x-l), & x > x' \end{cases}. \tag{15.123}$$

The joining conditions (15.122) lead to two linear equations for A and B. These are readily solved and we find

$$G(x,x') = \frac{1}{\sqrt{s}} \frac{1}{\sinh \sqrt{s}\, l} \begin{cases} \sinh \sqrt{s}x \sinh \sqrt{s}(x'-l), & x < x' \\ \sinh \sqrt{s}x' \sinh \sqrt{s}(x-l), & x > x' \end{cases}. \tag{15.124}$$

We note that $G(x, x') = G(x', x)$ is a symmetric function.

Substituting the differential equations (15.119) and (15.120) in Green's theorem,

$$\int_0^l dx \left[\widetilde{u}(x, s) \frac{d^2 G(x, x')}{dx^2} - G(x, x') \frac{d^2 \widetilde{u}(x, s)}{dx^2} \right] =$$

$$\left[\widetilde{u}(x, s) \frac{dG(x, x')}{dx} - G(x, x') \frac{d\widetilde{u}(x, s)}{dx} \right]_0^l, \qquad (15.125)$$

and using the boundary conditions (15.121) and the symmetry of $G(x, x')$, we find

$$\widetilde{u}(x, s) = - \int_0^l dx' \, G(x, x') u(x', 0) + \widetilde{u}(l, s) \frac{dG(x, x')}{dx'} \bigg|_{x'=l} - \widetilde{u}(0, s) \frac{dG(x, x')}{dx'} \bigg|_{x'=0}.$$
$$(15.126)$$

This formula provides, in principle at least, the solution of the problem. It gives the Laplace transform of the solution in terms of the known initial condition $u(x, 0)$ and the Laplace transforms of the known boundary conditions $u(0, t)$ and $u(l, t)$. Substitution of (15.126) in (15.116) leads to a formula for $u(x, t)$ itself. Of course the evaluation of the integrals for all these Laplace transforms and inverse Laplace transforms may not be easy. A simple special case of the formulas derived above is considered in a problem.

15.8   THE WIENER-HOPF METHOD

This method uses the analytically continued Fourier transform to solve certain types of boundary value problems with mixed boundary conditions. We need the analytic continuation properties of one-sided Fourier transforms. By the same type of argument given in (15.78)ff., we see that

$$F_+(k) = \int_{-\infty}^0 dx \, e^{-ikx} f(x) \qquad (15.127)$$

can be analytically continued to the upper half of the k plane, i.e., $\text{Im } k > 0$, where it is an analytic function with no singularities, and similarly

$$F_-(k) = \int_0^\infty dx \, e^{-ikx} f(x) \qquad (15.128)$$

is an analytic function with no singularities in the lower half k plane, i.e., $\text{Im } k < 0$. We use the subscripts $+$ and $-$ in what follow to designate the half of the k plane in which the function has no singularities. We consider two examples[4].

---

[4] Other examples are given in the book by B. Noble, Methods Based on the Wiener-Hopf Technique (Pergamon Press, New York, 1958). See also Carrier, Krook and Pearson, op. cit., Chap. 8, p. 376.

15.8.1 Potential Given on Semi-Infinite Plate

The first example is a two-dimensional analogue of the problem of the potential due to a thin charged disk. The disk problem is more interesting, but complicated, because it involves Fourier-Bessel transforms. Unfortunately, in the two-dimensional analogue the problem becomes trivial (or not well posed) for a sheet of constant potential, so we consider a semi-infinite sheet on which the potential is a given function $\neq$ const. Thus we wish to solve the two-dimensional Laplace equation

$$\frac{\partial^2 \phi(x,y)}{\partial x^2} + \frac{\partial^2 \phi(x,y)}{\partial y^2} = 0 \tag{15.129}$$

with the boundary condition

$$\phi(x,0) = e^{ax}, \quad x < 0, \tag{15.130}$$

where a is a positive constant as shown in Figure 15.15. We also impose the boundary condition at infinity

$$\phi(x,y) \to 0, \quad x^2 + y^2 \to \infty. \tag{15.131}$$

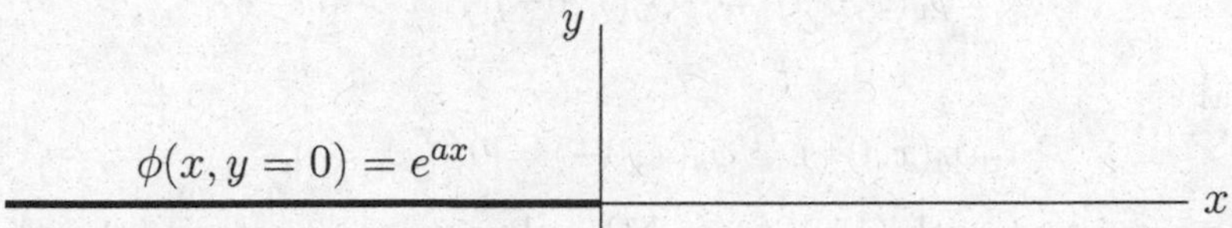

Figure 15.15: Semi-infinite sheet with potential $\phi(x,y)$

Fourier transforming in x

$$\phi(x,y) = \frac{1}{\sqrt{2\pi}} \int_{-\infty}^{\infty} dk \, e^{ikx} \widetilde{\phi}(k,y), \tag{15.132}$$

$$\widetilde{\phi}(k,y) = \frac{1}{\sqrt{2\pi}} \int_{-\infty}^{\infty} dx \, e^{-ikx} \phi(x,y), \tag{15.133}$$

Laplace's equation (15.129) becomes

$$-k^2 \widetilde{\phi}(k,y) + \frac{\partial^2}{\partial y^2} \widetilde{\phi}(k,y) = 0. \tag{15.134}$$

The solution of (15.134) consistent with (15.131) is

$$\widetilde{\phi}(k,y) = \begin{cases} A(k)e^{-|k|y}, & y > 0 \\ \\ B(k)e^{+|k|y}, & y < 0 \end{cases}. \tag{15.135}$$

Continuity of $\phi(x, y)$ at $y = 0$ implies $B(k) = A(k)$ in (15.135). We then find $\widetilde{\phi}(k, y) = \widetilde{\phi}(k, -y)$ and consequently also

$$\phi(x, y) = \phi(x, -y), \tag{15.136}$$

which is physically obvious anyway. We also need the partial derivatives with respect to y. Differentiating (15.132) with respect to y using (15.135) for $\widetilde{\phi}(k, y)$, and setting $y = 0+$ and $0-$, i.e., just above and below the plate, we find

$$\phi_y(x, 0+) = \left.\frac{\partial \phi(x, y)}{\partial y}\right|_{y=0+} = \frac{1}{\sqrt{2\pi}} \int_{-\infty}^{\infty} dk\ e^{ikx}[-|k|\,A(k)], \tag{15.137}$$

$$\phi_y(x, 0-) = \left.\frac{\partial \phi(x, y)}{\partial y}\right|_{y=0-} = \frac{1}{\sqrt{2\pi}} \int_{-\infty}^{\infty} dk\ e^{ikx}[+|k|\,A(k)]. \tag{15.138}$$

Thus we have

$$\phi_y(x, 0+) = -\phi_y(x, 0-) \tag{15.139}$$

for all x. On the other hand, for $x > 0$, outside the plate, continuity implies

$$\phi_y(x, 0+) = +\phi_y(x, 0-), \quad x > 0. \tag{15.140}$$

Thus we find

$$\phi_y(x, 0+) = \phi_y(x, 0-) = 0, \quad x < 0, \tag{15.141}$$

which is also consistent with (15.136) and the physical symmetry of the problem. We can now make a table, see Table 15.1, of the known and unknown boundary values of $\phi(x, y)$ and $\phi_y(x, y)$ at $y = 0$.

	$x < 0$	$x > 0$
$\phi(x, 0)$	e^{ax}	?
$\phi_y(x, 0)$	?	0

Table 15.1: Potential for ranges of x for semi-infinite plate

This is the kind of mixed boundary value problem for which the Wiener-Hopf method is effective.

Writing out (15.133), using (15.135), for $y = 0$ we find

$$A(k) = \frac{1}{\sqrt{2\pi}} \int_{-\infty}^{\infty} dx\ e^{-ikx} \phi(x, y)$$

$$= \frac{1}{\sqrt{2\pi}} \int_{-\infty}^{0} dx\ e^{-ikx} e^{ax} + \frac{1}{\sqrt{2\pi}} \int_{0}^{\infty} dx\ e^{-ikx} \phi(x, y)$$

$$= \frac{1}{\sqrt{2\pi}} \left(\frac{i}{k + ia}\right)_+ + F_-(k). \tag{15.142}$$

Here we have explicitly calculated the Fourier transform of the part of $\phi(x,y)$ which is known. This gives a $+$ type function; the only singularity is at $k = -ia$ in the lower half-plane, so the function is analytic in the upper half-plane. The Fourier transform of the unknown part of $\phi(x,y)$ for $x > 0$ is given a subscript $-$ to indicate that we know it is analytic in the lower half-plane. Similarly the inverse of (15.137) gives

$$- |k|\, A(k) = \frac{1}{\sqrt{2\pi}} \int_{-\infty}^{\infty} dx\; e^{-ikx} \phi_y(x, 0+)$$

$$= \frac{1}{\sqrt{2\pi}} \int_{-\infty}^{0} dx\; e^{-ikx} \phi_y(x, 0+) + \frac{1}{\sqrt{2\pi}} \int_{0}^{\infty} dx\; e^{-ikx} \phi_y(x, 0+)$$

$$= G_+(k) + 0. \tag{15.143}$$

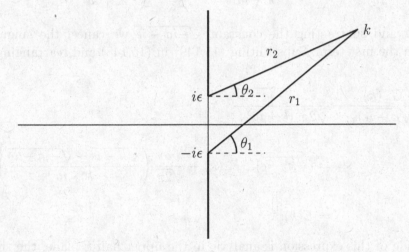

Figure 15.16: Geometry for treating $|k|$ as an analytic function with cuts in the imaginary plane for equation (15.144)

Before proceeding we must write $|k|$ as an analytic function. This can be done by introducing a parameter $\epsilon > 0$ which is allowed to approach 0 at the end of the calculation. Thus take

$$|k| = \sqrt{k^2 + \epsilon^2} = \sqrt{(k + i\epsilon)(k - i\epsilon)} = (r_1 r_2)^{\frac{1}{2}} e^{i(\theta_1 + \theta_2)/2},$$

$$-\frac{3\pi}{2} < \theta_2 < \frac{\pi}{2}, \quad -\frac{\pi}{2} < \theta_1 < \frac{3\pi}{2}, \tag{15.144}$$

with cuts as indicated in Figure 15.16. It is easy to check that $\sqrt{k^2 + \epsilon^2} \xrightarrow[\epsilon \to 0]{} |k|$ for negative real k. We thus have

$$|k| = (\sqrt{k + i\epsilon})_+ (\sqrt{k - i\epsilon})_- \tag{15.145}$$

in the notation which tells us where the singularities are <u>not</u>. Substituting (15.145) in (15.143) we find

$$A(k) = -\frac{G_+(k)}{(\sqrt{k+i\epsilon})_+} \frac{1}{(\sqrt{k-i\epsilon})_-}. \tag{15.146}$$

Eliminating $A(k)$ from (15.142), (15.146) we have

$$-\frac{G_+(k)}{(\sqrt{k+i\epsilon})_+} \frac{1}{(\sqrt{k-i\epsilon})_-} = \frac{1}{\sqrt{2\pi}} \left(\frac{i}{k+ia}\right)_+ + F_-(k) \tag{15.147}$$

or

$$-\frac{G_+(k)}{(\sqrt{k+i\epsilon})_+} = (\sqrt{k-i\epsilon})_- \, F_-(k) + \frac{i}{\sqrt{2\pi}} \frac{(\sqrt{k-i\epsilon})_-}{(k+ia)_+}. \tag{15.148}$$

We can rewrite the explicit function on the right of (15.148) as a sum of $+$ type and $-$ type functions:

$$\frac{(\sqrt{k-i\epsilon})_-}{(k+ia)_+} = \left(\frac{\sqrt{k-i\epsilon}-\sqrt{-ia-i\epsilon}}{k+ia}\right)_- + \left(\frac{\sqrt{-ia-i\epsilon}}{k+ia}\right)_+. \tag{15.149}$$

By adding and subtracting the constant $\sqrt{-ia-i\epsilon}$ we cancel the singularity at $k = -ia$ in the first term. Substituting (15.149) in (15.148) and rearranging slightly we find

$$-\frac{G_+(k)}{(\sqrt{k+i\epsilon})_+} - \frac{i}{\sqrt{2\pi}} \frac{\sqrt{-ia-i\epsilon}}{(k+ia)_+}$$

$$= (\sqrt{k-i\epsilon})_- \, F_-(k) + \frac{i}{\sqrt{2\pi}} \left(\frac{\sqrt{k-i\epsilon}-\sqrt{-ia-i\epsilon}}{k+ia}\right)_-$$

$$= E(k). \tag{15.150}$$

The left side of this expression is analytic in the upper half k plane, the right-hand side analytic in the lower half-plane. The two sides are the analytic continuations of each other. Together they constitute a function $E(k)$ which is analytic throughout the whole complex plane and thus an entire function. Solving (15.150) for $G_+(k)$ in terms of $E(k)$,

$$G_+(k) = -\frac{i}{\sqrt{2\pi}} \sqrt{-ia-i\epsilon} \left(\frac{\sqrt{k+i\epsilon}}{k+ia}\right)_+ - (\sqrt{k+i\epsilon})_+ E(k), \tag{15.151}$$

we see that if $E(k) \neq 0$, $G_+(k) \to \infty$ as $k \to \infty$ in the upper half-plane [Recall Liouville's theorem, Problem 7 from Chapter 10]. This contradicts our knowledge that $G_+(k)$ is analytic throughout the entire upper half-plane. Thus we find

$$E(k) = 0. \tag{15.152}$$

From (15.146), (15.151) we thus find $A(k)$:

$$A(k) = \frac{i}{\sqrt{2\pi}} \frac{\sqrt{-ia-i\epsilon}}{(k+ia)\sqrt{k-i\epsilon}}$$

or

$$A(k) = \begin{cases} \dfrac{\sqrt{a}}{\sqrt{2\pi}} e^{i\pi/4} \dfrac{1}{\sqrt{k}\ (k+ia)}, & k > 0 \\[4mm] \dfrac{\sqrt{a}}{\sqrt{2\pi}} e^{-i\pi/4} \dfrac{1}{\sqrt{-k}\ (-k-ia)}, & k < 0 \end{cases} \qquad (15.153)$$

Here we have used $\sqrt{k-i\epsilon} \xrightarrow[\epsilon\to 0]{} \sqrt{k}$ for $k > 0$, $\sqrt{k-i\epsilon} \xrightarrow[\epsilon\to 0]{} -i\sqrt{-k}$ for $k < 0$, and $\sqrt{-ia-i\epsilon} \xrightarrow[\epsilon\to 0]{} e^{-i\pi/4}\sqrt{a}$. The solution of the boundary value problem is thus given by (15.132), (15.135). For $y > 0$ we find

$$\phi(x,-y) = \phi(x,y)$$

$$= \frac{\sqrt{a}}{2\pi} \left[e^{i\pi/4} \int_0^\infty dk\, \frac{e^{ikx-ky}}{\sqrt{k}(k+ia)} + e^{-i\pi/4} \int_{-\infty}^0 dk\, \frac{e^{ikx+ky}}{\sqrt{-k}(-k-ia)} \right]$$

$$= \frac{\sqrt{a}}{\pi} \operatorname{Re} \left[e^{i\pi/4} \int_0^\infty dk\, \frac{e^{ikz}}{\sqrt{k}(k+ia)} \right], \quad z = x + iy. \qquad (15.154)$$

The integral here can be rewritten as an error function by finding a differential equation for the function

$$F(z) = \int_0^\infty dk\, \frac{e^{ikz}}{\sqrt{k}(k+ia)}. \qquad (15.155)$$

Thus, differentiating, we have

$$\frac{dF(z)}{dz} = \int_0^\infty dk\, \frac{e^{ikz}}{\sqrt{k}(k+ia)} i[k+ia-ia]$$

$$= aF(z) + \frac{\sqrt{\pi}}{\sqrt{z}} e^{i3\pi/4}, \qquad (15.156)$$

with the boundary condition

$$F(0) = \int_0^\infty dk\, \frac{1}{\sqrt{k}(k+ia)} = \frac{\pi}{\sqrt{a}} e^{-i\pi/4}. \qquad (15.157)$$

Solving the differential equation (15.156) and substituting the result in (15.154) we find

$$\phi(x,y) = \operatorname{Re}\left\{ e^{az}[1 - \operatorname{erf}(\sqrt{az})] \right\}, \qquad (15.158)$$

where the error function is defined by

$$\operatorname{erf}(u) = \frac{2}{\sqrt{\pi}} \int_0^u dv\, e^{-v^2}. \qquad (15.159)$$

It is easy to check that (15.158) satisfies the original boundary condition (15.130) since the erf of a pure imaginary argument is pure imaginary.

15.8.2 Diffraction by a Knife Edge

The second example we wish to consider is the diffraction of a plane wave by a semi-infinite plate as illustrated as Figure 15.17. The incident wave is taken to be

$$\psi_i = e^{-i\lambda(x\cos\theta + y\sin\theta)}. \qquad (15.160)$$

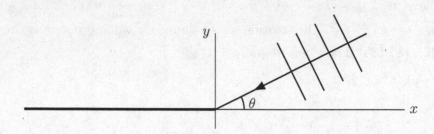

Figure 15.17: Plane wave incident upon a knife edge

Here λ is a positive constant, the wave number of the incident wave, and the time dependence $e^{-i\omega t}$ is dropped from (15.160) and all subsequent formulas. The presence of the plate induces a scattered wave ψ so that the total wave is

$$\psi_t = \psi_i + \psi. \qquad (15.161)$$

All three waves, in particular ψ, satisfy the Helmholtz equation

$$(\boldsymbol{\nabla}^2 + \lambda^2)\psi = 0. \qquad (15.162)$$

We shall use the boundary condition appropriate for a sound wave at the surface of the rigid plate, namely

$$\left.\frac{\partial \psi_t(x,y)}{\partial y}\right|_{y=0} = 0, \quad x < 0 \qquad (15.163)$$

or equivalently

$$\left.\frac{\partial \psi(x,y)}{\partial y}\right|_{y=0} = \psi_y(x,0) = -\psi_{iy}(x,0)$$

$$= i\lambda\sin\theta e^{-i\lambda x\cos\theta}, \quad x < 0. \qquad (15.164)$$

In addition we impose the radiation condition on ψ, i.e., ψ is to contain only outgoing waves ($\sim e^{ikr}$, $r = \sqrt{x^2 + y^2}$, as $r \to \infty$). As discussed in connection with (15.18) this is achieved by taking λ to have a small positive imaginary part,

$$\lambda \to \lambda + i\epsilon, \qquad (15.165)$$

in what follows. This will push certain singularities, which otherwise would lie on the integration contour, to one side or the other, and so give a definite meaning to integrals which otherwise would be ambiguous.

Fourier transforming $\psi(x,y)$ with respect to x,

$$\psi(x,y) = \frac{1}{\sqrt{2\pi}} \int_{-\infty}^{\infty} dk\, e^{ikx} \widetilde{\psi}(k,y), \tag{15.166}$$

and substituting in the differential equation (15.162) leads to

$$\frac{\partial^2 \widetilde{\psi}(k,y)}{\partial y^2} - (k^2 - \lambda^2)\widetilde{\psi}(k,y) = 0. \tag{15.167}$$

The solutions of this which, fall off with increasing $|y|$, are

$$\widetilde{\psi}(k,y) = \begin{cases} A(k)e^{-y\sqrt{k^2-\lambda^2}}, & y > 0 \\[2mm] B(k)e^{+y\sqrt{k^2-\lambda^2}}, & y < 0 \end{cases}, \tag{15.168}$$

where the cuts in $\sqrt{k^2 - \lambda^2}$ are chosen, with reference to the geometry illustrated in Figure 15.18, as follows (compare the corresponding choice in the previous example):

$$\sqrt{k^2 - \lambda^2} = (r_1 r_2)^{\frac{1}{2}} e^{i(\theta_1+\theta_2)/2}, \quad -\frac{3\pi}{2} \le \theta_2 \le \frac{\pi}{2}, \quad -\frac{\pi}{2} \le \theta_1 < \frac{3\pi}{2}. \tag{15.169}$$

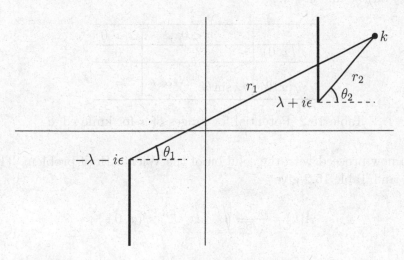

Figure 15.18: Geometry configuration for equation (15.169)

With this choice $\sqrt{k^2 - \lambda^2} > 0$ for $|k| > \lambda$ and both expressions in (15.168) fall off with increasing y; for $|k| < \lambda$ the expressions in (15.168) are oscillatory functions of y.

The continuity of $\psi_y(x,y)$ at $y = 0$ relates $A(k)$ to $B(k)$. According to (15.164) $\psi_y(x,0)$ has the same value on both sides of the plate, i.e., for $x < 0$, and for $x > 0$ the continuity of $\psi_y(x,y)$ follows from general considerations. Thus we have

$$\psi_y(x,0+) = \psi_y(x,0-), \tag{15.170}$$

which implies the same relation for the Fourier transforms, i.e.,

$$\tilde{\psi}_y(k, 0+) = \tilde{\psi}_y(k, 0-), \tag{15.171}$$

or, using (15.166), (15.168),

$$-\sqrt{k^2 - \lambda^2}\, A(k) = +\sqrt{k^2 - \lambda^2}\, B(k),$$
$$A(k) = -B(k). \tag{15.172}$$

Substituting (15.172) in (15.168) and the result in (15.166) we find

$$\psi(x, 0+) = -\psi(x, 0-) = \frac{1}{\sqrt{2\pi}} \int_{-\infty}^{\infty} dk\; e^{ikx} A(k). \tag{15.173}$$

Since we also know from general continuity considerations that for $x > 0$

$$\psi(x, 0+) = \psi(x, 0-), \tag{15.174}$$

we find

$$\psi(x, 0+) = \psi(x, 0-) = 0, \quad x > 0. \tag{15.175}$$

We can now make a table, Table 15.2, of the known and unknown values of $\psi(x, y)$ and $\psi_y(x, y)$ on the line $y = 0$.

	$x < 0$	$x > 0$
$\psi(x, 0)$	?	0
$\psi_y(x, 0)$	$i\lambda \sin\theta e^{-i\lambda x \cos\theta}$	?

Table 15.2: Potential for ranges of x for knife edge

We can now proceed with the solution of the Wiener-Hopf problem. The inverse of (15.173) and Table 15.2 give

$$A(k) = \frac{1}{\sqrt{2\pi}} \int_{-\infty}^{\infty} dx\; e^{-ikx} \psi(x, 0+)$$

$$= \frac{1}{\sqrt{2\pi}} \int_{-\infty}^{0} dk\; e^{-ikx} \psi(x, 0+)$$

$$= F_+(k), \tag{15.176}$$

where the $+$ notation indicates that $F_+(k)$ can be analytically continued to yield a function without singularities in the upper half-plane. On the other hand, the inverse transform for $\psi_y(x, 0)$ and the table above imply

$$-\sqrt{k^2 - \lambda^2}\, A(k) = \frac{1}{\sqrt{2\pi}} \int_{-\infty}^{\infty} dx\; e^{-ikx} \psi_y(x, 0)$$

or

$$-\sqrt{k^2 - \lambda^2}\, A(k) = \frac{1}{\sqrt{2\pi}} \int_{-\infty}^{0} dx\, e^{-ikx} i\lambda \sin\theta e^{-i\lambda x \cos\theta} + \frac{1}{\sqrt{2\pi}} \int_{0}^{\infty} dx\, e^{-ikx} \psi_y(x,0)$$

$$= -\frac{\lambda \sin\theta}{\sqrt{2\pi}} \left(\frac{1}{k + \lambda \cos\theta} \right)_+ + G_-(k). \tag{15.177}$$

In order to obtain an integral of the plane wave which is convergent at $x \to -\infty$ we must interpret $\lambda \cos\theta \to \lambda \cos\theta + i\epsilon$ so that in (15.177), $(k + \lambda \cos\theta)^{-1} = (k + \lambda \cos\theta + i\epsilon)^{-1}$ is a $+$ type function. Eliminating $A(k)$ from (15.176), (15.177) we find

$$-\sqrt{k^2 - \lambda^2}\, F_+(k) = -\frac{\lambda \sin\theta}{\sqrt{2\pi}} \left(\frac{1}{k + \lambda \cos\theta} \right)_+ + G_-(k). \tag{15.178}$$

The function $\sqrt{k^2 - \lambda^2}$ can be decomposed into $+$ and $-$ parts by referring to the diagram above:

$$\sqrt{k^2 - \lambda^2} = \left(\sqrt{k - \lambda} \right)_- \left(\sqrt{k + \lambda} \right)_+. \tag{15.179}$$

Thus (15.178) can be rewritten

$$-\left(\sqrt{k + \lambda} \right)_+ F_+(k) = \frac{-\lambda \sin\theta}{\sqrt{2\pi}} \frac{1}{(k + \lambda \cos\theta)_+ (\sqrt{k - \lambda})_-} + \frac{G_-(k)}{(\sqrt{k - \lambda})_-}. \tag{15.180}$$

The known function on the right of (15.180) can be decomposed by inspection[5] into $+$ and $-$ contributions:

$$\frac{\lambda \sin\theta}{(k + \lambda \cos\theta)_+ (\sqrt{k - \lambda})_-} =$$

$$\left(\frac{i\sqrt{\lambda(1 - \cos\theta)}}{k + \lambda \cos\theta} \right)_+ + \left(\frac{\sqrt{\lambda(1 - \cos\theta)} \left[\sqrt{\lambda(1 + \cos\theta)} - i\sqrt{k - \lambda} \right]}{(k + \lambda \cos\theta)\sqrt{k - \lambda}} \right)_-. \tag{15.181}$$

Substituting this into (15.180) we find

$$-\left(\sqrt{k + \lambda} \right)_+ F_+(k) + \frac{i\sqrt{2\lambda}}{\sqrt{2\pi}} \frac{\sin\theta/2}{(k + \lambda \cos\theta)_+}$$

$$= \frac{G_-(k)}{(\sqrt{k - \lambda})_-} - \frac{\sqrt{2\lambda}}{\sqrt{2\pi}} \sin\frac{\theta}{2} \left(\frac{\sqrt{2\lambda} \cos\theta/2 - i\sqrt{k - \lambda}}{(k + \lambda \cos\theta)\sqrt{k - \lambda}} \right)_-$$

$$= E(k). \tag{15.182}$$

---

[5]Evidently, it is not always easy to make this type of decomposition.

The left side here is analytic in the upper half-plane, the right side in the lower half-plane. The two sides are the analytic continuations of each other and together describe an entire function $E(k)$. Were $E(k)$ not zero, $G_-(k)$ would go to infinity as $k \to \infty$ in the lower half-plane in contradiction to our knowledge that it is analytic throughout the entire lower half-plane. Setting $E(k) = 0$ and using (15.176) we thus find

$$A(k) = i\sqrt{\frac{\lambda}{\pi}} \sin\frac{\theta}{2} \frac{1}{(k + \lambda\cos\theta)_+ (\sqrt{k+\lambda})_+}, \tag{15.183}$$

and substituting this in (15.168), (15.166) we find the solution of the problem. For $y > 0$ we have

$$\psi(x, y) = -\psi(x, -y)$$

$$= \frac{i}{\pi}\sqrt{\frac{\lambda}{2}} \sin\frac{\theta}{2} \int_{-\infty}^{\infty} dk \, \frac{e^{ikx - y\sqrt{k^2 - \lambda^2}}}{(k + \lambda\cos\theta + i\epsilon)\sqrt{k + \lambda + i\epsilon}}. \tag{15.184}$$

The integral here can be evaluated in terms of Fresnel integrals. First, divide the integration region into three parts

$$\int_{-\infty}^{\infty} dk = \int_{-\infty}^{-\lambda} dk + \int_{-\lambda}^{\lambda} dk + \int_{\lambda}^{\infty} dk. \tag{15.185}$$

In these three regions, make the following changes of variables, with $x = r\cos\varphi$, $y = r\sin\varphi$:

$$-\lambda < k < \lambda : \quad k = \lambda\cos\alpha, \quad 0 < \alpha < \pi$$
$$\sqrt{k^2 - \lambda^2} = -i\lambda\sin\alpha$$
$$kx + i\sqrt{k^2 - \lambda^2}\, y = \lambda r\cos(\varphi - \alpha)$$
$$\sqrt{k + \lambda} = \sqrt{2\lambda}\,\cos\frac{\alpha}{2}$$

$$\lambda < k < \infty : \quad k = \lambda\cos\alpha = \lambda\cosh\sigma, \quad \alpha = i\sigma, \quad 0 < \sigma < \infty$$
$$\sqrt{k^2 - \lambda^2} = \lambda\sinh\sigma = -i\lambda\sin\alpha$$
$$kx + i\sqrt{k^2 - \lambda^2}\, y = \lambda r\cos(\varphi - \alpha)$$
$$\sqrt{k + \lambda} = \sqrt{2\lambda}\,\cosh\frac{\sigma}{2} = \sqrt{2\lambda}\,\cos\frac{\alpha}{2}$$

$$-\infty < k < -\lambda : \quad k = \lambda\cos\alpha = -\lambda\cosh\sigma, \quad \alpha = \pi - i\sigma, \quad 0 < \sigma < \infty$$
$$\sqrt{k^2 - \lambda^2} = \lambda\sinh\sigma = -i\lambda\sin\alpha$$
$$kx + i\sqrt{k^2 - \lambda^2}\, y = \lambda r\cos(\varphi - \alpha)$$
$$\sqrt{k + \lambda} = i\sqrt{2\lambda}\,\sinh\frac{\sigma}{2} = \sqrt{2\lambda}\,\cos\frac{\alpha}{2} \tag{15.186}$$

One can check that the signs of the square roots have all been chosen in the way agreed on in the derivation of (15.184). Making these transformations in the integral (15.184), it becomes

$$\psi(x, y) = \frac{i}{\pi} \sin \frac{\theta}{2} \int_C d\alpha \, \frac{\sin \alpha/2}{\cos \alpha + \cos \theta} e^{i\lambda r \cos(\varphi - \alpha)} \tag{15.187}$$

with the contour C shown in Figure 15.19. In order to decide the contour around the pole in $(k + \lambda \cos \theta + i\epsilon)^{-1} = \lambda(\cos \alpha + \cos \theta + i\epsilon)$ we used $\cos(\theta - i\epsilon) = \cos \theta \cos i\epsilon + \sin \theta \sin i\epsilon \xrightarrow[\epsilon \to 0]{} \cos \theta + i\epsilon'$; with $\sin \theta$ positive in the range $0 < \theta < \pi$, we have $\epsilon' > 0$.

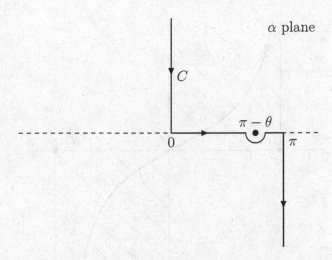

Figure 15.19: Contour for use with equation (15.187)

Manipulating the trigonometric functions in (15.187), we find

$$\psi(x, y) = \frac{i}{4\pi} \int_C d\alpha \left[\frac{1}{\cos \frac{1}{2}(\alpha + \theta)} - \frac{1}{\cos \frac{1}{2}(\alpha - \theta)} \right] e^{i\lambda r \cos(\alpha - \varphi)}$$

$$= \frac{i}{4\pi} \int_{C'} d\alpha \left[\frac{1}{\cos \frac{1}{2}(\alpha + \theta + \varphi)} - \frac{1}{\cos \frac{1}{2}(\alpha - \theta + \varphi)} \right] e^{i\lambda r \cos \alpha}$$

$$= I(\theta) - I(-\theta), \tag{15.188}$$

where the integral $I(\theta)$ is given by

$$I(\theta) = \frac{i}{4\pi} \int_{C'} d\alpha \, \frac{1}{\cos \frac{1}{2}(\alpha + \theta + \varphi)} e^{i\lambda r \cos \alpha}. \tag{15.189}$$

Using $\text{Re}(i\lambda r \cos \alpha) = \lambda r \sin \alpha_1 \sinh \alpha_2$ for $\alpha = \alpha_1 + i\alpha_2$ and remembering that we have $\varphi > 0$ for $y > 0$, as assumed in (15.184), we see that the contour C', which in the first instance is C of Figure 15.19 displaced a distance φ to the left, can in fact be located anywhere in the shaded regions as indicated in Figure 15.20. A C' of this form goes into itself under the transformation $\alpha \to -\alpha$, so that we also have from (15.189)

$$I(\theta) = \frac{i}{8\pi} \int_{C'} d\alpha \left[\frac{1}{\cos \frac{1}{2}(\alpha + \theta + \varphi)} + \frac{1}{\cos \frac{1}{2}(\alpha - \theta - \varphi)} \right] e^{i\lambda r \cos \alpha}$$

$$= \frac{i}{2\pi} \int_{C'} d\alpha \, \frac{\cos \frac{1}{2}\alpha \cos \frac{1}{2}(\theta + \varphi)}{\cos \alpha + \cos(\theta + \varphi)} e^{i\lambda r \cos \alpha}. \tag{15.190}$$

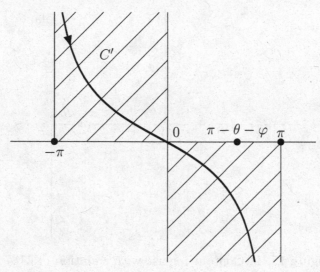

Figure 15.20: Shaded regions indicate allowable contours for equation (15.189)

If we now make a further change of variables,

$$\tau = \sqrt{2} e^{i\pi/4} \sin \frac{\alpha}{2}, \tag{15.191}$$

the contour C' is transformed into a straight line along the real τ axis from $-\infty$ to $+\infty$ and the integral (15.190) becomes

$$I(\theta) = \frac{e^{-i\pi/4}}{2\pi} \eta e^{i\lambda r} \int_{-\infty}^{\infty} d\tau \, \frac{e^{-\lambda r \tau^2}}{\tau^2 - i\eta^2}, \tag{15.192}$$

where η is given by

$$\eta = \sqrt{2} \cos \frac{1}{2}(\theta + \varphi). \tag{15.193}$$

The integral in (15.192) can be reexpressed by multiplying the known integral

$$\int_{-\infty}^{\infty} d\tau \, e^{-\xi \tau^2} = \sqrt{\frac{\pi}{\xi}} \tag{15.194}$$

by $e^{i\eta^2\xi}$ and integrating ξ from λr to ∞:

$$e^{i\lambda r\eta^2}\int_{-\infty}^{\infty} d\tau \, \frac{e^{-\lambda r\tau^2}}{\tau^2 - i\eta^2} = \sqrt{\pi}\int_{\lambda r}^{\infty} d\xi \, \frac{e^{i\eta^2\xi}}{\sqrt{\xi}}$$

$$= \frac{2\sqrt{\pi}}{|\eta|}\int_{|\eta|\sqrt{\lambda r}}^{\infty} d\mu \, e^{i\mu^2}. \qquad (15.195)$$

Thus if we define the complex Fresnel integral by

$$F(a) = \int_{a}^{\infty} d\mu \, e^{i\mu^2}, \qquad (15.196)$$

we find

$$\eta \int_{-\infty}^{\infty} d\tau \, \frac{e^{-\lambda r\tau^2}}{\tau^2 - i\eta^2} = \pm 2\sqrt{\pi}e^{-i\lambda r\eta^2} F(\pm\eta\sqrt{\lambda r}), \qquad (15.197)$$

where the upper sign holds for $\eta > 0$, the lower for $\eta < 0$.

From (15.188) and (15.197) we then find

$$\psi(x, y) = \frac{-1}{\sqrt{\pi}} e^{-i\pi/4} \times$$

$$\left\{ e^{-i\lambda r \cos(\varphi-\theta)} F\left[\sqrt{2\lambda r}\,\cos\frac{1}{2}(\varphi - \theta)\right] \mp e^{-i\lambda r \cos(\varphi+\theta)} F\left[\pm\sqrt{2\lambda r}\,\cos\frac{1}{2}(\varphi + \theta)\right] \right\}. \qquad (15.198)$$

In choosing the signs in this expression we note that $y = r\sin\varphi > 0$, so that $0 \le \varphi \le \pi$ and by assumption $0 \le \theta \le \pi$. Thus $-\pi/2 \le (\varphi-\theta)/2 \le \pi/2$, so that the η calculated with this combination of angles is positive. In the second term of (15.198) choose the upper or lower sign depending on whether $\theta + \varphi \lessgtr \pi$.

In fact, in deriving (15.198) it has been tacitly assumed that $\theta + \varphi < \pi$. For the case $\theta + \varphi > \pi$ the pole appears on the other side of the origin in Figure 15.20, and when the contour is distorted to C' one picks up the residue at this pole. This is easily calculated from (15.187) to be

$$e^{-i\lambda r \cos(\varphi+\theta)}, \qquad \varphi > \pi - \theta, \qquad (15.199)$$

and is just the reflected wave of geometrical optics.

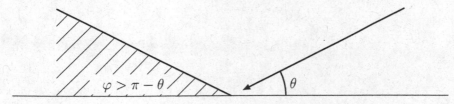

Figure 15.21: Reflected plane wave from knife edge at angle φ

The scattered wave $\psi(x, y)$ is thus the sum of (15.198), with the upper or lower signs depending on whether $\varphi \lessgtr \pi - \theta$, and the reflected wave (15.199) if $\varphi > \pi - \theta$. The formulas can be written in a neater form which covers all special cases. From the definition (15.196) we see that

$$F(a) + F(-a) = 2F(0) = 2 \int_0^\infty d\mu \, e^{i\mu^2} = \sqrt{\pi} e^{i\pi/4}, \tag{15.200}$$

according to (12.61). Using this to transform the second term in (15.198) for $\varphi > \pi - \theta$ and at the same time cancel the extra contribution (15.199), we obtain

$$\psi(x,y) = \frac{-i}{\sqrt{\pi}} e^{-i\pi/4} \times$$

$$\left\{ e^{-i\lambda r \cos(\varphi-\theta)} F\left[\sqrt{2\lambda r}\, \cos \frac{1}{2}(\varphi-\theta)\right] - e^{-i\lambda r \cos(\varphi+\theta)} F\left[\sqrt{2\lambda r}\, \cos \frac{1}{2}(\varphi+\theta)\right] \right\} \tag{15.201}$$

If we add in the incident wave $\psi_i = e^{-i\lambda r \cos(\varphi-\theta)}$ to obtain the total disturbance, we find, again using (15.200),

$$\psi_t(x,y) = \psi_i(x,y) + \psi(x,y) = \frac{1}{\sqrt{\pi}} e^{-i\pi/4} \times$$

$$\left\{ e^{-i\lambda r \cos(\varphi-\theta)} F\left[-\sqrt{2\lambda r}\, \cos \frac{1}{2}(\varphi-\theta)\right] + e^{-i\lambda r \cos(\varphi+\theta)} F\left[\sqrt{2\lambda r}\, \cos \frac{1}{2}(\varphi+\theta)\right] \right\}. \tag{15.202}$$

This expression has been derived for the range $0 \leq \varphi \leq \pi$, $0 \leq \theta \leq \pi$. Note, however, that the condition $\psi(x,y) = -\psi(x,-y)$ [see (15.184)] is satisfied by (15.201) so that (15.202) remains valid for y negative, i.e., φ negative.

The formula (15.202) is a famous exact result, due originally to Sommerfeld, for the diffraction of waves by a knife edge. An extensive discussion of the result with many graphs of numerical evaluations is given in the book by Born and Wolf.[6]

---

[6]M. Born and E. Wolf, <u>Principles of Optics</u>, Fourth edition, (Pergamon Press, New York, 1970), p. 565ff.

PROBLEMS

15.1 Use the Fourier transform technique to find the appropriate Green's function for all space for the equation

$$\nabla^2 u(\mathbf{r}) - \lambda^2 u(\mathbf{r}) = -4\pi\rho(\mathbf{r}),$$

and use your Green's function to give the solution of this equation in terms of the source term $\rho(\mathbf{r})$.

In solving for the Green's function by the Fourier expansion technique, what happened to the analogue of the ingoing wave Green's function (for the Helmholtz equation)?

15.2 Derive the Laplace transforms given in Table 15.3.

$f(t)$	$F(s) = \int_0^\infty dt\, e^{-st} f(t)$
1	$\dfrac{1}{s}$
t^n	$\dfrac{n!}{s^{n+1}}$
e^{-at}	$\dfrac{1}{s+a}$
$\sin at$	$\dfrac{a}{s^2 + a^2}$
$\cos at$	$\dfrac{s}{s^2 + a^2}$
$J_0(at)$	$\dfrac{1}{(s^2 + a^2)^{\frac{1}{2}}}$

Table 15.3: Table of Laplace transforms for Problem 15.2

For the last entry, use the integral representation (14.134). For a longer table see Erdélyi, Magnus, Oberhettinger and Tricomi, Tables of Integral Transforms, Vols. 1 and 2 (McGraw-Hill, New York, 1954).

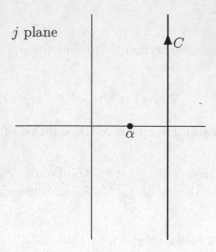

Figure 15.22: Reference contour C and pole α for Problem 15.3

15.3 Evaluate the Mellin transform integrals

a) $F(s) = \dfrac{1}{2\pi i} \displaystyle\int_C dj \; s^j \dfrac{1}{j - \alpha}$,

b) $G(s) = \dfrac{1}{2\pi i} \displaystyle\int_C dj \; s^j \dfrac{1}{\sqrt{j - \alpha}}$.

In these integrals s = real number > 1, α = real number. The contour C is parallel to the imaginary axis and to the right of α as shown in Figure 15.22. In (b) take the cut in $\sqrt{j - \alpha}$ along the negative real axis.

15.4 Three radioactive nuclei have a parent-daughter relationship described by the differential equations

$$\frac{dN_1}{dt} = -\lambda_1 N_1,$$

$$\frac{dN_2}{dt} = \lambda_1 N_1 - \lambda_2 N_2,$$

$$\frac{dN_3}{dt} = \lambda_2 N_2 - \lambda_3 N_3.$$

Use Laplace transforms to find $N_1(t)$, $N_2(t)$, and $N_3(t)$. The initial conditions are $N_1(0) = N$, $N_2(0) = 0$, $N_3(0) = 0$.

15.5 As a special case of the development (15.113)-(15.126) consider a semi-infinite region, $l \to \infty$, with boundary conditions

$$u(0, t) = T_0 = \text{const.},$$
$$u(l, t) = u(\infty, t) = 0,$$

and the initial condition

$$u(x,0) = 0.$$

Show that for this case

$$\tilde{u}(x,s) = \frac{T_0}{s} e^{-\sqrt{s}\,x},$$

$$u(x,t) = \frac{T_0}{2\pi i} \int_{c-i\infty}^{c+i\infty} ds\, \frac{1}{s} e^{st} e^{-\sqrt{s}\,x}.$$

Show that the integral for $\partial u/\partial x$ can be evaluated,

$$\frac{\partial u}{\partial x} = -\frac{T_0}{\sqrt{\pi t}} e^{-x^2/4t},$$

and that $u(x,t)$ is given by the formula

$$u(x,t) = T_0 \left[1 - \mathrm{erf}\left(\frac{x}{2\sqrt{t}} \right) \right],$$

where

$$\mathrm{erf}(u) = \frac{2}{\sqrt{\pi}} \int_0^u dy\, e^{-y^2}$$

is the error function.

15.6 The integral equation

$$A(s) = c \left(\frac{s}{s_0} \right)^\alpha + \lambda \int_{s_0}^s \frac{ds'}{s'} \left(\frac{s}{s'} \right)^\alpha A(s')$$

is important in high-energy particle theory. Use Mellin transforms to find the solution. Calculate explicitly $A(s)$ and its Mellin transform.

Bibliography

The following list of books is representative but in no sense complete. It is meant to provide a starting point for the student who wants more information or a more elementary or more advanced treatment than is to be found in the text. Many of the books listed below have extensive bibliographies of their own, which can be used to branch out further into the literature. A rough attempt was made to list the books in order of increasing difficulty and to list separately those books which provide a more rigorous mathematical treatment.

GENERAL REFERENCES

H. Margenau and G. M. Murphy, The Mathematics of Physics and Chemistry (Van Nostrand, New York, 1943).

R. V. Churchill, Fourier Series and Boundary Value Problems (McGraw-Hill, New York, 1963).

G. Arfken, Mathematical Methods for Physicists (Academic Press, New York, 1970).

E. Butkov, Mathematical Physics (Addison-Wesley, Reading, Mass., 1968).

J. Mathews and R. L. Walker, Mathematical Methods of Physics (Benjamin, New York, 1970).

A. Sommerfeld, Partial Differential Equations in Physics (Benjamin, New York, 1949).

I. N. Sneddon, Elements of Partial Differential Equations (McGraw-Hill, New York, 1957).

G. F. D. Duff and D. Naylor, Differential Equations of Applied Mathematics (Wiley, New York, 1966).

H. Jeffreys and B. Jeffreys, Methods of Mathematical Physics (Cambridge, London, 1956).

G. F. Carrier, M. Krook and C. E. Pearson, Functions of a Complex Variable (McGraw-Hill, New York, 1966).

G. F. Carrier and C. E. Pearson, Partial Differential Equations (Academic Press, New York, 1976).

COMPREHENSIVE TREATISE

P. M. Morse and H. Feshback, Methods of Theoretical Physics, Vols. I and II (McGraw-Hill, New York, 1953).

PHYSICS TEXTS

W. R. Smythe, Static and Dynamic Electricity, Third Edition (McGraw-Hill, New York, 1969).

W. K. H. Panofsky and M. Phillips, Classical Electricity and Magnetism (Addison-Wesley, Reading, Mass., 1955).

J. D. Jackson, Classical Electrodynamics, Second Edition (John Wiley, New York, 1975).

MORE RIGOROUS MATHEMATICS

F. B. Hildebrand, Methods of Applied Mathematics (Prentice Hall, Englewood Cliffs, New Jersey, 1965).

P. Dennery and A. Krzywicki, Mathematics for Physicists (Harper and Row, New York, 1967).

F. W. Byron and R. W. Fuller, Mathematics of Classical and Quantum Physics, Vols. I and II (Addison-Wesley, Reading, Mass., 1969).

R. Courant and D. Hilbert, Methods of Mathematical Physics, Vols. I and II (Interscience, New York, 1953).

I. Stakgold, Boundary Value Problems of Mathematical Physics, Vols. I and II (Macmillan, New York, 1967).

E. L. Ince, Ordinary Differential Equations (Dover, New York, 1956).

S. L. Sobolev, Partial Differential Equations of Mathematical Physics (Pergamon, Oxford, 1964).

F. John, Partial Differential Equations (Springer-Verlag, New York, 1971).

I. G. Petrovsky, Lectures on Partial Differential Equations (Interscience, New York, 1954).

F. Riesz and B. Sz.-Nagy, Functional Analysis (Ungar, New York, 1955).

S. G. Mikhlin, Mathematical Physics, An Advanced Course (North Holland, Amsterdam, 1970).

SPECIAL FUNCTIONS

E. T. Whittaker and G. N. Watson, A Course of Modern Analysis (Cambridge, London, 1958).

G. N. Watson, A Treatise on the Theory of Bessel Functions (Cambridge, London, 1958).

MATHEMATICAL HANDBOOKS

E. Jahnke and F. Emde, Tables of Functions (Dover, New York, 1945).

W. Magnus, F. Oberhettinger and R. P. Soni, Formulas and Theorems for the Special Functions of Mathematical Physics (Springer-Verlag, New York, 1966).

M. Abramowitz and I. A. Stegun, Handbook of Mathematical Functions, National Bureau of Standards Applied Mathematics Series 55, Washington, D. C. 1964.

A. Erdélyi, W. Magnus, F. Oberhettinger and F. G. Tricomi, Bateman Manuscript Project: Higher Transcendental Functions, Vols. I-III, Tables of Integral Transforms, Vols. I and II (McGraw-Hill, New York, 1953).

ADDITIONAL REFERENCES

Books on integral equations – see page 258

Books on complex variables – see page 274

Index

Aerodynamics, 7, 8
Analytic continuation, 313–317
    of Fourier integrals, 414–419, 424
    of gamma function, 316–317
    of hypergeometric function, 377–380
Analytic function, definition, 275
Asymptotic expansions, 368–370
    for Bessel functions, 92, 150,
        393–399
    for Bessel functions of imaginary
        argument, 93
    for exponential integral, 369–370
    for gamma function, 370–372
    for Hankel functions, 92, 150,
        393–399
    for hypergeometric function, 379
    for Neumann functions, 92, 150,
        393–399
    for $P_l(x)$, 56
    for $Q_l(x)$, 62
    for spherical Bessel, Neumann, and
        Hankel functions, 127–128, 151
Ausstrahlungsbedingung, see radiation
        condition

Bernoulli numbers, 318
Bessel's equation, 25–26, 85, 150, 386
Bessel functions, 25, 26, 85–87, 150–151,
        387
    asymptotic expansions, 92, 93, 151,
        395–399
    generating functions for, 104, 386
    identities for, 104
    integral representation for, 392
    of imaginary argument, 92–93
    series for, 86, 88
    small argument expansions, 91, 150
    Sommerfeld's integral representation
        for, 389

spherical Bessel functions, 124–128,
        151–152
Beta function, 366–367
Boundary conditions, see also initial
        conditions, radiation conditions
    for Bessel functions, 425
    for dielectrics, 162–163
    for diffusion equation, 146, 232, 235,
        357
    for diffusion equations, 233
    for drumhead, 112
    for Helmholtz equation, 108, 147
    for inhomogeneous ordinary
        differential equations, 190–192,
        196
    for Laplace's equation, 76–78,
        147–148
    for Legendre polynomials, 52
    for magnetic media, 177
    for Poisson's equation, 198–199
    for spherical Bessel functions
        equation, 128
    for spherical harmonics, 65
    for Sturm-Liouville equation, 34
    for wave equation, 109, 147, 221, 223
Branch point, 288, 292, 294, 296, 297

Cauchy Riemann equations, 287
Cauchy's integral formula, 305–307
Cauchy's theorem, 301, 303–305
Čerenkov effect, 231
Circle chain method, 314–316
Completeness relations
    for Fourier Bessel transforms, 149
    for Fourier integrals, 41, 148
    for Legendre polynomials, 57
    for spherical Bessel functions, 149
    for spherical harmonics, 67
    for Sturm-Liouville problem, 39

Contour integrals, 298–301, 323–335
Convergence
    absolute, 281
    Cauchy criterion, 280
    of power series, 283–284
    of series, 280
    ratio test for, 281
    uniform, 307
Cut in complex plane, 288, 290–293,
        295–297, 350, 379, 381, 383–384,
        427, 431

Differential equations, ordinary, 27–30
    indicial equation, 31, 52, 86
    ordinary points, 28
    second solution, 29, 33, 60, 87
    series solution, 30–34, 52–55, 85–87
    singular points, 28, 52, 85, 372–373
    with constant coefficients, 419–420
    with three regular singular points,
        372
    Wronskian, 28–30, 33, 104, 207
Differential equations, partial, see
        diffusion equation, Helmholtz
        equation, Klein-Gordon
        equation, Laplace's equation,
        Poisson's equation, wave
        equation
Diffusion equation, 3–5, 22, 107,
        114–115, 146, 232–235
Dipole moment, 72–73
Dirac's formula, 345
Discontinuity problem, 345–346
Dispersion relation, 347–354
    subtracted, 353

Eigenfunctions, eigenvalues
    for angular part of Laplacian, 65
    for Helmholtz equation, 108–109,
        147
    for integral equations, 245–247, 252,
        256–257, 260–264
    for Legendre's equation, 56, 63
    numerical methods for, 43–44, 252
    for particle in cylindrical box, 115

for spherical Bessel functions,
        128–130
    for Sturm-Liouville problem, 34–39
    for vibrating drum head, 112–114
    for vibrating string, 111–112
Electric field, 11
    macroscopic, 157
Electrodynamics, 11–15, 219, 229–232
Electromagnetic field
    due to point source, 227–232
Essential singularity, 312
Euler's constant, 89, 368
Expansion theorems
    for associated Legendre polynomials,
        63
    for Bessel functions, 96–97, 149
    for Legendre polynomials, 370
    for plane wave in cylindrical waves,
        104, 390
    for plane wave in spherical waves,
        130–131, 153, 243
    for spherical Bessel functions, 129,
        149
    for spherical harmonics, 67
    for spherical waves about displaced
        center, 131–133, 153, 243
    for Sturm-Liouville problem, 37–39
    Fourier Bessel transforms, 102, 149
    Fourier integral theorem, 41, 148
    Fourier series, 40, 148

Fourier Bessel transforms, 101–102, 129,
        149
Fourier integral transforms, 40–41, 148
Fourier series, 40, 148
Fredholm's solution of integral equation,
        254–257
    conditions for validity of, 258–260

Gamma function, 88–90, 316–317,
        365–368, 370–372
Gauge transformations, 13–15
Generating functions
    for Bessel functions, 104, 390
    for Legendre polynomials, 58, 153

Green's function
    advanced, 226
    expansion in eigenfunctions,
        196–197, 208–209
    for bowed string, 193–195
    for diffusion equation, 232–235
    for Helmholtz equation, 204–208,
        404–407
    for ingoing waves, 216
    for Klein-Gordon equation, 409–414
    for ordinary differential equations,
        190–193
    for outgoing waves, 216, 242
    for Poisson's equation, 198–199
    for scattering problem, 242
    for wave equation, 215–227, 407–409
    retarded, 219, 222–232
Green's theorem, 190, 198, 222, 223, 234

Hankel functions, 92
    asymptotic expansions, 92, 150,
        395–399
    Sommerfeld's integral
        representations for, 387
    spherical Hankel functions, 127, 128,
        151, 152
Heat conduction equation, see diffusion
        equation
Helmholtz equation, 22, 23, 25, 107–109,
        112, 116, 117, 122
    as eigenvalue problem, 108–109
    eigenfunctions for, 108
Hilbert transforms, 348
Hilbert-Schmidt theory, 260–264
Hydrodynamics, 7–8, 79–80
Hypergeometric functions, 285, 372–380
    confluent, 380
    Legendre functions expressed as,
        383, 386
Hysteresis, 181

Identity theorem for analytic functions,
        313–314
Images, method of
    electrostatics, 164

for Green's function, 200–202
    magnetostatics, 182
Initial conditions
    for diffusion equation, 107, 147,
        233–235, 237
    for Klein-Gordon equation, 409
    for vibrating membrane, 114
    for vibrating string, 112
    for wave equation, 109, 147, 221,
        223, 226–227, 408
Integral equations
    convolution equations, 248–249,
        420–422
    eigenvalue problem, 245–247, 252,
        256–257, 260–264
    first kind, 244
    for scattering problem, 242, 244, 250
    Fredholm formulas, 254–258
    Hilbert-Schmidt theory, 260–264
    numerical solution, 251–254
    second kind, 244–246, 248, 249, 252,
        254, 256
    separable kernels, 245–248
    singular, with Cauchy kernels,
        359–362
    Volterra, 244
Integral transforms
    Fourier, 40, 148, 403
    Fourier Bessel, 101–102, 129, 149
    Fourier, one-sided, 414–419
    Hilbert, 348
    Laplace, 418–420, 423–424
    Mellin, 421–422
Integrals, evaluation, 323–335
    angular, 333
    on range $(, \infty)$, 330–332
    rational functions, 325–326
    transformation of contour, 334–335
    with exponential factors, 326–330
Irrotational flow, 9, 79–80
Iteration of integral equation, see
        Liouville-Neumann series

Jordan's lemma, 326–330

Klein-Gordon equation, 409

Laplace's equation, 75–80, 148
  between two planes, 100–102
  cylinder in external field, 99–100
  exterior problem for cylinder, 98–99
  exterior problem for sphere, 77
  flow around sphere, 79–80
  for dielectric sphere, 167–169
  interior problem for cylinder, 94–95,
    97–98
  interior problem for sphere, 76
  sphere in external field, 78–79
Laurent expansion, 312
Legendre functions $P_l(z)$, 380–383
Legendre functions $Q_l(z)$, 27, 60–62,
    357, 383–386
  identity for, 400
Legendre polynomials $P_l(x)$, 27, 53–59
  formulas, 54, 55
  generating function for, 58
  identities for, 59, 62, 82
  normalization, 55, 57
Legendre polynomials, associated,
    $P_l^m(x)$, 62–65
  formulas, 63, 64, 152
  normalization, 65
Legendre's equation, 27, 51, 53, 60, 380
Liouville's theorem, 319
Liouville-Neumann series, 249–250
  convergence of, 258

Magnetic field, 11
  macroscopic, 172
Magnetic moment, 172
Magnetic pole density, 179
Magnetization, 175, 178, 179
  current, 172–175, 178–179
Maximum modulus theorem, 319
Maxwell's equations, 11
Mean value theorem, 319
Morera's theorem, 302–303
Multipole expansion
  for scalar potential, 71–75
  for vector potential, 169–172

Natural boundary in complex plane, 313

Neumann functions, 87–90, 150, 349,
    355, 389
  series for, 418
  spherical Neumann functions,
    123–128, 151
Normal modes
  for acoustic resonant cavity, 116–117
  for acoustic wave guide, 117–119
  for vibrating drumhead, 112–114
  for vibrating string, 111–112
Numerical solution
  integral equations, 251–253
  ordinary differential equations,
    41–44

Omnès's equation, 359

Partial fraction expansions, 335–339
Plasma dispersion function, 357–359
Plemelj formulas, 343–345
Poisson's equation
  scalar, 12, 71, 198–199
  vector, 13
Poisson's solution of Dirichlet problem
    for circle, 320–321
Polarization, 161
  charge, 158–163, 166, 169
Pole in complex plane, 28, 278, 312
Potential
  scalar, 12, 14, 71, 157
  vector, 12, 14, 169
Power series, 30, 53, 85, 282–285
Product expansions
  for $\Gamma(z)$, 368
  for $\sin \pi z$, 338

Quadrupole moment, 74, 160

Radiation condition, 135, 217, 227, 242,
    406–407, 430
Rational functions, 277
Regular function, see analytic function
Residue theorem, 323–325
Riemann P symbol, 374
Riemann sheets, 290–292, 294, 297, 298,
    354

Rodrigues's formula, 380
Runge-Kutta numerical integration, 42

Scattering
    in quantum mechanics, 241–244, 250
    of waves by sphere, 136–140
Schrödinger's equation, 5, 110, 115
Schwarz reflection principle, 350–352
Separation of variables, 21–27, 145–146
    Helmholtz equation
        cartesian coordinates, 22
        cylindrical coordinates, 23
        spherical coordinates, 25
    Laplace's equation, 25, 26
    spherical harmonics, 51
    time dependence, 22, 108, 111
Small argument expansions
    for Bessel functions, 91, 150
    for Bessel functions of imaginary
        argument, 93
    for Neumann functions, 91, 150
    for spherical Bessel and Neumann
        functions, 127, 151
Sommerfeld-Watson transformation, 338
Spectral representation, 348
Spherical harmonics, 65–67, 72, 75, 123,
        131, 133, 134, 137, 152
    addition theorem, 68–70, 153
    completeness, 67
    formulas, 65–66, 152–153
    normalization, 67
Steepest descents, method of, 370–372,
        393–399
Stirling's formula, 368–372
Sturm-Liouville problem, 34–39
    completeness of solutions, 37–39
    for associated Legendre polynomials,
        63–65
    for Bessel functions, 96–97
    for Legendre polynomials, 56–57
    for spherical Bessel functions,
        128–130
    orthogonality of solutions, 36
    relation to Green's function, 196–197

Taylor expansion, 309

Wave equation, 109, 146, 215–218
    acoustic, 8, 116–119
    for electromagnetic fields, 13
    for membranes, 6, 112–114
    for scalar potential, 14, 219
    for strings, 6, 111–112
    for vector potential, 15, 219
Wave guide, acoustic, 117–119
Waves
    acoustic, 8, 116–119
    diffraction by knife edge, 430–438
    due to a point source, 227–232
    emission of spherical wave, 133–136
    expansion in spherical coordinates,
        130–133
    scattering by a sphere, 136–140
Wiener-Hopf method, 424
    for diffraction by a knife edge,
        430–438
    for potential given on semi-infinite
        plate, 425–429

Printed in the United States
By Bookmasters